AF845275

# GOLD

## *Forgotten Histories and Lost Objects of Australia*

Throughout history, gold has been the stuff of legends, fortunes, conflict and change. The discovery of gold in Australia 150 years ago precipitated enormous developments in the newly settled land. Immigrants flooded in from Asia and Europe, and the population and economy boomed in spontaneous cities. The effects on both the environment and indigenous Aboriginal peoples have been profound and lasting. In this book, a team of prominent historians and curators have collaborated to produce a cultural history of gold and its impact on the development of Australian society. This fascinating book will add a new dimension to previously recorded Australian history, and provide a richer understanding of the foundations of Australian culture. Like a handful of tailings, *Gold* brings together a collection of stories that have been left out of standard Australian histories. It is an essential history.

**Iain McCalman** is Director of the Humanities Research Centre at the Australian National University and author of *Radical Underworld* (1988) and *An Oxford Companion to the Romantic Age* (1999).

**Alexander Cook** is a doctoral candidate at King's College, Cambridge, who specialises in Australian, Pacific and European cultural history.

**Andrew Reeves** is a historian who has published studies in labour history, museology and cultural policy. He previously served as the Director of the Western Australian Museum and is currently a member of the Council of the National Museum of Australia.

DURHAM
UNIVERSITY
LIBRARY

# GOLD

## *Forgotten Histories and Lost Objects of Australia*

*Edited by* Iain McCalman, Alexander Cook *and* Andrew Reeves

0 3 SEP 2008

CAMBRIDGE
UNIVERSITY PRESS

994
GOL

PUBLISHED BY THE PRESS SYNDICATE OF THE UNIVERSITY OF CAMBRIDGE
The Pitt Building, Trumpington Street, Cambridge, United Kingdom

CAMBRIDGE UNIVERSITY PRESS
The Edinburgh Building, Cambridge CB2 2RU, UK
40 West 20th Street, New York, NY 10011–4211, USA
10 Stamford Road, Oakleigh, VIC 3166, Australia
Ruiz de Alarcón 13, 28014 Madrid, Spain
Dock House, The Waterfront, Cape Town 8001, South Africa

http://www.cambridge.org

© Cambridge University Press 2001

This book is in copyright. Subject to statutory exception
and to the provisions of relevant collective licensing agreements,
no reproduction of any part may take place without
the written permission of Cambridge University Press.

First published 2001

Printed in Australia by Brown Prior Anderson

Typeset in New Baskerville 10.5/12 pt

*A catalogue record for this book is available from the British Library*

*National Library of Australia Cataloguing in Publication data*
Gold: forgotten histories and lost objects of Australia.
Bibliography.
Includes index.
ISBN 0 521 80595 3.
1. Gold mines and mining – Australia. I. McCalman, Iain.
II. Cook, Alexander. III. Reeves, Andrew.
622.18410994

2793301

ISBN 0 521 80595 3 hardback

NATIONAL MUSEUM OF AUSTRALIA

Produced with the assistance of the National Museum of Australia, Canberra.

# Contents

# Illustrations

PLATES

FIGURES

 


TABLES



MAP

# Acknowledgements

x We wish to thank Robert Edwards of Art Exhibitions Australia for encouraging us to undertake this publication, and for assisting us with some of the expenses. Dawn Casey, Director of the National Museum of Australia, and her excellent managers, Stephen Foster and Greg Andrews, also provided invaluable encouragement and financial help. Ian Bryson helped us to procure some rare photographs in the Tanami Desert region, and he and Derek Elias undertook a research trip there which went way beyond the ordinary expectations of scholarship or duty. On this last score, Lindy Shultz deserves our particular thanks: she has worked long, hard and cheerfully, often in her spare time, to prepare and deliver the manuscript at very short notice. Without her effort and expertise, this book would not have met its tough deadline. Katie Russell helped us to find some elusive photographic images. Numerous of the contributors assisted us with their private illustrations or gave us information on useful sources. Weston Bate and Bill Gammage gave us sage advice at early points in the process. Glen Barclay and Caroline Turner offered vital editing help when we most needed it. Leena Messina organised our contributors' conference at very short notice with her usual efficiency and good humour. Art Exhibitions Australia gave us access to useful data on goldfields paintings that they had commissioned in preparation for the forthcoming National Museum Gold exhibition. Finally, we would like to thank Sharon Mullins of Cambridge University Press for her enthusiastic support and excellent advice throughout the production of this book.

# Contributors



**Margaret Anderson** has researched and written on women's history, demography and the history of the family in nineteenth-century Australia. She also has a professional interest in public history and museology. This is her first foray into goldfields history.

**Patrick Bertola** lectures in history at Curtin University, Western Australia. He has researched and published extensively on the labour and working life of the West Australian goldfields.

**Bill Bunbury** works in the Oral History Unit of the ABC. He has published widely on aspects of Australian daily life and social experience, including a book on the 'Woodlines' in Western Australia.

**Anita Callaway** is a postdoctoral and Getty Fellow in the Centre for Cross-Cultural Research, Australian National University. She has recently published a book on varieties of popular visual culture in nineteenth-century Australia.

**Ian Coates** works in the Research Program of the National Museum of Australia. He recently completed a PhD at the Australian National University, examining exchanges between British museums, collectors and Australian Aboriginal people at the end of the nineteenth century.

**Alexander Cook** is currently undertaking a PhD on French and British history, at King's College, Cambridge. He has also written on Australian and Pacific cultural history in the eighteenth and nineteenth centuries.

**Carol Cooper** is an anthropologist who works as a curator and administrator at the National Museum of Australia. She has undertaken intensive research on indigenous art and culture in Australia.

**Ann Curthoys** is Professor and Head of History in the School of Humanities Faculty at the Australian National University. She has researched and published widely on Australia's race, gender and labour history.

**Graeme Davison** is Professor and Head of the Division of Historical Studies at Monash University. He has published extensively on Australian urban history and culture, including a classic study of the city of Melbourne.

**Derek Elias** was previously the Senior Anthropologist at the Central Land Council in Alice Springs, where he worked for seven years and continues to be employed on a contract basis. He is currently completing his PhD in anthropology at the Australian National University, focusing on the complex relationships between Warlpiri people, their places and gold exploration and mining.



**Dorothy Erickson** is a freelance art historian, writer, curator and goldsmith in Perth. Her doctoral thesis was on gold- and silversmithing in Western Australia.

**Charles Fahey** lectures in history at the Bendigo Campus of La Trobe University. He has researched extensively on goldfields history and land occupation, and is co-author of a book on local histories in Victoria.

**David Goodman** lectures in American and Australian history at the University of Melbourne. He is author of a major book comparing Australian and Californian gold-rush experiences.

**Tom Griffiths** is a Senior Fellow in the History Program of the Research School of Social Sciences, Australian National University.

**Suzanne Hunt** is a garden historian and consultant. She is currently developing resources for the study and exhibition of garden history with the State Library of Victoria.

**Susan Lawrence** lectures in archaeology at La Trobe University in Melbourne. She has recently published a book on the archaeology and material culture of the Dolly's Creek goldmining community in Victoria.

**Iain McCalman** is Professor and Director of the Humanities Research Centre at the Australian National University. He has published extensively on British and colonial cultural histories of the eighteenth and nineteenth centuries.

**Barry McGowan** is an environment and heritage consultant on Victorian and New South Wales goldfields sites. He has published numerous reports and is currently undertaking a PhD on New South Wales gold history at the Australian National University.

**Kim McKenzie** is an anthropologist, documentary film-maker, and museum curator. He is currently working with the National Museum of Australia and the Centre for Cross-Cultural Research, Australian National University, on a major multimedia project treating the kinship systems of indigenous communities in Arnhem Land.

**Hank Nelson** is a Professor of History and documentary film-maker at the Research School of Pacific and Asian Studies, Australian National University. He has published extensively on Australian and New Guinean history, including a book on the New Guinea gold experience.

**Paul Pickering** is an Australian Research Council Research Fellow at the Humanities Research Centre, Australian National University. He has written widely on Chartism in Britain and is currently working on radical and democratic political traditions in nineteenth-century Australia.



**Alan Platt** is a chartered engineer with a particular interest in the history of technology in the nineteenth century. He is author of *The Life and Times of Daniel Gooch* (Alan Sutton, Gloucester, 1987).

**David Raftery** is an anthropologist and historian who works with Aboriginal Land Councils in northern Australia. He is currently working on a study of indigenous–European relations in northern desert regions.

**Andrew Reeves** is a historian who has worked in a number of Australian museums, most recently as Director of the Western Australian Museum. His research has concentrated on trade unionism and the history of working-class communities. He has published studies in both labour history and in applied fields of museology and cultural policy. He is currently a member of the Council of the National Museum of Australia.

**Dianne Reilly** is the La Trobe Librarian at the State Library of Victoria. She has a longstanding research interest in Australian–French relations and is co-author of a photographic and historical study of the work of Richard Daintree and Antoine Fauchery.

## Conversions



LENGTH

1 inch = 2.54 cm
1 foot = 30.5 cm
1 yard = 0.9 m
1 mile = 1.6 km

MASS

1 ounce = 28.3 g
1 pound = 454 g
1 ton = 1.02 t

AREA

1 acre = 0.4 ha
1 square yard = 0.836 square metres
1 perch = 25.29 square metres

VOLUME

1 cubic yard = 0.7 cubic metres

CURRENCY

There were 12 pennies (d) in 1 shilling (s), and 20 shillings in £1. A guinea was worth 21 shillings. When Australia converted to decimal currency in 1966, $2 were equal to £1.

# Introduction

*Iain McCalman, Alexander Cook and Andrew Reeves*

The ancient tension between gold as a symbol of the sacred and gold as a tool for the profane has still not been resolved. It probably never will be. Perhaps it is the malleability of gold, as both material and metaphor, which has caused it to seize the imagination of countless men and women over the centuries. It has mobilised them – bond or free – to travel the globe, has driven them to lie and cheat, to suffer and speculate, to risk their lives, to move mountains and reshape the landscape, to create a myriad of objects as conspicuous for their beauty as for their ostentation. The production and consumption of gold stands at the heart of numerous cultures, and the history of gold seeking and goldmining in Australia is part of this long global epic.

During the 1850s, the population of the tiny Australian colonies nearly trebled as a direct result of the discovery of gold; Victoria alone rapidly came to produce almost one-third of the world's total output of the prized metal.[1] As gold discoveries moved north and west from the sites of the first finds, massive population movements and social upheaval followed. The 'gold rushes' to Queensland in the 1870s and 1880s, and to Western Australia in the 1890s, would reshape the social bodies of their respective colonies as immigrants flooded from all corners of the globe. These men and women carried with them their skills, their dreams and their prejudices, and they fertilised the culture of Australia for generations to come. Gold immigrants, in turn, were refashioned by their new environments, sometimes in strange and unpredictable ways.

Gold discoveries were instrumental in finally ending the British government's desire to transport convicts to the eastern colonies of Australia and they gave force to the campaigns for responsible self-government which achieved success in the 1850s. The quest for gold led prospectors into the remote interior of the continent, opening transport routes for other forms of settlement. It led to conflict between settlers and the indigenous population over watercourses and food, and it produced massive environmental upheaval. For a time, gold brought great financial prosperity to the colonies, although uneven distribution of the spoils generated significant social traumas.

The secondary economic effects were no less huge. Gold generated capital for other industries and the population boom created markets for other produce, fostering the development of whole sectors of the Australian economy. The lure of gold created labour shortages and raised wages widely across the spectrum. In Victoria's Western District, shearing rates rose by almost 50 per cent within the first year of the rushes, and average wages doubled by 1854, although inflation swallowed much of the increase in income.[2] Sometimes whole industries were pushed into temporary decline, and there was widespread fear that the attraction of gold digging would leave the colonies unable to produce anything else.

Although the social consequences of the generation of instant wealth were clearly substantial, there was sharp disagreement over their merit. Newspapers and parliamentary debates of the period featured anxious monologues on the imminent threats to law and order created by 'the rush to be rich'. The clergyman, the social engineer and the moralist lamented the unholy obsession with the 'lottery of gold digging' and warned of the consequences, both in this life and the next, of succumbing to gross materialism. Champions of the gold culture claimed that it fostered independence, ingenuity and 'manliness' and praised an environment in which the advantages of inherited privilege meant nothing to an individual's chances of success. For supporters and opponents of the gold culture alike, the perceived 'democracy' of the goldfields was an inexhaustible topic for discussion.



This book is about the material and cultural impact of gold seeking on Australian history. It explores the social, political, economic and environmental transformations wrought by the quest for gold and the practice of mining it – both in the past and today. From the 1850s to the present, gold has informed Australia's development. It remains one of the country's major export commodities, and the 'roaring days' of the gold rushes have been the inspiration for a substantial portion of its national mythology (Plate 1).

The subject is not a new one for students of Australian history. The gold rushes have attained an iconic status in the pantheon of epic nation-making events. The soaring figures of population growth and per capita income are cited again and again. The Eureka uprising is often claimed to have been a formative moment in the evolution of an Australian political consciousness. Indeed, for many the diggers stand between the convicts and the Anzacs as landmarks on the road towards a national self-image built on 'mateship' and a sceptical distrust of authority. Yet, for all the ink which has been expended in asserting the centrality of the rushes to the development of modern Australia, the specific nature of gold's impact remains a matter of fierce contention.

**Figure 1** Henry Marsh, *The Galop for Gold* [cover for piano music, 1852].
(Rex Nan Kivell Collection, National Library of Australia)

Contemporaries never doubted the importance of the discoveries, yet their assessments of the likely consequences ranged from utopian visions to predictions of pandemonium. When a Sydney geologist, W. B. Clarke, discovered a sample of ore near Lithgow in the Blue Mountains in 1841, the Governor Sir George Gipps, wary of the potential disturbance in a penal colony, is reported to have responded urgently: 'Put it away, Mr Clarke, or we shall all have our throats cut'.[3] The story is probably apocryphal, Clarke himself may have been the author of the statement,[4] but something of the sentiment would be echoed seven years later once the gold rushes had commenced in New South Wales. In the early days, the *Sydney Morning Herald* fervently hoped that 'the treasure does not exist in large quantities' and predicted 'calamities far more terrible than earthquakes or pestilence'.[5] Concerns for security, property and order were expressed by elements of the urban elites and the pastoral interest, and by others disturbed by the acquisitive materialism of the goldfields. The eminent squatter John Macarthur warned of the 'pernicious and demoralising results' to be expected from allowing the populace to ransack the country for gold. John Sherer lamented that 'plebeianism of the rankest and, in many instances, the lowest kind at present dwells in Australia' as a result of gold.[6]



Within a decade the political climate had changed to the extent that gold discoveries were to be fostered rather than suppressed. Edward Hargraves, confidence man, entrepreneur and spark for the first Australian gold rush, reported his personal response to his find near Bathurst in 1851: 'This', I exclaimed to my guide, 'is a memorable day in the history of New South Wales. I shall be a baronet, you shall be knighted, and my old horse will be stuffed, put into a glass-case, and sent to the British Museum.'[7] The comment reflects Hargraves' confidence in the positive social influence of goldmining, and the prospector certainly received much that he had hoped for by way of recompense from the grateful authorities – although his horse never achieved the glorious end he envisaged for it. We see another counterpoint to Gipps' pessimism in the triumphant declaration of the fiery minister John Dunmore Lang that gold was 'the divinely appointed means of ensuring, not merely political liberty, but national existence and a brilliant and glorious future for Australia'.[8] By 1859, William Kelly would marvel at the modern metropolis of Melbourne, built largely from gold wealth, and describe it as 'the overtopping wonder of the world'.[9]

These remarks are but the tip of an iceberg of commentary on the gold rushes. Debates concerning the best techniques of goldmining, appropriate methods of administration, revenue collection, immigration, environmental degradation and the impact of the whole affair on the political and social climate of the country had begun even before news of the first find reached the metropolis. They would continue long after the dust had settled on the rushes. Indeed, the self-congratulation and recrimination about various aspects of the history of the gold rushes continue today.

Australian gold finds were watched keenly from abroad by would-be gold seekers, desperate to escape the economic hardship and political turmoil of

Europe in the 1840s, and by investors looking to turn a quick profit. In Britain, the Secretary of State acknowledged the folly of conveying offenders at the public expense 'to the immediate vicinity of those very gold fields which thousands of honest labourers are in vain trying to reach'.[10] In Norway, the young Henrik Ibsen had his characters in *Love's Comedy* (1863) dream wistfully of transportation to a 'Ballarat beyond the sands'.[11] In China, thousands of poverty-stricken farmers looked to Australia as a means of accumulating enough money to help struggling relatives back home and to save a little for their old age. In America, as the sources of alluvial gold in California were slowly exhausted and the costs of mining rose, veterans of the first gold rush looked to Australia as the next opportunity for the small man to make good. As the immigrants flooded in, cocksure locals announced that 'a voyage to Australia is the working man's only road to the future'.[12] For many, probably for most, the road to the goldfields would prove a dead end, although other opportunities might present themselves.

Even those who had no interest in the wealth to be won from the ground watched, fascinated, to see the social forces unleashed. The prior history of goldmining had seen it almost universally controlled by the state or by local potentates. In Great Britain, for example, all gold found within the kingdom was automatically property of the Crown. Yet in the Anglo-Saxon colonies, especially in America which set the model for Australia, the history of pioneer settlement encouraged the private development of mineral resources. The New World goldfields were seen by many as a huge laboratory, where diverse human ingredients were brought together free from the constraints of convention and community – they would either learn to cohere or they would combust. Karl Marx wrote to Friedrich Engels in 1852 that 'California and Australia are two cases which were not provided for in the Manifesto: creation of great new markets out of nothing', and he went on to predict that in the Antipodes 'Carlyle will see his aristocracy of rogues established in full glory'.[13] It was an expectation shared by many contemporaries of diverse political persuasions.

Yet, by the 1890s, the outlines of the goldfields legends were taking shape. Radical nationalists like Victor Daley and Henry Lawson were writing poetry idealising the gold digger's life and depicting the miners' Eureka uprising of 1854 as a stand for social justice and political freedom against the oppressive forces of the administration.[14] The socialist William Lane looked back nostalgically to a time when workers were able to work independently on shallow gold leads, free from the tyranny of wage-slavery.[15] These strands fed into a brand of historical scholarship which quarried the goldfields for signs of an emergent 'national type'. Vance Palmer in the 1930s and 1940s and Russel Ward in the 1950s were the most prominent of a series of left-wing nationalist historians who found in the gold rushes a crucible for forging the Australian character – an individual characterised by independence, ingenuity and a commitment to 'a fair go'.[16] It is this tradition, more than any other, which has shaped the heroic image of the gold digger in Australia. It is reflected in the tradition of calling Australian soldiers 'diggers'. It has proved a remarkably versatile image. Communists, Catholics and champions of laissez-faire liberalism, to name but a few, have all found something to admire in the idea of the digger.[17]

In 1930 W. K. Hancock remarked that Australians 'have acclaimed the diggers as their Pilgrim Fathers, the first authentic Australians, the founders of their self-respecting, independent, strenuous national life, the fathers of their soldiers', and he went on to claim that 'the legend of the Australian diggers does not greatly distort the fact . . . despite the generous sprinkling among them of outlandish foreigners, low-born rascals, and well-born ne'er-do-wells'.[18]

In his history of Victoria during the gold rushes, Geoffrey Serle emphasised, rather, the quality of the immigrant population to the diggings, claiming that gold 'peopled it . . . with men of more diverse talents, skills and backgrounds, and perhaps more vigour, than Australia had yet seen'.[19] It was, on the whole, a more subtle picture of the goldfields population than that espoused by the romantics. While it praised the mettle of the population of gold seekers, it acknowledged diversity rather than the monolithic culture of the digger icon. But Serle, too, highlighted the remarkable number of subsequently eminent Australians who came to Victoria during the gold rushes or were descended from those who did. His catalogue of Victorian political notables included Higinbotham, Berry, Syme and Duffy. His national tally included the ancestors of Alfred Deakin, John Monash, Robert Menzies and Arthur Calwell.[20]



The gold-rush myths have not been without their critics. Manning Clark noted that 'just as some have erred in seeing the convict ships as Trojan horses for planting the ideals of the French Revolution in Australia so others have seen the gold diggers as evangelists for Chartism rather than men in pursuit of material gain'.[21] As early as the 1930s, A. G. L. Shaw was keen to downplay the role of the gold rushes in Australian political development, preferring to suggest that 'it was the rate rather than the nature of development that the gold discoveries influenced'. Yet even Shaw had to acknowledge that 'no country can have its population almost trebled in a decade without obvious effects'.[22]

In the early 1960s, Geoffrey Blainey weighed into the debates about gold with his history of Australian mining, *The Rush That Never Ended*. Intellectually unsympathetic to the class-based analyses of goldfields politics made by some earlier scholars, Blainey believed the diggers were individualists to the core. It was something of a simplification, just as some of the earlier interpretations had been, and his evidence was based heavily on the first phases of shallow alluvial mining when individual diggers had their best chance of success. Yet Blainey's work on mining history made important contributions to our knowledge of the subject. He was the first to pinpoint the temporal dynamics of Australian gold booms, to notice the way in which gold rushes seemed to come at times when the rest of the economy was in crisis, because limited job prospects encouraged individuals to try their hand at the hard life of the independent goldminer.[23] Blainey also had some basis for his claim that 'gold checked and for a time reversed Australia's tendency to become a land that favoured the big man', although the durability of its effects in this regard is questionable.[24]

In the 1970s, Humphrey McQueen turned his attention to the gold rushes in his neo-Marxist reassessment of Australian history, *A New Britannia*. He accepted the claim of his left-wing predecessors that gold had permanently shaped the political culture of the country, and particularly its radical tradition. He differed from them, however, in seeing this as no cause for celebration. Where prior historians had admired the diggers for their lack of deference to social superiors and their determination to secure political privileges for the working man, McQueen blamed the gold rushes for co-opting the working classes into an acceptance of the over-arching hegemony of the capitalist system. Discussing what he referred to as 'the mytho-poetic properties' of gold, he claimed that 'they sustained the belief that advancement was possible, that the small man could make good, that capitalism was not without promise'.[25] In some ways his analysis of the gold rushes echoed the individualist emphasis of Blainey. He saw the independent digger as a would-be capitalist and he blamed the gold-driven prosperity of the second half of the nineteenth century for the quietist, compromising nature of the Australian labour movement.

A number of other critics since that time have emphasised the limitations of the much-vaunted 'diggers democracy'. Some, after Manning Clark, have suggested that the gold rushes fostered a philistine acquisitiveness alongside the culture of resistance to hierarchy. Others have suggested that the independent, manly digger, in so far as he existed at all, tended also to be a sexist, racist digger, who left a legacy of distrust of women and hostility to non-Anglo-Saxon immigrants which this country is still working its way through. The use of a variant of the Eureka flag for a banner against the Chinese during the brutal riots at Lambing Flat in 1861 does a great deal to symbolise the limitations of goldfields democracy.



These debates about the virtues and vices of gold-rush society highlight our ongoing sense of the importance of our colonial past. Yet they reflect a preoccupation with the teleology of national development, and a conversation with the narratives of the 1890s, which sometimes overlooks the question of how gold and gold society were understood at the time, and how they affected the lives of those who lived through them. These are some of the things we have encouraged contributors to attempt in various ways in this book. We believe it is instructive to study the dynamics which shaped gold-rush society and the cultural lens through which they have been understood.

As its sub-title suggests, then, this book is not intended to be a comprehensive and systematic revision of the classic works of Australia's pioneering gold historians. Our intentions are more modest – to illustrate, amplify, complicate or update the central gold themes that have been so well traversed, and to suggest some relevant new themes for exploration. We have chosen to do this in the main through the medium of twenty diverse 'stories' or microhistories. Why take such an approach?

In the first place we want to celebrate the return of the story, or narrative history, to mainstream importance in the practice of academic history writing. For too long, professional scholars have misguidedly neglected the historical narrative as seemingly too humble a means of demonstrating our expertise, compared with the sophisticated theoretical paradigms, analytical models and specialist vocabularies of our counterparts in other disciplines. For the general public, however, who still read and experience history in growing numbers, the narrative has never gone away: it is seen as the natural and axiomatic preserve of the historian. Fortunately, a multitude of independent writers, journalists, critics, novelists, biographers, screenwriters, film-makers and museum curators have more than made up for our desertion. In 1989 the distinguished Harvard historian Simon Schama published what he mischievously called a 'chronicle or annal' of the French Revolution, by which he meant a racy general history written in narrative form and intended to appeal to the mass public engaged in celebrating the bicentenary of the revolution.[26] This provocative action sent a frisson of outrage through the ranks of his academic colleagues – still more so when they learnt how much he had been paid for writing such a regressively popular and accessible history. Since then, and not only for mercenary reasons, the historical narrative has been making a steady comeback in university circles. We hope to continue this trend, and we have faith in the possibility of combining a commitment to narrative with an engagement with broader theoretical issues and analytical approaches.



Of course, we cannot hope in this book to emulate Schama's financial good fortune, but we can aspire similarly to using the narrative as a way of reaching out to a broad public who may be moved to re-explore stories of Australian gold by the advent of a new national museum which is mounting a major exhibition on gold in the year of the Centenary of Australian Federation. Our book contains twenty, relatively short historical narratives written by a range of professional historians, cultural anthropologists, journalists, museum curators, heritage consultants and art specialists. The stories they offer cannot, and do not, pretend to offer a comprehensive coverage of Australia's entire gold history, with its huge geographical spread across the continent and its equally broad temporal sweep from the mid-nineteenth century to the present. But gold is also good to think – to adapt the words of a famous anthropologist[27] – because it offers a peculiarly effective window into the diversity of Australian culture over the last two centuries. One way and another gold has touched so many of Australia's peoples, places and things. Our book offers a representative and entertaining selection of these, including case studies of what we believe to be comparatively neglected or forgotten aspects both of gold's history and of Australian culture.

As a form of writing the historical story also has the enormous advantage of accessibility. Although it is not fashionable to talk about human nature these days, storytelling must come pretty close to being a universal human impulse. Casting information into ordered, comprehensible sequences and patterns is almost a reflex. Stories surround us, informing the multiple engagements and transactions of daily life. Without them, we could make no sense or meaning out of the barrage of images, information and sensation that are an inescapable part of the contemporary condition. Perhaps it is the very ubiquity of historical stories, and their associated appearance of artlessness, that has led scholars to devalue them. Yarns, tales, chronicles and sketches are seen as the domain of the unsophisticated or the amateur, not of the credentialled expert. This fallacy – and it is one – makes it all the more important that a book such as this, intended equally for scholar, student and general reader, should work through the medium of historical stories. As the good conductor that it is, gold may help us to reopen connections between the interested public and the specialist researcher, as well as to bring some less familiar methodologies into wider circulation. Our microhistories might even enable us to bridge the larger gulfs that have customarily divided the work of different university specialists.

Anyone who has written historical stories will know that they are never easy to craft, but some stories are undoubtedly more difficult than others. The historical evidence needed can present itself for potential capture in myriad forms, including the oral, visual, aural, tactile and material. Photographs, for example, which have been extensively used in this collection, are a case in point. Cameras were to be found on all but the earliest fields and became increasingly common as the nineteenth century progressed. Photography served many purposes – as an alternative to landscape art, capturing the broad social sweep of the goldfields; or, alternatively, the passing moment of a mine's workforce collected at the pithead to mark a special occasion, or prior to going on shift. Yet, as an instrument of historical interpretation, photographs capture particular moods or moments like no other medium. To discover a nugget was the height of mining achievement for many, and the carefully posed photograph of the discovery, often re-created, generally with the miner in fresh, clean clothes, is surely goldmining's equivalent of the wedding photograph. Similarly, while many photographs reflect the routines of mining or those elusive moments of success, there is also a minority which preserve fugitive images of the physical cost of mining – deep lines etched into the faces of miners, demonstrating beyond words the pitiless demands made by the pursuit of gold. At their best such photographs stand with the classic images that document industrial life in nineteenth-century Europe or the faces of the American Midwest during the Great Depression.



Moreover, to extract hidden meanings from seemingly mute images or objects requires impressive decoding skills. Many of our stories are based on just such unfamiliar forms of evidence, modes of inquiry and methods of narration. Museum curators Kim McKenzie and Carol Cooper, for example, have to interpret the subtle, often ambiguous, messages inscribed in the coloured pencil drawings of an illiterate Aboriginal boy from the Palmer River goldfields. Art historian Anita Callaway has to recover the motifs and meanings of popular panoramas and theatrical tableaux from the Victorian goldfields. Visual anthropologist Ian Coates has to compare the pictorial encodings of ethnographic photographs and ink cartoons from the Pilbara goldfields. Artist and art historian Dorothy Erickson must unravel the cultural symbols and life experiences fashioned into material objects such as West Australian goldfields brooches. Garden historian Suzanne Hunt must uncover the cultural aspirations lovingly cultivated in goldfields gardens. Archaeologist Susan Lawrence must disinter and analyse the life of a goldfields community, using the buried detritus of lost villages like Dolly's Creek. In order to write their stories of 'lost places', anthropologists Derek Elias and David Raftery have had to travel into the remote Central Australian desert and spend long hours learning how gold and its European seekers have intruded into the lives of indigenous peoples over many years (Plate 2).

Another benefit of producing such a series of microhistories is that of plurality – the ability to evoke multiple voices, different angles of vision and diverse disciplinary frameworks. Such case studies are naturally rich in variety because of their embeddedness in specifics of time, character and place. At the same time we must contend with the fact that gold seems to have particularly fissile properties when it comes to generating imagined stories. Tantalising myths have gone hand in hand with gold seeking all over the world, and Australia is no exception. We too have our equivalents of Jason's golden fleece or of the lost treasures of El Dorado. David Raftery here unravels one of the most persistent and protean of these in his account of a 1930 expedition to search for Lasseter's Reef, a local myth which he shows has entangled itself with the Dreaming of peoples whose stories have animated the continent for thousands of years.



In considering the big picture we need also to assess how our microhistories relate to those grand assumptions, the larger narratives of national progress or regress that have become inextricably associated with Australian gold. Perhaps the foremost of these narratives is the idea that gold rapidly and utterly transformed Australia into the basic form of 'modern' society we know today. Perhaps because of its age-old associations with instant alchemical transmutation, gold seems to attract such a language of revolution. True, the bald figures of, say, Victoria's economic growth between 1851 and 1853 make this language seem appropriate, as does the dramatic martial imagery of the Eureka Stockade of 1854 (Plates 3 and 4), notwithstanding the cautions of historians like Shaw that even in Victoria gold acted as a catalyst to accelerate already existing social, political and economic trends. This debate over the revolutionary or stimulant effects of gold will probably never be resolved. All we can do is to underscore through the geographical and temporal variety of our inclusions the need to assess gold's impact both in its early dramatic phases and incrementally, across a spread of time extending to the present.

Many popular histories and cultural representations also offer various versions of gold history as the prime source of our remarkable progress towards nationhood. By this account gold created modern, liberal Australia. Or, to put it another way, the infusion of gold wealth into the colonies is seen to have initiated a dramatic and progressive movement towards contemporary patterns of Australian life such as our federated, democratic political system,

**Figure 2** *Sly Grog Selling on the Goldfields*, in *Victoria and Its Metropolis*, Sutherland, Melbourne, 1888. (National Library of Australia)

our liberal, capitalist industrial economy, our multi-ethnic and multi-cultural population, and our sophisticated urbanised mode of living with its associated transport, communications, technological and welfare infrastructures. Perhaps the only caveat to this hymn of progress is a concern about the environmental costs of goldmining, one of the acknowledged sources of ongoing Australian habits of squandering and destroying natural resources.

We can guess that such a generally upbeat story speaks to our deep and continuing needs and fantasies, which is not to say that the narrative of progress is without foundation. Many of these twenty microhistories do, however, serve to complicate or modify its tenor in important ways. David Goodman's opening plea for 'an edgier history of gold' points out that, even at the climax of the early gold mania in Victoria and New South Wales, there were

**Figure 3** J. A. Gilfillan, *Gaol and Commissioner's Station*, wood engraving. (Rex Nan Kivell Collection, National Library of Australia)

many who doubted the link between material and social progress that gold seemed to promise. Dreams of democratic progress brought an equal and opposite conservative reaction of fear and resistance. Existing cultural elites and religious groupings, in particular, were worried that gold would skew organic patterns of social development in favour of lawlessness and reckless ambition. Hard-won colonial achievements in agriculture, commerce and industry would give way before the lure of the instant buck. In the place of sound morality would flow the violence, disorder, crime, drunkenness and profligacy that had supposedly characterised the previous gold-rush society of California.

Australia's foremost urban historian, Graeme Davison, also subtly modifies what is often, and many ways rightly, seen as a modern gold-fuelled triumph: the astounding growth of 'Marvellous Melbourne'. Here, surely, was an example of gold's revolutionary energies, the creation in half a decade of a metropolitan global wonder, a city prophetic of the future, which exhibited modernisation in all of its facets. In a few short years from 1851 it had grown to become a hub of commerce, banking and industry, of multi-ethnic migration and global communication, serving its torrent of new inhabitants with sophisticated infrastructures of transport, housing, communication and transport. Though Davison does not refute Melbourne's glory story – indeed, he is its pre-eminent historian – he does show that such modernisation came at considerable social cost and in forms far removed from current assumptions.

Gold's sonic boom created an 'instant city', certainly, but by telescoping years of ordinary growth into frantic weeks it subjected the young urban complex to devastating social strains. Indeed, we might see the challenges created by such an instant accession of wealth as akin to those engendered by disasters such as war and famine rather than triumph. Davison himself offers the more sober comparison of the instant cities of the 'Asian tigers', where social provision often falls far behind explosive industrial, commercial and financial growth. Hank Nelson's chapter surveying the broad sweep of New Guinea gold history furnishes comparisons even closer to home, for here the social dislocations created by mushrooming gold towns are very much a thing of the present. Whether in colonial Melbourne or contemporary Bougainville, such instant cities not only create slums and canvas towns with all the attendant social problems but also give succour to potentially lethal urban developers. While it might be slightly harsh to put the Victorian gold-rush miner-engineer Edward Snell into that category, Tom Griffiths and Alan Platt here show that, having made a fortune as a mining-boom engineer, Snell beat a hasty retreat to England ahead of mounting colonial criticism that he had been more interested in gold profits than gradient curves. Arguably, such instant gold cities create psychic conditions more akin to the fragmented consciousness we associate with postmodernity than with modernity, echoing the claim of Bernard Smith that early twentieth-century Australian painting became postmodern before it was modern.



Another of Australian gold-history's most cherished progress tales involves the mineral's modernising political role, enabling the triumph of democratic individualism against the elitist aspirations of a wool-based squattocracy. This in turn is supposed to have prepared the ground for national political unification and independence in 1901. Appropriately in the Federal Centenary year, Paul Pickering's chapter focuses on gold's reputation as a national unifier by exploring the impact of the early rushes on the democratic republican aspirations of New South Wales. Initially, to the jubilation of radicals and the anguish of conservatives, the levelling magic of gold seemed destined to create the first independent Chartist state through the combined effects of increased British working-class migration, local wealth redistribution and the power of the ballot box. Thereafter, the Australian colonies under the leadership of New South Wales would, it was believed, develop both the economic capacity and the political incentive to achieve immediate republican independence.

But such golden dreams turned quickly to dross when the rush shifted to Victoria, generating a cocky challenger for economic and political primacy. Blainey has shown that this kind of pattern was to be replicated throughout the nineteenth and twentieth centuries as successive rushes broke out in an anticlockwise sequence around the country. Though such modernising processes might indirectly have aided long-term trends towards democratic unification, there seems little doubt that the short-term effects were negative. Everywhere that gold surfaced, it invigorated state rivalries and local chauvinisms, often stimulating dreams of separation, generating paralysing competitiveness and an increasing dependence on the Empire as neutral broker.

It is not clear, either, whether Australian gold seeking led to individualist democracy at an experiential level in the manner that is so often claimed of California's 'forty-niners'. True, many of the early alluvial mining methods used in the Australian colonies (including New Guinea) did enable individuals or small groups to obtain varying quantities of gold with relatively little capital cost. As a result, it is often argued that this 'democratic mineral', a term coined by the historian Weston Bate,[28] stimulated a spirit of individualist enterprise which militated against class and industrial conflict. Not necessarily, argues Charles Fahey, in a fifty-year survey of relationships between labour and capital in Bendigo that offers us a plausible alternative version of industrial modernisation in the Australian goldfields. Between 1861 and 1915 this archetypal Australian gold town generated, not digger individualism and class harmony, but a pugnacious legacy of employer recalcitrance, union militancy and state-based industrial arbitration.

Some indirect light on why Australian gold may have been a less democratic mineral than we think also emerges from Barry McGowan's meticulous analysis of the environmental impacts of alluvial mining in Victoria and New South Wales. For unlike quartz crushing and deep-lead tunnelling, whose destructive environmental effects have been well analysed by historians such as Blainey,[29] alluvial mining is widely assumed to have been both individualistic and environmentally benign. Images of the solitary digger armed only with gold pan or cradle have fashioned an enduring impression of the alluvial phase as a small-scale personal activity. But it was not always so. McGowan shows us that the technology and capital cost of alluvial extraction escalated rapidly to the point where huge hydraulic hoses were being used to wash away entire cliff faces and factory-like dredges were gouging vast areas of earth and forest before spewing their wastage back into already damaged and polluted watercourses. What gold began, erosion and overstocking completed, so that large portions of both states remain environmental disaster areas in the wake of gold seeking. If any consolation can be drawn from McGowan's judicious account, it is that throughout the nineteenth and early twentieth centuries state bodies, liberal newspapers, and environmental pressure groups mounted a vigorous and partly successful resistance to this destruction of our natural heritage. Inquiries, press campaigns and state regulation, as well as the healing long-term effects of bush regrowth, have done something to soften an ugly face of the gold modernisation story.

This bleaker note marks an appropriate point at which to shift our attention from the master theme of gold-based progress in Australia to an equally powerful contemporary narrative which associates gold with the emergence of modern Australian racism. This is the sorry, and at many levels undeniable, story that depicts gold as the catalyst for a new phase of legally institutionalised racial discrimination, primarily through the attempts of digger communities and colonial legislatures to exclude Asians (especially Chinese) from the goldfields. Gold is also seen to have extended and intensified the earlier pastoralist persecution of the country's indigenous peoples through increased land expropriation, physical violence and brutal interference in family, sexual and personal life.

With respect to anti-Chinese violence and discrimination, neither Ann Curthoys' history of European–Chinese encounters on the New South Wales goldfields nor Patrick Bertola's analysis of Western Australia's history of anti-Asian goldmining legislation reverses the sad tenor of this story. Indeed, both accounts show a disturbing commonality of discriminatory response across state boundaries and temporal periods. Curthoys tells mainly of direct and violent action against Chinese miners by inflamed gatherings of European diggers; Bertola reveals a long history of union and mine-worker pressure for discriminatory legislation in concert with the state legislature. While not neglecting the importance of prevalent racial theories or ingrained Anglo-Celtic prejudice, each chapter illustrates the value of giving due weight to the increasingly unfashionable conception of class interest in Australian race history. Both in New South Wales and Western Australia, the limited defence of Asian interests derived not from any more humane or enlightened attitudes to  race on the part of conservatives, but from the powerful incentive of providing cheap labour for the pastoral and mining industry. In turn, such clear expressions of capitalist self-interest gave a fillip to the paranoid and exclusionary tendencies of mine workers. Within each state, too, a clear diminution in the economic threat posed by Chinese immigration produced an immediate lessening, or abandonment, of discriminatory pressures. On the other hand, economic competitiveness seems to have played little part in regulating the prejudice experienced by European migrants on the isolated 'woodline' railway which fed the voracious needs of Western Australia's eastern goldfields for sixty years from the 1890s. Bill Bunbury's lively evocation of the deprivation and mistreatment of Italian and Slavic migrant workers makes sober reading. His story offers only the small and bitter consolation that Asians were not the only targets of goldminer prejudice and that the external goads of war-time chauvinism worsened the local situation.

Notable in all of these chapters is the active resistance and agency of the oppressed, whether Chinese goldminers, Japanese prospectors or Italian woodline workers. The same holds true within all of our chapters analysing widely different aspects of indigenous goldmining experience in Australia and New Guinea. Whether we are considering Kim McKenzie and Carol Cooper's gripping history of the Aboriginal boy, Oscar, stolen from the Palmer River goldfields area in the 1870s, or Ian Coates' interrogation of wry 1890s sketches of Aboriginal–European interactions on the Pilbara, or Derek Elias' interviews with traditional Warlpiri men ruminating on the changes wrought to their sacred places by goldmining in the Tanami, or Hank Nelson's evocation of tiny Kukukuku bowmen ambushing European prospectors on the Lakekamu River, nowhere do we encounter Indigenes as passive victims of gold. True, they were all too often the targets of victimisation: we meet stolen children and dispossessed elders; see depictions of lynch-mob reprisals and of Aboriginal women being bought for a tin of salmon; hear how an ancient desert people could be grievously damaged because of their accidental contiguity with a pointless, glittering mineral.

Yet we equally encounter examples of extraordinary sagacity, agile resourcefulness and wily adaptability in the face of great odds – occasions where European compulsions for gold are harnessed by shrewd Indigenes so as to defend or improve their circumstances. Encouraged to take up drawing by a well-disposed European, the Palmer River boy Oscar records his brutal life experiences, with no holds barred. The sympathetic cartoonist Hodgson's most compelling drawing captures the sardonic gaze of an Aboriginal woman as she watches her sometime European lover piously writing home to his wife. On the Tagula island off New Guinea, Sudest people take up mining for themselves, undercutting Australian diggers by their frugality and becoming so indispensable to European storekeepers that these same diggers dare not 'stop them from mining the gold on their own land'. Warlpiri peoples on the Tanami make sense of the European gold lust by likening the underground veins of gold to Dreaming tracks (*jukurrpa*) – except that in the white man's dreamings there is a pot of gold at the end of the rainbow serpent.



We also take away from these stories a strong sense that Australian history cannot be properly understood without the realisation that European and indigenous peoples have been and remain deeply entangled in the processes and consequences of gold extraction. There are in these stories some uncomfortable moments for modern liberal sensibilities. We must confront the certainty that the Native Police were widely used by European mining and pastoral interests in order to discipline and dispossess other indigenous peoples, and the possibility that Oscar, torn from family and place, found in these powerful, uniformed blacks a source of solace and racial pride. But there are also moments of harmony and exchange across cultures, even occasionally of mutual tenderness. Pastoral manager Gus Glissan goes to extraordinary lengths to encourage Oscar's artistry and to preserve his tiny anonymous sketchbook, so that it survives for us today. Hodgson's pencil catches a rough white digger clowning gently with a small Aboriginal boy. The titan of the Tanami, NormandyNFM goldmine, has evolved a process of consultation and compensation with Aboriginal miners and custodians that is far in advance of anything yet developed by governments. Could goldmining, the origins of so much oppression of indigenous peoples over the centuries, yet turn out to be a beacon for future reconciliation?

Our book also suggests the possibility that gold played a more positive role in the formation of European-Australian domestic and aesthetic sensibilities than is often assumed. It is easy to see how the feckless promiscuity, violence and drunkenness of the early alluvial diggings, so well captured in the diaries of English engineer Snell and French artist Antoine Fauchery, could seem to have instantiated the culture of rough masculinity that was already a legacy of Australian convictism. At one level, Margaret Anderson's poignant account of the experiences of artisanal wives Sarah Davenport, Lucy Hart and Emily Skinner provides detailed confirmation of this culture of misogyny. But Anderson's dazzling analysis of women's lives and experiences on the early Victorian goldfields also does much to question conventional assumptions. Above all, she stresses the need for us to appreciate the crucial differentiator of class: even within the tent life of the diggings, women like Elizabeth Ramsay-Laye were able to carve out ways of living that retained or re-created conventional markers of gentility and distinction. Moreover, irrespective of their class or education, women asserted an energetic and powerful presence within digger life – whether they were rearing or burying babies, operating general stores or running sly-grog shops, riding the hunt or, like the celebrated Lola Montez, flaunting unconventionality with a sizzling spider dance. All over the Australian diggings, we seem to confront the strength and resilience of the home-making reflex within digger life.

Susan Lawrence's fascinating uncovering of the domestic life of subsistence miners at Dolly's Creek in the mid-nineteenth century not only demonstrates the high incidence of married families in the settlement but also reveals their strenuous efforts to imbue canvas dwellings with the practical comforts and aesthetic pleasures of permanence. The latter imperative also lay behind the plots and pleasure gardens that Suzanne Hunt has painstakingly traced within the central Victorian goldfields towns of Ballarat, Bendigo and Castlemaine. Here, too, inscribed in flowers, trees, shrubs and borders, we find a full microcosm of the diversity of the gold experience. Whether to mask the scars of mining, feed the family, remind one of home, bring health and knowledge to the community, give pleasure to the public or attract tourists to the town, the goldfields gardening impulse ran deep. Here, surely, is a positive entry on the ledger to set alongside the sorry record of environmental destruction traced by Barry McGowan.



Neither should we overlook the powerful impulse towards respectability that so many of these gold histories reveal. Against Snell's depiction of the ceaseless firing of guns in the night or Fauchery's record of the presence of an infamous Parisian courtesan on the diggings, we need to set Lawrence's accounts of Dolly's Creek women who daily carried white clay from the diggings to ornament their fireplaces or who spent hard-won earnings on fragile porcelain rather than cheap, practical tinware. For all their raffish appearance and reputation, the early Victorian goldfields seem hardly to have loosened conventional codes of feminine (and even masculine) respectability; indeed, it is possible that the reputation of the diggings provoked an unusually zealous policing of the boundaries of propriety. We need to remember, too, that the ensemble of values associated with respectability had its pleasurable sides – even to modern sensibilities habituated to scorning the self-improving ethics of the nineteenth century. One of self-improvement's most appealing aspects was its attention to art, aesthetics and sensibility, none of which is conventionally associated with goldfields culture.

**Figure 4** Walter Mason, *A Scene on Sunday at Oakenville Creek Diggings, New South Wales*, wood engraving. (Rex Nan Kivell Collection, National Library of Australia)

But we should not assume that the diggings were cultural deserts, just because conditions on the goldfields were not conducive to the practice and exchange of high art. Both Dorothy Erickson in her study of West Australian goldfields jewellery, and Anita Callaway in her analysis of spectacle and performance on the Victorian and New South Wales goldfields, show that original and important forms of popular art evolved to suit the rough, transient conditions of mining life. Gold jewellery could be fashioned with simple tools by highly skilled itinerant European jewellers, and there were women on the diggings to wear these brooches or wealthy men to purchase them for sweethearts across the seas.

So, too, with the revolutionary new popular art form of the panorama: enterprising artists/showpeople like Edwin Stocqueler of London or Gus Peirce of Massachusetts evolved a spectacular, educative art that could be transported on dray or riverboat, and that anticipated the cinema in its capacity to generate illusion and grip its viewers. Theatrical spectacles of this sort could convey

PLATE 1 *Gold Mining at Ballarat*, in *Voyage Autour Du Monde*, Le Comte de Beauvoir, 1878. (National Library of Australia)

PLATE 2 *Australian Life: A Party of Diggers Joining a Native Encampment*, tinted lithograph.
(Rex Nan Kivell Collection, National Library of Australia)

PLATE 3 Charles Alphonse Doudiet, *Eureka Riot, 17th October*, watercolour, 1854. (Ballarat Fine Art Gallery)

PLATE 4 Charles Alphonse Doudiet, *Swearing Allegiances to the Southern Cross, December 1st 1854*, watercolour, 1854. (Ballarat Fine Art Gallery)

PLATE 1.1 Eugène von Guérard, *Aborigines Met on the Road to the Diggings*, 1854.
(Rex Nan Kivell Collection, National Library of Australia)

PLATE 1.2 Theodore King, *The First Parliamentary Election, Bendigo*, 1855. (Bendigo Art Gallery)

PLATE 2.1 T. E. Balcolme, *Mr E. H. Hargraves, the Gold Discoverer of Australia, Feb. 12th 1851, Returning the Salute of the Gold Miners on the 5th of the Ensuing May,* wood engraving, hand-coloured, in *Sydney Morning Herald,* 1851. (Rex Nan Kivell Collection, National Library of Australia)

Plate 2.2 Frederick William Hulme, *Ophir Diggings, Summer Hill Creeks*, lithograph, 1851. (National Library of Australia)

Plate 2.3 Marshall Claxton, *The Government Escort and Mail Containing the Gold, Arrival at the Treasury, Sydney*, pen and wash drawing, 1851. (National Library of Australia)

comical musical fantasy or didactic surveys of colonial progression towards nationhood and prosperity. Here it was difficult, as always, to draw a sharp distinction between popular commercial art and the kind of high art revered in European connoisseur circles.

Dianne Reilly shows us that Antoine Fauchery, friend of Murger, Baudelaire and de Nerval, and the original of Marcel in *La Bohème*, honed his skills as a writer on the Victorian diggings where he also developed the innovative appreciation of technology that later helped to make him one of Australia's most celebrated early portrait photographers. In Fauchery's writings and his documentary images, we find the goldfields as they appeared to an outsider eager to report his conclusions to a fascinated public beyond the seas. They remind us of the global stage on which the Australian gold rushes were set.

In the tiny goldfields brooch, inscribed with 'swag and swan', and the gigantic panoramas which celebrated the rise to nationhood and the conquest of a continent, we find nascent expressions of national pride and confidence – personal and public symbols of how deeply gold had become ingrained in the Australian experience.



## Notes

1 The population rose from 437,665 to 1,168,149, according to official figures. The growth in some of the individual colonies was even more remarkable: Victoria increased from 97,459 to 539,764 people in the period 1851–61, New South Wales from 197,265 to 357,362, and South Australia from 66,538 to 130,812. Tasmania, Western Australia and the future Queensland (which did not become a separate colony until 1859) also experienced substantial population growth but on a smaller scale, primarily due to the absence of a gold industry at this stage. During the period 1851–60, Victoria produced more than twenty million ounces of gold, and the Australian colonies together produced 39 per cent of the world's output. These figures are taken from M. Clark, *A Short History of Australia*, 2nd edn, 1969, Heinemann, London, p. 127, and from G. Davison, J. Hirst and S. Macintyre (eds), *The Oxford Companion to Australian History*, 1998, Oxford University Press, Melbourne, p. 284.

2 These figures are taken from H. McQueen, *A New Britannia*, 1970, Penguin, Victoria, p. 144.

3 For an account of this exchange, see E. Scott, *A Short History of Australia*, 6th edn, 1936, Oxford University Press, London, p. 210. The story comes from W. B. Clarke's *Researches in the Southern Goldfields of New South Wales*, 1860, Reading & Wellbank, Sydney.

4 G. Blainey takes this view in his general history of Australian mining, *The Rush That Never Ended*, 3rd edn, 1978, Melbourne University Press, Melbourne, p. 8.

5 Quoted in A. G. L. Shaw, *The Story of Australia*, 1954, Faber, London, p. 124.
6 J. Macarthur to E. Deas Thompson, 10 June 1851. Reproduced in G. R. Quaife, *Gold and Colonial Society, 1851–1870*, 1975, Cassell, Sydney, p. 18. J. Sherer, *The Gold-Finder of Australia: How He Went, How He Fared, and How He Made His Fortune*, 1853, Clarke & Beeton, London, p. 10.
7 E. H. Hargraves, *Australia and Its Gold Fields*, 1855, London. Extract in M. Clark (ed.), *Select Documents in Australian History*, 1955, Angus & Robertson, Sydney, p. 4.
8 J. D. Lang, sermon at Sofala, 5 October 1851. Reproduced in Quaife, *Gold and Colonial Society*, p. 18.
9 W. Kelly, *Life in Victoria, or Victoria in 1853 and Victoria in 1858*, 1859, Chapman & Hall, London, vol. 2, p. 43.
10 Cited in Clark, *Short History of Australia*, p. 116.
11 This reference is discussed in Scott, *Short History of Australia*, p. 213.
12 Cited in R. White, *Inventing Australia: Images and Identity 1688–1980*, 1981, Allen & Unwin, Sydney, p. 41.
13 Marx to Engels, 21 August 1852. Extract quoted in McQueen, *A New Britannia*, pp. 141–2.
14 For a discussion of Daley's 'Ballad of Eureka' and Lawson's 'Eureka', see Clark, *Short History of Australia*, pp. 122–3.
15 See W. Lane, *The Workingman's Paradise: An Australian Labour Novel*, 1892, Edwards, Dunlop & Co., Brisbane.
16 See V. Palmer, *The Legend of the Nineties*, 1954, Melbourne University Press, Melbourne, and R. Ward, *The Australian Legend*, 1958, Oxford University Press, Melbourne. For a discussion of their role in the cultural debates of the inter-war and post-war period, see White, *Inventing Australia*, pp. 153–4.
17 For a discussion of the diverse uses of the 'digger myth' in different political contexts, see White, *Inventing Australia*, p. 157.
18 W. K. Hancock, *Australia*, 1945, E. Benn Ltd, London, pp. 43–4.
19 G. Serle, *The Golden Age: A History of the Colony of Victoria 1851–1861*, 1963, Melbourne University Press, Melbourne, p. 381.
20 Serle, *The Golden Age*, p. 381.
21 See his introduction in M. Clark (ed.), *Select Documents in Australian History, vol. 2, 1851–1900*, 1955, Angus & Robertson, Sydney, p. 2.
22 Shaw, *Story of Australia*, p. 136.
23 G. Blainey, 'A Theory of Mineral Discovery: Australia in the Nineteenth-Century', *Economic History Review*, vol. 23, 1970, pp. 298–313.
24 Blainey, *Rush That Never Ended*, p. 62.
25 McQueen, *A New Britannia*, p. 142.
26 S. Schama, *Citizens: A Chronicle of the French Revolution*, 1989, Knopf, New York.
27 Claude Lévi-Strauss.
28 W. Bate, *Victorian Gold Rushes*, 1988, McPhee Gribble/Penguin, Victoria, p. 61.
29 Blainey, *Rush That Never Ended*.

# Part I Gold and Modernity

THE IMPACT OF THE GOLD RUSHES ON THE DEVELOPMENT OF THE AUSTRALIAN COLONIES

# Making an Edgier History of Gold

Chapter 1

*David Goodman*

> The discovery of the Victorian Gold Fields has converted a remote dependency into a country of world-wide fame; it has attracted a population extraordinary in number, with unprecedented rapidity; it has enhanced the value of property to an enormous extent; it has made this the richest country in the world; and, in less than three years, it has done for this Colony the work of an age and made its impulses felt in the most distant regions of the earth.[1]



This was the considered judgement of a Select Committee of the Victorian Legislative Council in 1854, concluding a report with a summary of the effects of the discovery of gold upon the colony. The committee spoke the language of political economy – there is a sense of a global economy, of economic stimuli being felt and responded to by people around the world, and of a general satisfaction with the miraculous way in which gold had fostered progress and prosperity.

The *Economist* expressed such a view from London: 'There is happily but one opinion as to the ultimate effect of the discovery of great quantities of gold, that, like all things not of human contrivance, it cannot fail to promote the civilisation of the species, and contribute to the progress and power of man'.[2] The *Economist* had even maintained that the consequent 'great increase in the business of the world' might lead to an increase in the price of gold.[3]

Historians have also tended to look back on the gold rushes with general satisfaction. It is hard not to be impressed by the extraordinary figures of population increase, of gold exported, of cities and civic cultures being built. The story of the sleepy pastoral colony transformed into a booming world centre is an attractive one. Gold created cities, industries, a nation. 'Before the gold era', wrote G. V. Portus in 1933, 'Australia was regarded in the main as a kind of outlandish suburb of Britain at best; at worst, as a place of exile for those Britons who had to live there. After that time there is apparent in Australia the consciousness of a distinct national identity.'[4]

It is true that this fundamental story of progress is usually accompanied by stories about temporary dislocation, filled with colourful characters who waste their wealth, with comic stories of a world turned upside down, in which the distinctions of social class were muddled and the poor man at last had his day. But the outcome is never in doubt, the political economy of gold – the sure transition from the discovery of enormous golden wealth to the growth of the new metropolis – seems unquestionable.

We would make a mistake if we wrote our histories of gold only in that voice. People at the time did not always know or believe that progress would be the outcome of the gold discoveries. Writing hopes and fears into the history of the gold-rush era will more accurately convey the uncertainty and anxiety which the mid-nineteenth-century gold rushes provoked in Australia.

The language of political economy was only one of the ways in which people thought about gold in the 1850s. There were many who were yet to be persuaded that the market would provide the only regulation and discipline necessary, that a single-crop gold economy would work out satisfactorily in the long run, or that the society created around such an economy would be worth living in. For many contemporaries, gold posed problems of government. How could any kind of order be brought to a society in which each was preoccupied only with his own rapid enrichment? This was not a freak problem in the 1850s – the debate between laissez-faire and government was everywhere in the intellectual life of the time.



We tend to forget how concerned most contemporaries were about gold, and we have underestimated the extent to which most saw the events of the gold rush in anxious and moral terms. Frances Perry, wife of the Anglican Bishop of Melbourne, wrote in a letter of 'the subject of gold – that hateful, fearful, baneful subject. All wise and good men look with great alarm on this new feature in our history.'[5] Both those who viewed the disruptions of gold first-hand, and those who speculated on the situation from afar, often concluded that there was something troubling about the prospect that gold-rush society held out to them. When a public meeting was held in June 1851 to offer a reward to anyone who could discover gold in Victoria, merchant Asher Hyman Hart announced that he believed gold to be 'one of the greatest curses with which they could be afflicted'. At the same meeting, 'A Working Man' declared that the working classes 'should not leave their work and run away after gold finding' – 'health was always gold to the working man', he argued, and the authorities should improve conditions in the city rather than encourage gold seeking away from it.[6] Such dissenters have not figured prominently in our histories.

Even at the level of political economy itself, there were some prominent dissenters. The Romantic essayist Thomas de Quincey, for example, thought it quite implausible that the demand for gold would increase to match the enormous new supplies: 'As to fancying that the Australian gold will continue to force a market for itself, you cannot seriously suppose that a man, who never thought of buying a gold watch or other trinket when such articles were made of Uralian gold, will suddenly conceive a fancy for such an article, simply because the gold is raised in an English colony'.[7] It was by no means self-evident that prosperity would be the result of an increase in the supply of gold.

The concern about a concentration on gold production was debated most sharply around the question of food. Melbourne lawyer Thomas a'Beckett argued forcefully in 1851 that gold was merely the sign of wealth: 'Wealth consists in the abundance of things and not in the accumulation of the signs by which they are represented'.[8] A'Beckett, like many other critics of the time,

**Figure 1.1** *Great Meeting of Gold Diggers Dec. 15th 1851,* drawn by D. Tulloch, engraved by Thomas Ham, in *The Gold Diggings of Victoria, In Five Views Taken on the Spot by D. Tulloch,* Thomas Ham, Melbourne, 1854. (La Trobe Picture Collection, State Library of Victoria)

drew from this premise the conclusion that the pursuit of gold might impoverish rather than enrich a country:

> The sudden discovery of Gold, scattered in bewildering profusion, over the face of our adopted country, has caused a simultaneous rush, by all but a small minority, from their ordinary occupations and employments, which, if long continued, must render the Gold acquired all but valueless to its finders, and injurious in proportion to its extent to those who have made no effort to obtain a share of the treasure.[9]

A'Beckett contended that only cultivation of the soil and the production of food rendered a people genuinely prosperous. That theme pervades the writings of 1850s Victoria. 'The danger is', warned historian George Rusden, 'lest the absorbing pursuit of gold should dull the finer energies of our adventurous colonists; lest their gathering together on the soil of Victoria should be but for a time – lest the soil be disturbed with no purpose to plant, and to reveal those treasures which are far richer than gold, and which will enable it to maintain happy millions in plenty'.[10] So many commented on the richness of the land, 'this land of Australia, a land flowing if not with milk and honey, unquestionably with milk, beef, mutton, flour, and fruits of every description'.[11] In this context gold digging seemed a destructive activity, one which plundered rather than nurtured the land. 'The diggers', observed the once-radical journalist William Howitt, 'seem to have two especial propensities, those of firing guns and felling trees'. 'Every tree is felled', he wrote of the diggings, 'every feature of Nature is annihilated'.[12]



The concern about the concentration on gold digging rather than food production was widespread. A Mr Hellicar warned a Special Committee of the Melbourne Chamber of Commerce that, if the large importations of food were to cease, 'they would be left, with their ships laden with gold, to die of starvation'.[13] No country, Hellicar insisted, should be dependent on another for its subsistence.[14] The committee concluded that the effect of the gold discoveries and the influx of population had been to 'diminish the resources of the Colonists and impoverish the Colony generally, by causing a yearly importation of bread-stuffs and vegetable productions to the value of two millions, all of which we have to pay to the neighbouring colonies, to Chile, the United States, and the Mother Country'.[15] To many, dependence on imported food was a sign of national vulnerability. Only the surest believers in political economy could be confident that the free market would supply all their wants in the best possible way. The area under cultivation in Victoria fell by 40 per cent between 1851 and 1853, and the importation of food rose to very high levels.

There was also something very troubling to many about the unreflective nature of gold seeking. As the Anglican Dean of Melbourne observed, 'It is not easy to imagine emigration conducted in more reckless thoughtlessness'.[16] Life in gold-rush society seemed too fast and too obsessively focused on one object. William Howitt's son Alfred wrote of the diggers as 'wild looking men', and observed that 'no one here seems to care about any other pace than a hard gallop over break neck ground'.[17] A letter to the editor of the *Argus* in 1852 complained that the '*sole* object of life amongst us is to get gold': 'Should Gold, Gold, Gold, be our only desire? Should gold absorb everything else? Was man made merely to acquire glittering metal – *filthy lucre.* Is there no higher object, no more noble aim?'[18] Gold digging seemed an addiction, provoking a level of excitement which meant that men lost any sense of other goals or values. Chief Justice Sir William a'Beckett wrote:

> For e'en hatred can keep cold
> Sooner than stop work for Gold,
> Digging for it, as if God
> Had hid salvation 'neath the sod![19]

There was a great deal of drinking on the goldfields. 'Many of the diggers have reduced themselves to mere savages from the effects of drink', wrote Elizabeth Ramsay-Laye, 'to which they are tempted by disappointment to resort'.[20] Samuel Lazarus met a man whose 20-year-old brother had died from the effects of drunkenness at Bendigo: 'It is painful to contemplate the horrible havoc which drunkenness makes on the diggings'.[21] Despite the obviously self-serving role of the frequent comments about Aboriginal drunkenness (blaming the dispossessed for the symptoms of their dispossession), it is clear that the remaining Aboriginal people around the goldfields did suffer badly from the effects of alcohol. Excitement, drunkenness, the frenzied pursuit of individual wealth and the neglect of all else – these were some of the dominant images of the goldfields at the time.



Gold ominously seemed to be removing the link between work and wealth. A society in which wealth threatened to become completely detached from work – and in which hard work no longer necessarily met any reward – was profoundly disturbing to the Victorian work ethic. Sir William a'Beckett wrote that 'the cause of our evils is not in the gold, but in the fatal facility of its acquirement'.[22] Robert Ramsay observed:

> Land of the gold where wealth is spread
> Around our feet wher'ver we tread
> Where gold is scarce to labor wed...[23]

Repeatedly, gold digging was compared with a lottery. And no one could seriously argue that a society in which the central principle of wealth distribution was a lottery was likely to be peaceable, industrious, progressive – any of the things valued in mid-nineteenth-century social thought.

The Australian (and Californian) gold rushes were major literary events in Britain. As early as August 1852, the *Athenaeum* commented archly on 'the number of strange, racy, and crazy books devoted to the New World, and the new miracle, which appear month by month and week by week'.[24] The social reversals of gold were endlessly interesting to those at home in Britain. Travellers' books made much of this. Lucky diggers smoking or eating £10 notes, dressing themselves or their new wives in clothing well above their station in life, became staple figures in the literature. But the reverse was also undeniably a part of the gold-rush experience. Dr John Milton told a Legislative Council committee in 1854:

> We can scarcely tell who are below the middle classes. The diggers appear in labourers' garb, but many of them have a great deal of money. I have known clergymen, medical men, gentlemen, gentlemen's sons, men of education, in the City Court cells; many of them in rags.[25]

Few saw through the comedy or tragedy of this to the possibilities it might hold for the future. As the Anglican Dean of Melbourne put it, 'The extraordinary and important fact that the lowest offices of our society are filled by persons of education, intelligence and refinement, is utterly overlooked'.[26] The popular memory of the gold rushes emphasises the comic disruptions of gold, but with the happy reassurance or grim reconfirmation of hindsight that the events *were* comedy rather than tragedy – that the norms of social and moral life did reassert themselves, that the scarcity of wealth again became the great regulator of life, in Australia as elsewhere. For communities unaccustomed to accepting the efficacy of market forces as social regulators, those (reassuring or disappointing) outcomes were by no means clearly visible.



Many found that the reality of life on the diggings did not at all conform to the narratives they had read back in Britain. One gold-rush immigrant, George Ross Reid, wrote home:

> I think I may as well inform you that very few that I am acquainted with have seen or heard of the strange things you read of in the London papers. There is no Diggers ordering butts of Port and requesting passers by to drink their health; no knocking men's eyes in or out with nuggets; no giving half crowns to have their teeth picked for them, etc, that I can hear, and I can find no one knowing anything of the young *Lady* who dressed in the boy's rig, the case made so much of in the London papers.[27]

The theme echoes through letters. 'Indeed, you must believe nothing you hear of it at home', a young Scot wrote home from Forest Creek.[28]

On the other hand, correspondence published in British local newspapers gave those at home a chance to read about how people from their community were faring at the diggings. The *Brighton Gazette*, for example, carried numerous letters from Brightonians at the Victorian diggings – both encouraging and discouraging accounts. 'Tell Fred not to think of coming out yet till he hears more from us', two Brighton men wrote to their father from

Melbourne.[29] We need more research on how information about gold spread in Britain and Europe and China at a local level, but the information carried through personal networks in this way was obviously more credible than the published experiences of unknown authors. At least one well-known published account of Australian adventures – John Sherer's *The Gold-Finder of Australia: How He Went, How He Fared, and How He Made His Fortune* (1853) – has been shown to have been written in London.[30]

Almost from the beginning there were observers looking for signs of the emergence of normality. Governor La Trobe wrote to London in 1852 to calm fears that the social order might entirely collapse at the diggings:

> In such a crowd, one half utter strangers to the other and to the colony, met together in a wild tract of broken forested country, full of secluded hollows, honeycombed with hundreds or thousands of ready-made graves, under such strong inducements to cupidity, disorder and crime, the imagination is free and unrestrained to picture the extent to which crime may, however improbable, prevail in secret without the probability of discovery or the chance of detection.

But, La Trobe assured Earl Grey, 'no such general disorder and rejection of law  and constituted authority . . . has ever been observable'.[31] The survival of social order at all seemed remarkable to many contemporaries.

The London *Morning Chronicle* reported in 1854 that society in Australia was 'beginning to assume a healthier tone and aspect':

> Thanks to the love of order inherent in the Anglo-Saxon character, and to the prompt vigilance both of the home and local Governments, there is reason to hope that the disturbing effects of the gold discoveries will prove less pernicious and less protracted than had been apprehended, and that the greater part of that vast country will, at no distant day, recover from a state of feverish excitement alike perplexing to the moralist and the statesman.

It was not of course, the paper made clear, that Australia had attained 'an English, or even a European, standard of civilization', since 'social refinement is the last thing to be looked for in a newly colonised country'. To the radicals at the *Morning Chronicle*, though, there was something plainly disturbing about the gold discoveries, a sense that a counterbalancing force would need to be found to the acquisitive desires which gold had unleashed, and which had led so many men to seek their fortunes in Australia.[32]

To most, that force was government. Despite the emphasis of much of the historiography on the inadequacy of La Trobe's response to gold, the extent of government of the goldfields population was actually remarkable, given the rapidity of the arrival of the new population and the way diggers 'rushed' unpredictably across the countryside. Despite the gold-rush conditions, people appealed to government for judgement on all kinds of things. One man wrote to La Trobe, saying that for some years he had been trying to obtain a 'rational and satisfactory answer' to the question of why the sun and moon appear largest when on the horizon – he appealed to 'Your Excellency's mature judgement and Experience in such matters'.[33]

A system of goldfields commissioners was established, each reporting to the Chief Goldfields Commissioner. Their weekly reports were then sent to La Trobe, who read them carefully. He expected information at his desk which would allow governmental oversight down to the level of individuals. The Acting Resident Commissioner at Castlemaine wrote in his weekly report to the Chief Commissioner that 'a young man named Thomas Floyd was stopped in the Road between Saw Pitt Gully and Friar's Creek, on the 14th instant, by two armed men, one of whom discharged a Pistol at the youth, and wounded him severely in the face with small shot'.[34] La Trobe read such reports from all the goldfields and annotated them with scrawled minutes down the side, usually on matters of procedure, sometimes urging action with terse directions: 'This must be looked to', and 'Enquire what the facts may be'.[35]

The information which flowed to the top of government also provided important statistical information. The commissioners' weekly reports contained figures on the goldfields population considered as a whole – their productivity in terms of ounces of gold and numbers of licences issued, their health, and usually a comment that the population had remained 'orderly and well conducted'. The statistics maintained in 1850s Victoria were remarkable – in part due to the work of W. H. Archer, who was Victorian Registrar-General from 1859 to 1874. Archer saw in gold-rush Victoria a chance to advance statistical science: 'The novelty of the condition of the population with respect to climate, age, proportion of the sexes, and general prosperity, will develop laws that could not come under scientific observation in any other part of the world'.[36]



Comparisons with California have often led to a stress on the regulated nature of the Australian gold rushes. Government reached out across the colony to license and survey. Goldfields commissioners and their camps established bases of government presence. Local magistrates took on social responsibilities such as some provision of welfare for the poor.[37]

Two of the most important themes in the history of the mid-nineteenth-century gold rushes, then, are the fears and anxieties which gold provoked about disorder and disorganisation, and the ambitions of government for the maintenance of order and the establishment of some kind of social contract among wealth seekers. These, in turn, were provoked and challenged by the spectre of democracy within the colonies, and led to debates concerning the character of the populace and the nature of its attachments to place and position.

Many observers stressed the manly independence which the occupation of gold prospecting seemed to offer to those working-class diggers who had escaped poverty in the Old World. On the Victorian goldfields, the English labourer

> is no longer the abject, cringing half-starved pauper, whose air and carriage denote his broken spirit and humiliation . . . he is now bold, independent, and

> fully conscious of the importance of his position: his step is firm, his bearing manly and bold, his eye bright, his countenance cheerful, every movement of the man proves that his valuable services to the community are fully known to himself, and duly appreciated by others, and that he may now be justly called his country's pride, a 'bold peasant' . . .[38]

Contemporaries commonly linked this much remarked-upon masculine independence at the diggings with the political movements which arose there. La Trobe put it more analytically than most, despite being out of sympathy with the changes he described:

> We live in times of restlessness and desire for political change, and there is nothing in the character and prospects of the mass of the Colonists of these appendages of the Crown which may induce us to believe that they can remain unaffected by the spirit of the age: or that the growing sense of importance and independence arising from unexampled prosperity, emancipation from old ties and obligations, and powers of self-support, and self-government, should not influence the multitude.[39]



The British government also recognised that new emigrants to Australia would expect self-government and some democratic rights. There were no doubt many, wrote the Duke of Newcastle to the governors of New South Wales and Victoria, who would maintain that the 'peculiar character of the population' of the colonies, its 'as yet slight connexion with the soil, and its addiction to the engrossing pursuits of material progress', rendered it unfit for self-government:

> But even if that opinion were well founded . . . all will agree as to the extreme difficulty of withholding political privileges from bodies of men to whom the maxims now prevalent in British domestic policy afford so strong a right to claim them and of keeping our fellow subjects in Australia on a different political footing from those to whom those rights have been fully conceded in America.[40]

The diggers would develop a political consciousness of their shared interests and their common exclusion from full citizenship in Victoria (Plate 1.2).

Historians have long recognised the advance towards white male democracy as one of the most important themes of the mid-nineteenth-century gold rushes, but it would be good to know more about how democratic politics came into being. How did the language of rights spread through the diggings? How were the great democratic instruments of mass meetings and petitions brought into use? How did an infectious and gradually formed sense of the possibilities of life in this new world inform political movements? There is something remarkable in that process, something which matters more than all the debate about the factual details of the events at the Eureka Stockade.

Some of the irritation at the licence fee system arose from its operating as a tax on successful and unsuccessful alike. Edward Kemp, a Ballarat digger, told an 1853 Legislative Council inquiry into the goldfields that he looked upon the licence fee as a kind of robbery: 'Labor is the poor man's capital, and if it were expended in any other occupation he would at all events get something for it'.[41]

I am struck always by how lyrically ordinary men spoke about their future in the colonies – and almost always in relation to the land which they saw awaiting their ownership and cultivation. Digger Thomas Kennedy proclaimed to a Ballarat meeting in 1857: 'Let the lands be opened in order that we may settle down in peace and quiet, and make the land flow with wine and milk, and become lords of the soil'.[42] At Ararat, a Mr Wood told a public meeting of four or five thousand: 'We want our equal rights and must assert them, our right to the soil and the land we tread on; God made the land, and the land for the people, which was their just inheritance'.[43] In the longer term, taking possession of the land excited the imaginations and fantasies of the men who rushed to seek gold even more than the gold itself. The gold was a means to an end. It was only in late nineteenth-century nostalgia for the 'roaring days' that the passing of gold-rush conditions became a matter for lament. The gold-rush diggers were no more prone than were the governing classes to imagine gold seeking as the permanent basis for a society.



In the democratic movements of the 1850s, the diggers sought both the right to vote and access to ownership of a piece of the land occupied by the squatters. The diggers smarted under claims by the squatters that they were only 'wandering vagabonds', with no genuine claims to the land.[44] In these debates, the memory of one dispossession informed thoughts about another:

> The same right by which these squatters first grasped the soil from the thriftless and improvident aborigines, now in the present posture of affairs, holds equally good against themselves . . . The claims of the natives were as valid against the squatters, as those of the squatters against the superior demands of an agricultural population. But, as the original possessors were obliged to yield, so must in their turn those who supplanted them.[45]

The democratic politics of the 1850s thus placed the issue of right to the land at the very centre of public discussion.

Gold, coming so soon after the rapid pastoral invasion of the Victorian countryside, was a cruel blow to the Aboriginal people of the region. Nothing but a gold rush could have sent so many of the invaders so rapidly up-country (Plate 1.1). Few of those who came could have been unaware of the fate of the indigenous people, and in the intellectual circles of the colony it was a matter for serious discussion. Pamphleteer R. L. Milne wrote that 'the aboriginal inhabitants, with a title from Heaven . . . are plundered, dispossessed and slain, forming a chapter in British history that will not bear inspection'.[46] 'Tell me not', he warned, 'that the greater part of you are innocent, merely because ye have not done the act with your own hands. Accessories, by social connexion, and participation of the spoil, are, in the eye of the law, and most assuredly in the eye of Heaven, murderers, as well as the actual perpetrators of the deed.'[47]

Daniel Bunce published in 1851 his *Language of the Aborigines of the Colony of Victoria, and Other Australian Districts*, which offered 'Specimens of Dialogue, as a Guide to Protectors of Aborigines and others, desirous to assist in ameliorating their condition'.[48] He had a strong sense, in June 1851, of the imminent disappearance of 'a people who are evidently fast sharing the fate of their savage prototypes of the adjoining islands'.[49] In his introduction to the 1859 second edition, eight years into the gold era, Bunce again called attention to the 'present helpless condition' of the Victorian Aboriginal people, and attributed that state to an over-indulgence in 'Ardent Spirits'. The original inhabitants, he argued, needed 'a large area of country, and if possible an island' where they would be removed from the temptations of alcohol and might 'be induced to fall into better habits of industry by cultivating patches of land'.[50]

Sympathetic commentators in the 1850s thus often mixed humanitarian concern with a diagnosis which blamed the Aboriginal people for their own displacement. They argued that, if only the Aborigines could become a little more sober and industrious, they would succeed in colonial society – quietly cultivating patches of the land which had once been all theirs. These tensions marked much of the discussion of the plight of the indigenous population during the gold era in Victoria. The 1858 Select Committee of the Victorian Legislative Council, set up to inquire into the condition of the Aborigines of the colony, heard some bleak testimony, and concluded that a 'great injustice has been perpetrated upon the Aborigines'.[51] From the goldfields came reports that there were few Aboriginal people left in those districts. The Police Magistrate at Castlemaine wrote: 'There exists no regular tribe in the Castlemaine districts. Some few natives hang about the diggings and are employed by dairymen and slaughtermen, and occasionally a straggler comes in from beyond the Loddon.'[52] Committee member C. W. Sherard reported from Ballarat: 'The mortality rate has been very great the last ten years more especially since the discovery of gold'.[53] A former Protector of Aborigines, Edward Stone Parker, testified: 'The remnants of the native tribes on the Loddon . . . pick up a scanty subsistence on the goldfields . . . Their present condition is most disastrous.'[54] The Select Committee, casting about for a solution, repeatedly sought opinions from witnesses as to the merits of taking away the children of the remaining Aboriginal population.

Until we see the vigorous, masculine, democratic politics of the 1850s gold-rush period, with its insistent calls for the land to be distributed among 'the people', as part of the same story as the taking of Aboriginal land and the breaking up of Aboriginal families and communities – until we see, that is, that the 'black armband' history of Australia and the history of democratic progress in Australia tell the same story from different perspectives – we will not have fully acknowledged the conflict of historical understandings which reconciliation aspires to resolve.

The gold rushes have been mythical events in Australian history – uproarious, larger than life – but they have been tamed in historians' prose into a comforting story of gold as a maker of cities and nations. We need more than a summary account of winners and losers (it was better to be a shopkeeper than a miner, better to be English than Chinese). We need to recover a sense of the gold rushes as dangerous, edgy events with unpredictable outcomes.

## Notes



1 Victoria, *Report from the Select Committee of the Legislative Council on the Claims for the Discovery of Gold in Victoria*, 1854, p. 6.
2 *Economist*, vol. 10, no. 461, 26 June 1852, p. 712.
3 *Economist*, vol. 10, no. 450, 10 April 1852, p. 393.
4 G. V. Portus, 'The Gold Discoveries 1850–1860', in E. Scott (ed.), *Australia: Cambridge History of the British Empire*, 1933, Cambridge University Press, Cambridge, vol. vii, pt. 1, pp. 270–1.
5 A. de Q. Robin (ed.), *Australian Sketches: The Journals and Letters of Frances Perry*, 1983, Queensberry Hill Press, Melbourne, p. 155.
6 *Argus*, 11 June 1851.
7 T. de Quincey, 'Final Catastrophe of the Gold Digging Mania', in *Littell's Living Age*, xxxvi, 1853, p. 599.
8 T. a'Beckett, *The Gold and the Government*, 1851, Daniel Harrison, Melbourne, p. 7.
9 A'Beckett, *The Gold and the Government*, p. 3.
10 G. Rusden, *Gathering Together for the Good of Work and Learning: A Lecture on the Occasion of Opening the Exhibition Building in Melbourne*, 1857, George Robertson, Melbourne, p. 22.
11 'Communications with the Port', *Illustrated Australian Magazine*, vol. 4, no. 19, January 1853, p. 28.
12 W. Howitt, *Land, Labour and Gold, or, Two Years in Victoria: With Visits to Sydney and Van Diemen's Land*, 1855, Longman, Brown, Green & Longmans, London, pp. 176, 252.
13 *Proceedings of the Melbourne Chamber of Commerce and the Special Committee Appointed to Consider and Report on the Best Means of Promoting Agriculture and Settling the Waste Lands of the Colony*, 1855, Goodhugh & Trembath, Melbourne, p. 7.
14 *Proceedings of Melbourne Chamber of Commerce*, p. 27.
15 *Proceedings of Melbourne Chamber of Commerce*, p. 12.

16 HBM, 'Canvas Town', *Church of England Messenger*, vol. 4, 1853, p. 131.
17 A. Howitt letter to Mary Howitt, 30 September 1852. Alfred Howitt papers, La Trobe Australian Manuscripts Collection, State Library of Victoria, MS 9356.
18 *Argus*, 17 April 1852.
19 W. a'Beckett, 'The First Gold Rush at Melbourne', in *The Earl's Choice and Other Poems*, 1863, Smith, Elder, London, p. 178.
20 E. Ramsay-Laye, *Social Life and Manners in Australia*, 1861, Longman, Green, Longman & Roberts, London, p. 37.
21 S. Lazarus, diary 1853–55, La Trobe Australian Manuscripts Collection, State Library of Victoria, MS 11484.
22 Colonus [W. a'Beckett], *Does the Discovery of Gold in Victoria, Viewed in Relation to Its Moral and Social Effects, as Hitherto Developed, Deserve to be Considered a National Blessing or a National Curse?*, 1852, Benjamin Lucas, Melbourne, p. 36.
23 R. Ramsay, 'Victoria's Anthem', in 'Poems Written by Robert Ramsay 1853–9', La Trobe Australian Manuscripts Collection, State Library of Victoria, MS 11021.
24 *Athenaeum*, 21 August 1852, p. 891.
25 *Report from the Select Committee of the Legislative Council on Intemperance*, 1854, p. 26.
26 HBM, 'Canvas Town', p. 132.



27 G. R. Reid letter, 1853, La Trobe Australian Manuscripts Collection, State Library of Victoria, MS 12430.
28 David Matthew Watson letter from Forest Creek, 18 January 1853, La Trobe Australian Manuscripts Collection, State Library of Victoria, MS 8297.
29 Henry and Herbert Humphrey letter, 21 November 1852, in *Brighton Gazette*, 24 March 1853. I am indebted to Dr Joyce Collins of Brighton for this reference.
30 See J. Molony and T. J. McKenna, 'All That Glisters', *Labour History*, 32, May 1977, pp. 33–45.
31 Despatch from La Trobe to Earl Grey, Secretary of State for the Colonies, 2 March 1852, in *Further Papers Relative to the Recent Discovery of Gold in Australia*, 1853, London, p. 170.
32 Reprinted in *Argus* (Melbourne), 7 January 1854.
33 Peter Robertson letter to La Trobe, 5 July 1853, C. J. La Trobe correspondence, La Trobe Australian Manuscripts Collection, State Library of Victoria.
34 Acting Resident Gold Commissioner, Castlemaine, to Chief Goldfields Commissioner, 17 January 1853, Victorian PRO VPRS 1189/P, Box 84.
35 Gold Commissioner, Castlemaine, to Colonial Secretary, 3 March 1853, Victorian PRO VPRS 1189/ P, Box 84.
36 W. H. Archer letter to editor of Melbourne *Morning Herald*, 10 January 1853. Cited in M. Beever, 'W. H. Archer, Civil Servant', MA thesis, University of Melbourne, 1971, p. 29.
37 C. Twomey, ' "Without Natural Protectors": Histories of Deserted and Destitute Colonial Women in Victoria 1850–1865', PhD thesis, University of Melbourne, 1996, ch. 6.
38 W. H. Hall, *Practical Experiences at the Diggings of the Gold Fields of Victoria*, 1852, Effingham Wilson, London, p. 52.

39 C. J. La Trobe letter to Sir John Pakington, 1 June 1853, La Trobe General Correspondence 630/11 Letter 501, La Trobe Library.
40 Copy of despatch from Secretary of State, the Duke of Newcastle, to Governor Charles Fitzroy, 4 August 1853, in Confidential Despatches from the Secretary of State 1852–60, vol. 1, part 1, Victorian PRO, VPRS 1090–1.
41 Victoria, *Report of the Select Committee of the Legislative Council on the Gold Fields*, 1853, p. 66.
42 *Ballarat Star*, 15 June 1857.
43 *Mount Ararat Advertiser*, 28 May 1858.
44 See, for example, editorial in *Ballarat Star*, 24 May 1858.
45 Editorial, *Ballarat Star*, 10 June 1857.
46 Aristides [R. L.Milne], *A Glance at the Forerunners of the Crisis*, Morning Herald Office, Melbourne, 1853, p. 15.
47 Aristides, *A Glance at the Forerunners*, Appendix, p. 14.
48 Advertised in the *Age*, 14 April 1852.
49 D. Bunce, *Language of the Aborigines of the Colony of Victoria and Other Australian Districts*, 2nd edn, 1859, Thomas Brown, Geelong, p. iii.
50 Bunce, *Language of the Aborigines*, p. xii.

51 Report of the Select Committee on the Aborigines, in *Votes and Proceedings of the Victorian Legislative Council 1858/9*, p. iv.
52 Report of Select Committee on Aborigines, p. 26.
53 Report of Select Committee on Aborigines, p. 28.
54 Report of Select Committee on Aborigines, p. 20.

# 'The Finger of God'

hapter 2

## GOLD'S IMPACT ON NEW SOUTH WALES

*Paul A. Pickering*

The tenth of May 1852 was a public holiday in New South Wales. It was a day when some Sydneysiders enjoyed a 'pleasure jaunt to Waverley, or to the South Head'; went to Botany Bay to see the elephant that was the highlight of the zoological gardens there; or paid five shillings to witness M. and M. Prieste 'SKATE THE LIQUID WAVES' off Circular Quay. Others took part in events that were intended to evoke the particular spirit of the holiday. At Malcom's Royal Australian Circus some witnessed Master Ord perform an 'Act of Horsemanship as the GOLD DIGGER in miniature', and, for the more active, there was a prospecting expedition to the Bark Huts.[1] The occasion was the commemoration of the first anniversary of the announcement that gold had been discovered by Edward Hargraves outside of Guyong, near Bathurst (Plate 2.1). Although it was not the first public holiday in the colony, it was notable because it was proclaimed by popular demand without the formal approval of the Governor-General, Sir Charles Fitzroy. 'The good people of Sydney', wrote Henry Parkes in his liberal newspaper, the *Empire*, 'have determined to have a holiday on the strength of the colony's golden prosperity'.[2]



Stiffly, Parkes went on to regret that more formal celebrations of the 'gala day' had not been organised, and some correspondents quibbled about the date, believing that 12 February, the actual day of Hargraves' discovery, was more appropriate. Sounding a similar note of caution the leading newspaper in the colony, the *Sydney Morning Herald*, felt that 'five or six years' would have to elapse before 'the actual working of the gold discovery is seen', but few doubted that there was something to celebrate.[3] As an 'Old Colonist' predicted, 'the importance of the discovery' would be even 'more generally appreciated by the next anniversary'.[4]

The prevailing mood was captured by the colony's leading politician, William Charles Wentworth. The discovery of gold, he observed in August

1851, 'must in a very few years precipitate us from a colony into a nation'. Quoted by friend and foe alike from the day he uttered the words, Wentworth's sentiment has gone on to become one of the clichés that provide most Australians with an understanding of their own past. By the time he composed his contribution to the *Cambridge History of the British Empire* in the early 1930s, G. V. Portus could even render Wentworth's 'rotund' prediction in the past tense: 'must precipitate' became 'had precipitated'.[5] Wentworth had seemingly been vindicated by history and it might have been really so had the Victorian 'golden triangle' been in Wentworth's home colony of New South Wales. Even as Sydneysiders were gazing amazedly at the elephant, the balance of the gold rush was shifting inexorably in favour of Victoria. Although few knew it at the time, by 1853 there was little left for Sydneysiders to celebrate.

Within a few short years, the young colony of Victoria, so recently separated from New South Wales, had supplanted Wentworth's 'golden colony' and established itself as a rival base for a nation. For a brief while before the opening of the Victorian fields late in 1851, however, the discovery of gold in New South Wales had transformed debates in Sydney and in London about responsible government and transportation. For radicals and democrats gold promised to provide national independence, even separation from Britain, and greater equality of opportunity. For liberals such as Parkes the lustre of gold illuminated a vision of the future based on liberal democracy, industrialisation and commercial prosperity. In both cases the assumption was that the putative Australian nation, populated by throngs of willing migrants, would be governed from Sydney: New South Wales had been picked out by 'the finger of God'.[6] The purpose of this chapter is to rescue these visions by exploring the impact of gold on New South Wales before it became clear that God's finger was actually pointed further south.



The story of Hargraves' discovery beyond the Blue Mountains is well known.[7] Its initial impact was both economic and demographic. The California rush two years previously, when shiploads of 'adventurers' had headed for the Yuba, had given Sydneysiders a taste of what was to come, although it quickly became clear, as the *Herald* commented, that 'that mania, compared with the one with which we are now menaced . . . [was] nothing'.[8] Following Hargraves' announcement, the 'greatest excitement' gripped the colony and Sydney soon resembled 'a deserted village'.[9] Within days the columns of the Sydney press were teeming with advertisements promising to equip diggers with everything from tents to 'Gold Magnets' tried and tested in California, and offering to purchase the fruits of their labour at premium prices.[10] The convulsion was not confined to Sydney. From Maitland one correspondent wrote to complain that prices had risen by 30 per cent and to predict that the town 'will literally be drained of its population' if things continued as they were.[11]

By mid-1851 it was estimated that as many as 6000 diggers were at work on the fields opening up at Ophir, Sofala (Turon) and Bathurst,[12] exacerbating the colony's already substantial labour shortage (Plate 2.2). A month prior to Hargraves' announcement, one commentator had referred to the 'universal acknowledgement' that '*some means* for securing copious immigration to this colony' was needed to address the crisis in the labour market. So concerned was the *Herald* that, initially, it hoped the rush would be short-lived, and it called for draconian restrictions on access to the fields in order to prevent interruption to the regular business of the colony: 'Should our gold prove to be abundant in quantity, rich in quality, of easy access – let the inhabitants of New South Wales and the neighbouring colonies stand prepared for calamities more terrible than earthquakes and pestilence'.[13]

In the *Empire*, Parkes' response was optimistic and tinctured with humour. Publishing an extract from a play entitled 'Gold! Gold!' in the wake of the *Herald*'s most dire predictions, Parkes ridiculed the concerns of the pastoralists with tongue firmly in cheek:[14]

> MESSENGER: Here's news! another lump as big as your hat –
>
> [WORKING MEN]: Hooray!!
>
> MESSENGER: All the shepherds have bolted –
>
> [WORKING MEN]: Hooray!!
>
> MESSENGER: All the sheep are dead –



In fact, Parkes might have been excused a complaint, as the production of the *Empire* was hampered by the departure of his compositors in search of gold.[15] Other liberals were similarly affected. James Pemell, a future Sydney councillor and liberal MP, for example, had delivered bread to Sydney homes for a decade but, in the aftermath of Hargraves' discovery, his horses and carts were soon up for sale since he no longer had any drivers for them.[16]

Despite these inconveniences Parkes was, by August, reassuring his readers that the depopulation of Sydney would be a temporary aberration: 'The greatest gainer by the gold discovery will be the City of Sydney'.[17] Here was the 'magnet' for immigration that would ultimately resolve the colony's intractable labour problem: 'The greater proportion of those who shall emigrate to New South Wales, even with the avowed object of digging for gold, will eventually settle down to the ordinary avocations of life, or become the producers of some new source of wealth'.[18] By this stage the *Herald* had abandoned its negative stance and had begun to muse over some of the potential benefits that gold promised to bring. The first of these was a predicted population increase of 100,000 'meritorious sons of toil' by the end of 1852.[19]

Radicals and liberals were concerned, however, by the short-term political implications of the exodus from Sydney. 'Party politics are forsworn in favour of digging parties', lamented Parkes. Gold had 'cast a mantle of obliviousness' over the candidates who had promised a 'warm little contest'.[20] The election, set to take place in September, was to be the first contested under the aegis of the Australian Colonies Government Act of 1850 that had separated Victoria from New South Wales, granted limited self-government to both colonies and halved the qualification for the franchise, bringing it into line with the £10 level that prevailed in Britain.[21] For all their continuing concerns arising from the implementation of this system – particularly the distribution of seats – radicals and liberals approached the election in Sydney with high expectations.

By the time of the gold rush, a number of candidates in the liberal-radical interest were already in the field, including Charles Cowper, a pastoralist and railway investor, John Lamb, a former naval officer and prosperous merchant, and the irascible Presbyterian minister and republican activist, the Reverend John Dunmore Lang.[22] Differences, principally between Cowper and Lang, prevented a joint campaign, although the respective committees contained many common members.[23] Unity was also encouraged by their common opposition to the conservative candidate, W. C. Wentworth. Wentworth made no secret of the fact that he did not expect to win Sydney under the extended franchise which he had bitterly opposed, using his platform appearances to rail against the 'spirit of democracy . . . which was almost daily extending its limits' and excoriating Chartists, Socialists and Republicans.[24]



That Wentworth did win one of the three seats at the poll (at the expense of Cowper) was due, in large part, to gold.[25] 'Register, Register, Register! Before you start for the Diggings, secure the right of going to the poll' had been Parkes' advice in June, but when the revised rolls were released in July it was clear that this had not occurred. Despite the substantial extension of the franchise, the Sydney electorate showed only a slight increase in total, with one ward – Bourke – actually registering a decline.[26] Moreover, when it came to election day less than 50 per cent of those on the roll voted (2499 out of 5715).[27] Lang's popularity had not been hampered by the fact that he spent most of the campaign languishing in Parramatta Gaol for libel – he topped the poll. The *Herald* thought that Lang had benefited from the low turnout, but at the declaration of the poll Lang suggested to Wentworth, amidst 'great cheers and laughter', that it was he who owed his election to the diggings: 'If the 1001 countrymen of his (Dr Lang) who had gone there, had remained to record their votes in Sydney, he felt pretty sure that they would have taught the honourable and learned gentleman the tune of "Ye'll gang no more to yon town" '. Wentworth agreed.[28]

In the longer term, the political implications of gold proved to be even greater than its demographic consequences. The deepening crisis in the labour market had its sequel in the developing political debates over transportation and immigration. One solution to the problem of labour was to bring in labourers from China and elsewhere in Asia under bonded arrangements that were tantamount to slavery. This nineteenth-century equivalent of what is nowadays called 'people trafficking' was extremely contentious, provoking an outburst of racist xenophobia that presaged the White Australia policy.

No part of the political spectrum had much to be proud of. In the colonial legislature the opposition to Chinese immigration was led by Dr

Henry Gratten Douglass, a graduate of Dublin University and an intractable opponent of reform in colonial politics, who is best known for his role in establishing Sydney University.[29] He attracted plenty of support among liberals. Parkes congratulated him for standing up against those involved in a 'shameless and wretched traffic in human flesh', not on humanitarian grounds, but because they were seeking to 'shovel in upon us an abhorrent mass of foreign Paganism, festering with all incestuous and murderous crimes'.[30] Even among radicals renowned for their internationalism, sympathies were rarely extended to include non-whites. Writing in the colony's only Chartist newspaper, radical journalist and agitator Edward Hawksley described the Chinese variously as 'cut-throat miscreants', 'accomplished thieves, adroit swindlers' and 'men unrestrained by any moral feeling . . . whose only God is the Devil'.[31] Although Hawksley was keen to point out that the Chinese and the 'coolies' were being exploited as part of a wider conspiracy to force down wages among the Sydney trades, his overall stand was unremittingly hostile.[32] Only John Lamb, a radical-liberal elected as one of the three members for Sydney behind Lang, stood out against the tide of racism. A retired sea captain who had visited a number of British colonies, Lamb insisted that the Chinese were not 'the horrible beings which they were said to be': 'In every British settlement that he had visited, he always observed them to be an orderly and quiet people'.[33]



Lamb had built his political career on the back of his opposition to the renewal of transportation and when he defended the Chinese he earned a rebuke from the *Empire* for inconsistency. For those facing a labour shortage, transportation was the other potential solution, but this was equally controversial. Indeed, many opponents of 'convictism' would have agreed with Douglass and Parkes that, if anything, two thousand British criminals were preferable to five hundred 'yellow slaves from a land of lies, infanticide and heathenism'.[34] Even before the news of gold reached Sydney, supporters of transportation – principally among the elite pastoralists – were basing their demand on the need for a 'constant supply of labour'.[35] By June a 'Separation Association' had been formed in the Moreton Bay district, with the twin objectives of separation from New South Wales and the 'introduction of exiles [convicts] and an equal number of free migrants'.[36] Almost two years previously Circular Quay in Sydney had been the scene of unprecedented crowds of protesters against 'convictism' and, early in 1851, Australia's first national political movement, the Australasian League against Transportation, had brought together New South Wales, Victoria, Tasmania and South Australia under the banner of the Southern Cross.[37] By the time Hargraves discovered gold, opposition to convict transportation had become a metaphor for the demand for responsible self-government. At times approaching the brink of open rebellion, anti-transportation was redefining the relationship of the Australian colonies to Britain.

For most commentators, gold was expected to put a swift end to any thoughts of continuing transportation. 'It would be a little too much even for Downing Street philosophy', argued the New South Wales branch of the

Australasian League, 'to transport a highwayman or a burglar to a gold mine!' The *Herald*, too, took it as a matter of course that the discovery of gold would put an end to transportation.[38] The continuing obstinacy of Colonial Secretary Earl Grey in the face of colonial opposition to transportation took on ever-increasing political significance and encouraged bolder claims for responsible self-government. Gold was an essential element to the developing case. 'If we are to become a rich and powerful people', editorialised the *Herald*, 'we must have the full exercise of our constitutional rights, and no Ministry can long keep us out of them'.[39] Although it had opposed him at every turn, the *Herald* even quoted Lang in support of this point: one of the 'certain results of this wonderful discovery', he wrote in 1852, 'is that Australia will, in spite of every effort to the contrary, very soon be free and independent'.[40]

A corollary to the vexed issue of transportation and self-government was the issue of control of the revenue that would arise from the production of gold. In the 'Gold, Gold!' play, Parkes' readers witnessed an exchange between the Australian and British flags that carried an unmistakable message:[41]

> [AUSTRALIAN FLAG]: Look, mother, at my gold! Have I not a fair share of the good things of the world! See how it glitters.
>
> [BRITISH FLAG]: Bless me! Let me look! The real metal as I live! Let me take care of it for you darling.
>
> [AUSTRALIAN FLAG]: No. Mother! I can take care of it myself. Let us make a bargain Mother. Let us part as friends.



The revenue issue was resolved in 1852 with a reaffirmation of the principle that the proceeds from public land were 'to be employed only in such ways as may be directly beneficial to the colony'. Revenue had proved to be an important test for the limits of self-government during 1851–52, and the victory of the New South Wales legislature was an important staging post on the road to greater self-government later in the decade.[42] A further institutional dividend from the gold revenue took the form of a successful campaign to establish a colonial branch of the Imperial mint in New South Wales in 1853.[43]

At a more ethereal level, gold illuminated a vision of the future for radicals and liberals alike. Hawksley regarded it 'as a peculiar providence, that the discovery of our auriferous treasures should have been made at the precise time that it was, just as one worn-out but mischievous Legislature was broken up, and a new, and we trust, far more vigorous one is about to be called into existence'.[44] With characteristic bluntness Lang put the case more succinctly: 'Our bane and antidote were thus before us within the course of one short week'.[45] Hawksley had long complained of apathy, ignorance and dissension in colonial politics, and, although the gold seemed to appeal to the worst traits of human nature, his optimism inspired a vision of New South Wales as a haven of prosperity, equality and mutuality: 'We have been too long a divided and self-seeking people; let us now become united, and exert ourselves for the common weal'. It was a chance for a 'revolution' not to be missed. 'Never before', Hawksley mused, 'was there such a glorious opportunity of upsetting, and for ever, the infamous oligarchical system which in this colony has too long tyrannized over the people'.[46]

Lang also welcomed gold as a platform for the foundation of a free and independent nation. 'The gold discovery has thrown the colony of New South Wales fifty years in advance of its previous position', he wrote. 'Divine Providence has virtually declared that these lands shall no longer be suffered to lie in waste – that Australia shall be inhabited – and has accordingly taken the work of Australian colonization into its own hands.'[47] Having resigned his parliamentary seat and hastily left the colony to avoid creditors, Lang was writing for a British audience, some of whom he delighted in tormenting. Many of the immigrants to New South Wales would flood in from California, he predicted, where they have 'not only been acquiring valuable experience in gold-mining . . . but have also been observing the working of free institutions among a people of kindred origin'.[48] The future was in the hands of the diggers themselves; military force would do no good across a front of a thousand miles and, 'at all events . . . [the troops] would soon discover that there is no better instrument for picking out nuggets from the crevices of Australian rocks than a Birmingham bayonet'.[49]



Lang's Australia was also to be a nation of equal opportunity. Gold would act as a leveller, ending 'the ascendancy of the Squatter' by placing 'the wealth and property of the country in the hands of men of nerve and sinew – men of industry and perseverance – men of honesty and integrity; who are perfectly willing to extend to others what they claim for themselves: a fair field and no favour'.[50] Although the 'relations of sovereignty and dependence between Britain and her Australian colonies' were doomed 'under the new order of things which the gold discovery has originated', Lang's vision was not bad for Britain. 'The United Provinces of Australia' must continue to have 'a thoroughly British population', he insisted, arguing that half of the revenue from gold for the next fifty years at least ought to be devoted to the immigration of good British Protestants who would accomplish a general development of Australia's vast potential resources such as wine, tobacco and cotton, the latter simultaneously ending Britain's dependence on America and striking a blow against slavery.[51]

For Henry Parkes, Hargraves was 'the greatest revolutionist that has yet appeared in Australian history' and gold itself was 'the true talismanic power to move us forward. Richard Cobden, or Daniel O'Connell himself, would be a snail agitator compared to this Sir Charmer Gold.'[52] Parkes had arrived at a similar conclusion to Hawksley and Lang: gold would initiate a colonial version of the nineteenth-century bourgeois revolution against a lingering *ancien régime*, represented in New South Wales by the squatters. The year 1851 in New South Wales was not 1848 in Europe, but it seemed to be. At the height of the rush Parkes reprinted an article from the *Economist*, the journal of the Manchester School that had been established with funds from Cobden's Anti-Corn Law League, which identified the decline of aristocratic power as the most significant development of the nineteenth century.[53]

Who would be the beneficiaries of this revolution? Here the reference to Cobden is also revealing. Following the model of Britain where Richard Cobden had been fêted around the nation in the aftermath of repeal of the Corn Laws in 1846, a group of the leading citizens of Sydney decided to honour Hargraves.[54] Formed in January 1852, the Hargraves Testimonial Committee brought together many of the leading merchants and businessmen in the colony, including a number directly involved in the gold trade and several of the founding members of the Sydney Stock Exchange and the Sydney Chamber of Commerce in 1851.[55] Typically, these were men who might have found a place in a colonial edition of Samuel Smiles' *Self-Help*, having arrived in New South Wales with virtually nothing and built business empires through hard work and perseverance.[56]

At first glance the Hargraves committee appears to have contained a cross-section of the political landscape in New South Wales, but this is misleading. Admittedly, there was one of the favourites of Government House, Thomas Icely, who had successfully prosecuted Lang for libel in 1850, and, at the same time, there was one of Lang's supporters, P. N. Russell.[57] The core of the committee, however, were moderate liberals, including more than a dozen who served on liberal election committees and the New South Wales Constitution Committee which would go on to lead the campaign for liberal amendments to the colony's constitution in 1853.[58] Reflecting a commitment to developing a liberal capitalist vision from Sydney, several of them had also assisted Parkes to establish the *Empire* which, as their oracle, responded to gold with a simple narrative of progress: 'Gold will prove the great progressionist. It will populate our country, construct railways for us, plant our cotton fields, keep our sheep walks within safe bounds, and make us a nation . . . It will do all, too, before Mr Hargraves himself is an old man.'[59]



The other elements of the *Empire*'s liberal capitalist vision of the future were a policy of free trade and the development of Australian manufacturing industries to complement Australia's natural resources. History was on their side. As Thomas Mort, a leading Sydney merchant and member of the Hargraves committee, enthused: 'Gold is the mainspring of commerce [and] commerce is the forerunner of civilization'.[60]

Radicals and liberals alike were correct when they believed that gold transformed the image of Australia in the eyes of the international community, especially in Britain. Up to this time colonial agitators felt, with some justification, that Australian affairs held little interest for British audiences. 'The British public have little political affinity to us, we are too distant for our grievances and struggles to do more than arouse in them a secondary interest', wrote Parkes within days of the news of Hargraves' discovery reaching Sydney.[61] Gold changed everything. The improvement in the tone of Parkes' editorials within a few short weeks was remarkable: 'The influence of the discovery will make us great and important in the eyes of all Europe', he boasted on 30 May. 'Australia will be the theme of all orators, the subject of all writers, and the study of all capitalists.' On the same day, the *Herald* rejoiced that Australia would no longer be 'a by-word of reproach': the 'land of sheep and cattle, wool and tallow, convicts and expirees' would now be 'respected and valued as the most precious, not only of all the colonies of Australasia, but of all the colonies subject to the British sceptre'.[62]

The news of gold arrived in Britain in September 1851 in the midst of tremendous excitement over two unconnected events: the Great Exhibition, then attracting massive daily crowds to the Crystal Palace, and the release and imminent British tour of Lajos Kossuth, the leader of the Hungarian separatist movement. Notwithstanding the exhibition and 'Kossuth fever', gold quickly captured the headlines. 'By this time', commented the *Globe* on 16 September, 'accounts of this newly explored region . . . have been eagerly perused by every newspaper reader from Lands End to John o'Groat's'.[63] An indication of the symbiotic relationship between the press in Britain and in New South Wales was the large number of newspapers that carried extracts from the colonial press. The *Empire* was aware of this market, producing a special edition for London. According to its own estimation at least 1000 copies of this edition, together with 'large numbers' of the *Herald*, were sent on the first available ship, the *Thomas Arbuthnot*.[64] Undoubtedly, these extracts contributed to a largely uniform response to the events in New South Wales across the political  spectrum represented by the British journals. From the 'Whig-ministerial' *Morning Chronicle* to the liberal journals, the *Daily News* and the *Spectator*, there was unanimity in the belief that the most important implications of gold would be political. It would necessitate a review of colonial policy, for instance, ending any plans to reintroduce transportation. 'Who knows but that the light-fingered gentry of the London streets', commented the *Chronicle*, 'may actually be glad to purchase a free passage to the gold country, at the price of seven or fourteen years' nominal detention on arrival'.[65]

This interpretation of events also owed much to the presence in London of John King and Robert Lowe. King, the former Town Clerk of Melbourne, was a delegate of the Australasian League and had just begun his mission to inform British public opinion about the evils of transportation; Lowe, a former member of the New South Wales parliament, had recently begun to write for *The Times* on colonial affairs.[66] The presence of both in the metropolis was propitious. According to his instructions, King was assiduously cultivating contacts in the press when the news of gold arrived and transformed his reception.[67] Similarly, the pro–New South Wales tone of Lowe's articles was particularly well suited to the anti-government stance of Britain's best-known newspaper.[68]

The response of the more conservative British journals was muted. The *Morning Advertiser*, for example, confined its comment to the comforting reflection that the discovery would reduce Britain's reliance on America.[69] In fact, some of the harshest criticism for the colonial policies of Lord John Russell's failing liberal administration came from the liberal press itself. For the *Spectator*, a widely respected liberal weekly, 'grave' consequences would arise from the population of New South Wales by 'new settlers hostile to dependence on England'. Even so, it sheeted the responsibility for this outcome to the home government: 'These evil omens might perhaps be averted . . . But neither the past conduct of our Colonial administrators nor their present temper, as far as can be gathered from the most recent indications, warrant any hope.'[70] It was left to the *Globe*, widely regarded as the mouthpiece of the government, to confirm the fears of many in Britain and Australia by hinting that Earl Grey had not been moved by Hargraves' discovery to abandon his plans for transportation. His recalcitrance pushed the Australasian League closer to the brink of open rebellion.[71]

The radical press was equally anxious to publicise the discovery of gold. Tales of fabulous wealth became the staple of weekly reports during 1852,[72] and these new riches were recognised as a potential platform for independence. 'Now they have gold', the *Weekly Dispatch* warned Earl Grey, 'it is time to give up this game . . . Give self government to a colony which will soon be in a position to exact it.' For Chartists such as Charles Nicholls, editor of the *Political Examiner*, in the wake of gold the Australian colonies were 'developing the most republican and democratic communities that now exist in the world'.[73] It is important to note that before the portentous discovery of gold the radical press had begun to present Australia in an entirely new light. When Lang opened his agitation for a federal republic, the leading Chartist newspaper, the *Northern Star*, anticipated a quick resolution in favour of the colonists, and urged reformers at home to contemplate the answers to the 'pregnant questions' that would arise following the imminent success of republicanism in Australia: 'Let the people at home make common cause with their brethren in Australia, and the downfall of aristocratic domination is certain. If Englishmen in America and Australia can conduct their affairs well, wisely and prosperously under a Republican form of government, why not at home?'[74]



Lang was well known to Chartists as a result of his lectures promoting emigration as well as his regular epistles in a leading nonconformist newspaper, the *British Banner*, which included a ringing endorsement of all the points of the Charter except annual parliaments.[75] The Chartist response to Lang's declaration became a regular reprise. Following the passage of the Australian Colonies Act, the *Northern Star* enthusiastically predicted the foundation of the 'United States of Australia'.[76] By the middle of 1852 the transformation was complete. The *Star* reprinted a 'cheery account' of conditions in New South Wales from the *Herald* and editorialised that 'no workman need starve at home while comfort and independence await him in Australia':

> The future Australian republic will be a refuge and a home for those of our workers in the cause of the people, whose souls shall yearn for liberty, should they ever be . . . compelled to abandon in despair the people of the British Islands, as debased wretches, hopelessly sunk in slavish degeneracy.[77]

The *Northern Star* was not alone when it incorrectly predicted that an independent federation of the Australian colonies would quickly emerge from the gold rushes. Wentworth, Parkes, Hawksley and Lang also believed that gold would create a nation, and their assumption was that the 'golden colony' of New South Wales would be at its head. Three months into the rush the *Sydney Morning Herald* felt confident enough about the future to boast at Victoria's misfortune: 'Our dear sister Victoria, who went mad with rejoicing at separation . . . will see that . . . there may be sterling value in what she despised. She has flared up in bonfires and lengthy processions, but the old colony will, nevertheless, take the shine out of her.'[78]

A few days later there was confirmation of the discovery of large deposits of gold at Deep Creek, north of Melbourne, and early in 1852 the *Empire* was publishing figures which showed that the balance of the gold rushes was shifting inexorably away from Sydney.[79] In the same month Hawksley had seen the writing on the wall. Now that news of Victorian gold, given an 'exaggerated colouring' by the 'poetical imaginations of our southern neighbours', was on its way to England, his fear was that 'but few of the industrious classes will come to New South Wales . . . the consequences of which will be that our trade must languish, our flocks and herds must remain untended, our fields uncultivated, and large quantities of our gold undug'.[80]



Rivalry between New South Wales and Victoria was not new in 1851. The campaign for separation that had culminated the year before had been a bitter one, but it had not been a struggle between equals. In 1850 Victoria was, as one veteran of the separation campaign put it, 'a small oppressed out-district of New South Wales'.[81] Although thriving, Melbourne was a relatively small town, anxious that its wealth and resources were being systematically siphoned off to pave Sydney streets. Gold changed everything. In terms of both immigration and wealth, Victoria was the main beneficiary of gold; by 1860 Melbourne was more prosperous and more populous than its northern rival. It was Victoria's turn to boast. By the end of 1852 it had become 'the first colony in the Southern Hemisphere'.[82]

Far from creating a nation, gold thus created the rival basis for a nation. By 1860 the orderly development of an Australian nation, dominated by its first colony, as envisioned by men as different as Lang, Parkes, Hawksley and Wentworth, was not secured, it was precluded. All that could rescue nationhood was an unwieldy compromise four decades later. Nearly one hundred years after Federation the effects of this compromise are still felt: in 1990 a container travelling by rail from Sydney to Perth was subject to three non-integrated rail systems, four changes in locomotives, five different safety systems, six different sizes of loading gauge, ten different engineering standards and more than twelve hours at sidings or junctions for crew changes.[83] The legacy of gold was not the nation but the ultimate compromise between Sydney and Melbourne: the national capital. The finger of God was pointed at neither Sydney nor Melbourne; it was pointed at Canberra.

## Notes

1 *Sydney Morning Herald*, 8 and 10 May 1852; *Empire*, 8 May 1852. I am grateful to Alex Tyrrell for his comments on a draft of this chapter.
2 *Empire*, 8 May 1852.
3 *Sydney Morning Herald*, 30 April, 5, 6 and 8 May 1852.
4 *Empire*, 8 May 1852.
5 *Sydney Morning Herald*, 12 August 1851; G.V. Portus, *Cambridge History of the British Empire*, 1933, Cambridge University Press, Cambridge, vol. 7, pt. 1, p. 271. Wentworth's election address was slightly misquoted by his nemesis, J. D. Lang (*Sydney Morning Herald*, 10 May 1853), and even borrowed by William Westgarth to describe Victoria in 1853; see *Personal Recollections of Early Melbourne and Victoria*, 1888, George Robertson & Co, Melbourne, pp. 142–3.
6 *Empire*, 5 February 1852. Historians generally pass quickly from New South Wales to tell the story of gold in Victoria. For example, a recent history devotes less than one sentence to gold in New South Wales: G. Davison, J. Hirst and S. Macintyre (eds), *Oxford Companion to Australian History*, 1998, Oxford University Press, Melbourne, p. 283.
7 See G. Blainey, *The Rush That Never Ended: A History of Australian Mining*, 4th edn, 1993, Melbourne University Press, Melbourne, pp. 13–27; B. Hodge, *Frontiers of Gold*, 1979, Cambaroora Star Publications, Penshurst.
8 *Sydney Morning Herald*, 17 May 1851.
9 *Sydney Morning Herald*, 23 May 1851; R. Therry, *Reminiscences of Thirty Years' Residence in New South Wales and Victoria*, [1863], 3rd edn, 1974, Sydney University Press, Sydney, p. 369.
10 *Daily News*, 13 September 1851.
11 *Sydney Morning Herald*, 2 June 1851.
12 Blainey, *The Rush That Never Ended*, p. 26.
13 *Sydney Morning Herald*, 17 and 19 May 1851; Portus, *Cambridge History of the British Empire*, p. 246.
14 *Empire*, 19 May 1851.
15 See A. W. Martin, *Henry Parkes: A Biography*, 1980, Melbourne University Press, Melbourne, pp. 81–2. Parkes would later complain about the high wages he had to pay them.
16 *Sydney Morning Herald*, 4 June 1851.
17 *Empire*, 27 August 1851.
18 *Empire*, 23 June 1851.
19 *Sydney Morning Herald*, 28 May 1851.
20 *Empire*, 14 and 30 June 1851.
21 See A. C. V. Melbourne, *Early Constitutional Development in Australia*, 2nd edn, 1963, University of Queensland Press, Brisbane, pp. 366–75.
22 See *Empire*, 17 July, 2 and 8 August 1851; See also A. Powell, *Patrician Democrat: The Political Life of Charles Cowper 1843–1870*, 1977, Melbourne University Press,

Melbourne; D. W. A. Baker, *Preacher, Politician, Patriot: A Life of John Dunmore Lang*, 1998, Melbourne University Press, Melbourne.

23 See *Empire*, 14, 22 and 24 July, 18 and 20 August 1851.

24 *Empire*, 16 and 18 September 1851.

25 The liberal vote was also divided by the first real outbreak of sectarianism in colonial politics.

26 *Empire*, 14 June, 1 July 1851.

27 *Sydney Morning Herald*, 19 September 1851.

28 *Sydney Morning Herald*, 18 and 19 September 1851.

29 See *Empire*, 24 November 1851; Martin, *Henry Parkes*, 1980, p. 91.

30 *Empire*, 24 November 1851, 21 February 1852.

31 *People's Advocate*, 3 and 10 March 1849.

32 *People's Advocate*, 29 November 1851; A. Messner, 'Contesting Chartism from Afar: Edward Hawksley and the People's Advocate', *Australian Colonial History*, no. 1, 1999, pp. 62–94. Messner does not make the point that this attitude can be seen as part of Hawksley's flawed inheritance from a particular form of British radicalism. See J. Hyslop, 'The Imperial Working Class Makes Itself "White": White Labourism in Britain, Australia and South Africa before the First World War', *Journal of Historical Sociology*, vol. 12, no. 4, December 1999, pp. 398–421. I am grateful to Neville Kirk for this reference; Parkes also made the point that the Chinese were being brought to New South Wales to be exploited. See *Empire*, 25 February 1852.



33 *Empire*, 24 November 1851. This point has been helped by a discussion with Nev Kirk.

34 *Empire*, 24 November 1851.

35 *Sydney Morning Herald*, 22 April 1851.

36 *Empire*, 6 June 1851.

37 *People's Advocate*, 28 April, 16 June 1849, 17 August 1850. I have work in progress on a broader study of this movement.

38 *Empire*, 13 June 1851; *Sydney Morning Herald*, 4 June 1851.

39 *Sydney Morning Herald*, 4 June 1851.

40 *Sydney Morning Herald*, 10 May 1853.

41 *Empire*, 19 May 1851.

42 *Sydney Morning Herald*, 6 June 1851; *People's Advocate*, 21 February, 6 March 1852; Melbourne, *Constitutional Development*, pp. 398–9.

43 See W. F. Morrison, *The Aldine Centennial History of New South Wales*, 1888, Belden Brothers, Sydney, vol. 1, p. 125; Portus, *Cambridge History of the British Empire*, pp. 267–8.

44 *People's Advocate*, 9 August 1851.

45 J. D. Lang, *A Historical and Statistical Account of New South Wales including a Visit to the Gold Region and a Description of the Mines*, 3rd edn, 1852, Longmans, London, p. 364n.

46 *People's Advocate*, 20 January, 29 September 1849, 24 May, 9 August 1851. See also J. Hirst, *The Strange Birth of Colonial Democracy*, 1988, Allen & Unwin, Sydney, pp. 98–9.

47 *Sydney Morning Herald*, 10 May 1853.

48 *Sydney Morning Herald*, 10 May 1853.

49 J. D. Lang, *Freedom and Independence for the Golden Lands of Australia*, 1852, Longmans, London, pp. 202–3. See also Baker, *Preacher, Politician, Patriot*, pp. 136–7.

50 Lang, *A Historical and Statistical Account*, pp. 425–6.

51 Lang, *A Historical and Statistical Account*, pp. 423, 425–6, 551, 555.

52 *Empire*, 3 June 1851.

53 See *Empire*, 1 July 1851. See also P. A. Pickering and A. Tyrrell, *The People's Bread: A History of the Anti-Corn Law League*, 2000, Leicester University Press, London.

54 *Empire*, 7 January 1852.

55 The committee comprised C. Cowper, Chair, S. Peek, Treasurer, R. S. Ross, Secretary, R. Campbell, J. Fairfax, G. R. Griffiths, T. W. Smith, G. A. Lloyd, J. S. Dowling, T. S. Mort, E. Tooth, J. Macnamara, A. B. Spark, J. S. Willis, H. Parkes, Dr Douglass, D. Lanarch, A. Campbell, T. Icely, Sir O. Gibbs, D. Cooper Jnr, J. Croft, P. N. Russell, W. Tucker, T. Hale, H. Fisher, J. O. Flavelle, S. H. Smyth, J. G. Cohen, and J. Gilchrist. See *Empire*, 2 August, 14 November 1851, 7 and 30 January 1852; *People's Advocate*, 1 and 15 November 1851, 13 March 1852. At least one member evidently knew little about it. A. B. Spark recorded in his diary for 24 December: 'Find my name in the *Herald* of this morning as one of a Committee for managing a testimonial to Mr Hargraves'. See G. Abbott and G. Little (eds), *The Respectable Sydney Merchant: A. B. Spark of Tempe*, 1976, Sydney University Press, Sydney, p. 209.

56 See, for example, the careers of John Fairfax, T. S. Mort, G. A. Lloyd, and (native-born) T. W. Smart in *Australian Dictionary of Biography*, vol. 5, pp. 95–6, 299–301, and in C. N. Connolly, *Biographical Register of the New South Wales Parliament 1856–1901*, 1983, Australian National University Press, Canberra, pp. 62, 100, 198, 305–6. S. Smiles, *Self-Help*, [1859] 1968, Sphere Books, London.

57 See *Empire*, 16 November 1850; Connolly, *Biographical Register*, p. 163; Baker, *Preacher, Politician, Patriot*, pp. 123, 126.

58 See *Empire*, 14 May, 24 July, 2 August, 16 October 1851; *People's Advocate*, 13 and 20 August, 3 September 1853, 29 April 1854. Anti-transportation was another common involvement.

59 *Empire*, 3 June 1851. Musing over this question of what gold would do for New South Wales, the *Herald* suggested that in addition to a 'great accession of working capital and working men' it would also bring railways and hasten steam communication with England. See *Sydney Morning Herald*, 2 and 4 June 1851.

60 See *Empire*, 30 June, 8 September 1851; Mort cited in T. Irving, '1850–70,' in F. K. Crowley (ed.), *A New History of Australia*, 1974, William Heinemann, Melbourne, p. 139.

61 *Empire*, 5 May 1851. See also *People's Advocate*, 22 June 1850; *Empire*, 17 November 1851.

62 *Empire*, 30 May 1851; *Sydney Morning Herald*, 30 May 1851. See also *People's Advocate*, 6 March 1852.

63 *Globe*, 16 September 1851.

64 *Daily News*, 13 September 1851. John Hirst has made the point that the leading newspapers in New South Wales relied heavily on British newspapers; see *Strange Birth*, pp. 48–9. The reverse was also true.

65 *Morning Chronicle*, 8 October 1851.
66 See *Empire*, 19 March, 16 June 1851.
67 *Empire*, 13 February 1852.
68 See *The Times*, 6 and 15 October 1851.
69 *Morning Advertiser*, 7 October 1851.
70 *Spectator*, cited in *Empire*, 5 February 1852.
71 *Empire*, 18 February 1852.
72 See, for example, *People's Paper*, 17 July, 21 August 1852; *Star of Freedom*, 26 June, 31 July, 14 August 1852.
73 *Weekly Dispatch* cited in *People's Advocate*, 21 February 1852; *Political Examiner*, 30 March, 11 July 1853.
74 *Northern Star*, 7 May, 14 September 1850. See also A. Beever, 'From a Place of "Horrible Destitution" to a Paradise of the Working Class: The Transformation of British Working Class Attitudes to Australia, 1841–1851', *Labour History*, no. 40, May 1981, pp. 1–15.
75 *British Banner*, 12 April 1848.
76 *Northern Star*, 19 April 1851.
77 *Star of Freedom* (the former *Northern Star*), 19 and 26 June 1852.
78 *Sydney Morning Herald*, 16 August 1851.
79 *Sydney Morning Herald*, 19 August 1851; *Empire*, 10 January, 20 February, 10 May 1852.
80 *People's Advocate*, 21 February 1852.
81 T. McCombie, *The History of the Colony of Victoria From Its Settlement to the Death of Sir Charles Hotham*, 1858, Sands & Kenny, Melbourne, p. 212.
82 McCombie, *History of the Colony of Victoria*, p. 212.
83 Australia, House of Representatives, *Debates*, 7 December 1994.

# Gold-Rush Melbourne

Chapter

*Graeme Davison*



For a golden moment in the 1850s Melbourne held the world record for urban growth. Between 1851 and 1861 the city more than quadrupled its population and grew from a small and unknown pioneer settlement into a proud metropolis. In London, Hamburg, Boston and Canton, Melbourne became a name synonymous with fame and fortune, and at the peak of its meteoric rise in 1852 and 1853 more British emigrants bought tickets to Melbourne than to any other destination in the world.

The secret of Melbourne's rapid rise can be told in a single word – gold. It was the lure of instant riches that drew the eager travellers so far across the world, the apparently boundless wealth beneath the soil that generated the frantic traffic in people and goods through the upstart metropolis. The pastoral land rush of the 1840s had given Melbourne a running start in the urban growth stakes – the town was gaining on its older rival, Sydney, even before gold had been discovered. But, in the 1850s, it quickly caught and overhauled the northern city, and by 1861 it had become the largest as well as the most glamorous city on the continent.[1]

Melbourne was perhaps the most spectacular example of a new type of city created by the advance of capitalistic enterprise in the New World. The American historian Gunther Barth describes them as 'instant cities'. Such cities usually emerged in regions of recent European settlement, such as the United States, Australia, New Zealand, South Africa or South America and often in the wake of major mineral discoveries. San Francisco, Denver, Johannesburg, Dunedin and Melbourne each grew in a matter of months from mere dots on the map to mighty commercial cities. The growth of Melbourne, wrote William Kelly, who compared its development in 1853 and 1858, 'was truly wonderful, not to be imagined or understood in the ordinary growth or progress of Old World cities'. Another visitor, the American merchant George Francis Train, compared Melbourne with Chicago, Cincinnati and San Francisco. 'Babes in the wood in the morning', they became 'full-grown men at night'.[2]

The bigger the mineral discovery and the more precious the metal, the faster the cities grew. Melbourne's nearest rival as an instant city in the 1850s was another golden gateway on the other side of the Pacific: San Francisco. The discovery of the Californian goldfields near Sacramento in 1847–48 transformed San Francisco, the nearest harbour to the fields, from a village of barely 300 people to a city of more than 50,000 in less than a decade. Melbourne began from a higher base (30,000), but ten years later it had reached a population of 125,000. Sydney's growth, from 54,000 in 1851 to 96,000 in 1861, though slow by Melbourne's standards, was faster than in any other decade of the nineteenth century.

Instant cities contrast with other New World cities in several interesting ways. Slower-growing commercial centres developed in an organic relationship with their agricultural or pastoral hinterlands. Their new urban industries used raw products and immigrant labour from the surrounding countryside. Instant cities, on the other hand, were often planted in remote and inhospitable regions and developed without a home-grown population, with a limited agricultural or industrial base and with only tenuous transport and communication links with the rest of the world. 'There are few sights more interesting to behold than vast cities rising out of the desert, and busy marts of trade opened in the midst of the forest', remarked Melbourne's leading newspaper, the *Argus*, in 1854. 'Tadmor and Damascus in ancient, and San Francisco and Melbourne in modern times, present interesting features for thoughtful contemplation of this sort.'[3] To be a Melburnian in 1852–53 was to experience the unusual combination of isolation and dependence known in our day perhaps only by Antarctic scientists and space explorers.



Instant cities were also remarkable in another way. They were settlements running perilously close to the limits of growth. They experienced all the usual problems of urban development – providing food and water, housing, sanitation, social security and public order – all at once. They were possibly the first cities in history to experience the kind of runaway growth and social dislocation now endemic in the rapidly urbanising societies of the Third World. Cities born under such extreme pressures often reacted in unusual ways, creating or improvising novel solutions to their problems. Sometimes they reverted to traditional ways once the emergency had passed; but sometimes, through inertia or expediency, these makeshift solutions became permanent features of their institutional life.

Melbourne was the natural gateway to the Australian goldfields. It lay directly in the path of the main route for sailing ships from Europe, around the Cape of Good Hope and through the 'roaring forties'. Sydney faced the Pacific and North America, an advantage which would one day tell in its favour; but Melbourne, as the English poet and adventurer Richard Horne noted in 1859, 'is right in the line of those westerly winds . . . down which all sailing ships coming from Europe must travel. On this line she is, I believe, at least a thousand miles nearer Europe than Sydney.' Like many of his contemporaries, Horne held great hopes for the opening of the continent by inland waterways. Melbourne's readier access to the Murray–Darling river system therefore placed it in a position analogous to that of New York in relation to the Great Lakes. 'Melbourne', he forecast, 'must become the New York of the southern hemisphere'.[4]

At the peak of the gold rush, Melbourne was more a way-station than a settled community, more a flooding river than a placid lake. Contemporaries struggled to orient themselves to a society whose 'enormous floating population' was 'always ebbing and flowing'.[5] Contemporary maps and panoramas reveal something of the perceptual geography of the city, highlighting landmarks associated with the traffic of people, goods and information through the city. Teale and Whittock's charming panorama of the city in 1854 views the gold-rush city, as an arriving gold seeker might have approached it, from the sea and the riverfront (Plate 3.1). It highlights the flotilla of ships jamming the lower reaches of the Yarra, the cluster of privately owned wharves and warehouses below Queen's Wharf, and the jumble of wooden houses and more permanent public buildings of the city's crowded western end. The thousands of newcomers who poured in during the previous few months were largely housed in the overcrowded pubs, boarding houses and cottages of the city grid. But signs of progress are also apparent in the newly opened Melbourne and Hobson's Bay Railway – the first in the colonies – joining the city to its ocean port, in the line of posts carrying the electric telegraph to the government observatory at Williamstown, and in the miniature Crystal Palace on the city's outskirts where the colonists were holding their own international exhibition.



In gold-rush Melbourne, three public buildings symbolised these sustaining flows of people, goods and information: the Immigration Depot, the Customs House and the Post Office. Melbourne's people were largely transients, passing through on the way to somewhere else. The city produced little of what it consumed, depending on the flow of commodities from elsewhere, and its fortunes were therefore influenced mightily for good or ill by news from elsewhere in the colony or abroad. The Immigration Depot was the gateway for newcomers. The Customs House was the official point of entry for imported goods. The Post Office was the place where news was received or despatched. Each was the centre of a distinctive zone or quarter of the city and the node of far-flung economic and social networks.

When news of the first gold discoveries reached Europe it generated a spontaneous rush of people and cargo towards Australia.[6] By 1852 ships were arriving in the port of Melbourne at the rate of two or three a day and in Port Phillip Bay there were often as many as 300 ships anchored at one time. Since there was only one pier for ocean-going vessels, the majority of new arrivals and most consignments of goods had to be unloaded into lighters and drawn by steam tugs upriver to Queen's Wharf where they were unloaded (Plate 3.2).

Here, on the city's threshold, bystanders could begin to gauge the staggering volume of people and things flooding into the colony. 'All down, near the wharves, it is a scene of dust, drays and carts hurrying to and fro, and heaps of boxes, trunks, bundles and digging tools', remarked Quaker journalist and radical William Howitt. 'Here you see ships unloading all kinds of goods and scores of drays fetching them away, making it almost impossible to pass among them without being crushed . . . It is everyman's business to take care of himself here.'[7] From the waterfront, the boxes, bundles and trunks were borne away by scores of drays, wagons and wheelbarrows into the city's adjoining commercial zone. Much of the south-western quarter of the city mile, especially along King Street and Flinders Lane and their narrow tributaries, was given over to warehouses, shipping agencies, chandleries and stevedores.

The gold rush revolutionised Melbourne's business life. Before the rush began, barely fifty merchants and commission agents were operating in Melbourne on behalf of the squatters of the interior. Most of them were conducting a simple import–export trade, largely on a commission basis. By the peak of the gold boom there were more than 300 merchants, and the number of persons employed in commercial occupations had more than doubled from 3445 (1851) to 7687 (1854). More than 1000 new buildings were erected in the central business district in the first six months of 1853, many of them improvised structures of wood or canvas. But among the more substantial were a growing number of bluestone banks and warehouses, like the handful which still stand at the south end of King Street.



Only a few of the new merchants were old colonists. Most were young men of capital, with previous experience in the commercial life of London, Glasgow, Boston or Berlin. George Train, a newly arrived merchant from Boston, noted their revolutionary impact on the town's commercial life:

> The legion of new claims [newcomers], real Simon Pure Yankees, solid go-ahead Englishmen, hard-working, quick-witted, close-fisted Germans . . . that are flocking into Melbourne act as a leaven upon the residents who never before dreamed of the possibility of going beyond their own office to transact business.[8]

The traditional outlook of the old commission agents, content to take a set percentage on the cargoes they imported, was supplanted by the more speculative behaviour of the new merchants, who aimed to maximise their profits by skilfully forecasting market conditions on both sides of the world.[9]

The growth of the goldfields had opened up an immense new consumer market. With most of the population preoccupied with the search for gold, almost everything they consumed – food, clothes, mining implements, vehicles – had to be imported. Gold-rush commerce was a risky line of business. In the early days of the rush everyone was mindful of the cautionary example of San Francisco, where thousands of tons of goods imported from the East Coast and Europe had had to be dumped either because they were unsuitable or in over-supply.[10] So Melbourne played a waiting game and European shippers, determined not to burn their fingers again, sent only modest consignments. Meanwhile, immigrants continued to arrive, supplies ran short and prices soared. Local merchants placed urgent new orders but there was at least a five- or six-month delay between the despatch of orders from Melbourne and the arrival of the cargoes themselves. The scarcity of 1852 was followed by the glut of 1853–54. Over the following months, crates of unwanted merchandise spilled out of the warehouses and into the streets. Perishable goods were sometimes left to rot on the wharves, and anything of value was sold off in liquidation sales. When the crisis was over, almost one-third of the city's merchants had been bankrupted.

The prudent or lucky survivors included some of Melbourne's richest and most powerful men. From the mid-1850s the merchant elite began to exert a strong counterbalance to the previously unrivalled power of the 'squattocracy'. 'Every shopkeeper became a merchant and every merchant a prince', recalled E. Carlton Booth.[11] During the early 1850s the squatters and their allies, the old commission agents, held aloof from the newly rich gold-rush merchants. While the squatters met at the Melbourne Club, now located in Leonard Terry's handsome new clubhouse at the fashionable eastern end of Collins Street, the merchants gathered at the newly formed Victorian Club.[12] Merchants not only dominated the political institutions of the city itself, such as the town council, but played an increasingly powerful role in both Houses of the colonial parliament, where commercial men such as James McCulloch, J. G. Francis and James Service would later head liberal-conservative ministries.



Gold-rush cities were cities of strangers. Most of their residents were birds of passage, ambitious individualists more intent upon making their fortunes and heading home to Europe than in putting down roots in Australia. The gold towns of California had attracted thousands of fortune-hunters from the East Coast, over the border from Mexico, across the Pacific from Australia and China, and around the Horn from Europe. The new residents of Melbourne and Sydney were less cosmopolitan than those of San Francisco, but more diverse than any generation of Australian immigrants before the 1960s. German, Chinese, American and Jewish settlers mingled with the Anglo-Irish majority.

Most newcomers saw the coastal cities merely as staging posts on the way to the goldfields themselves. In Melbourne, they sometimes entered the city through the Immigration Depot at the west end of Collins Street and found their first lodgings in one of the dozens of hotels, boarding houses or lodgings clustered in the west end of the town. 'The number of inns . . . would do credit even to London', William Howitt observed, noting the presence of American and German as well as English, Scottish and Irish establishments.[13] In the 1850s the main city grid remained largely a residential zone and housed more than two-thirds of the population of the metropolis. Assisted immigrants were often directed by government officials to up-country employments. But the great majority of unassisted immigrants were expected to fend for themselves. Even before their ship had arrived at Sandridge (Port Melbourne), new chums may have spied a likely lodging in an immigrants' guide or sought the advice of a shipmate. If they arrived without accommodation, they were likely to be accosted as they disembarked from the steam ferry at Queen's Wharf by one of the touts or 'runners' employed by the city hotels and boarding houses. Many bore names designed to exploit the nostalgia of the new immigrant, and we may imagine the homesick voyager knocking at the door of the 'Highlander' or the 'Kilkenny Inn' in search of comforting homeland company. Once inside, they were likely to stick with their fellow countrymen. In one Melbourne boarding house, diners were divided among three tables: the 'Britishers' table, the 'Yankeys' table and the 'Experienced Colonials' table. Thomas Pierson, a Scotsman with only six months 'colonial experience', sat at the last.[14]

In the early 1850s, when the immigrant tide was at full flood, the lodging house zone was crammed with thousands of young, mainly single, men. In 1854 the *Argus* noted that an overwhelming proportion of the houses of the city were being used as lodging houses. The average number of persons per dwelling had doubled within three years from four or five to about eight or nine. With every house seemingly bulging at the seams, newcomers were glad  to find even a couch, a table or a palliasse to sleep on. An indignant observer noted that many of the common lodging houses were not only 'overcrowded and filthy to a degree' but that some were notoriously 'scenes of extortion, drunkenness, riot and robbery, if not of murder'.[15]

The west end of the city, overflowing with restless young bachelors, was as turbulent as the American 'Wild West'. Energies and emotions bottled up in the crowded, masculine world of the lodging house exploded onto the streets in knuckle fights, drunken mêlées and impromptu horse races along the city streets. 'The Western Hill of the city', according to William Kelly, was known as the district of the 'Wild Sports of the West'. He claimed that 'a pedestrian who in those days ventured into the street would have had as little chance of escape as the amateur campaigner who hazarded a crossing during the Balaklava charge'.[16]

Gold-rush Melbourne was essentially a public city, a place in which people cooked, ate, drank, played, bathed and often slept away from home. Life was lived very largely under the eyes of others. One notable feature of the city was the number of eating and drinking establishments. A higher proportion of the city sites was occupied by cheap restaurants, hotels, taverns, sixpenny cafes, chophouses, oyster bars and cookhouses than at any other moment in its history. The streets themselves supported a colourful cast of itinerant hawkers, barrowmen, flower-sellers, street entertainers and food stalls.[17] Very few homes, in a city without piped water or underground sewage systems, had their own bathrooms. So people bathed, if they troubled to bathe at all, in public baths and bath-houses. Some private proprietors ran open-air baths by the sea at St Kilda or on the banks of the increasingly polluted Yarra. The unsavoury or unsanitary reputation of the private bath-houses stirred a demand among respectable citizens for the establishment of properly regulated premises under the auspices of the Melbourne Town Council. The Melbourne City Baths in Swanston Street still occupy the same site as the municipal baths first established in 1852, a tangible reminder of a time when bath night in Melbourne was also invariably a night out.[18]

Instant cities catered to the demand of young, spendthrift men for instant pleasures. Bourke Street, especially the middle section between Swanston and Russell Streets, was Melbourne's amusement parlour. There the lucky digger found a dazzling array of bars, theatres, billiard rooms, bowling saloons and rifle galleries ready to part him from his new-won wealth. Its glittering centre was, as Margaret Anderson's chapter shows us, George Coppin's Theatre Royal, where Lola Montez, Charles Thatcher and other musical stars of the day performed. Tucked away behind the facades of Bourke Street, in the narrow courts and alleys connecting Great and Little Bourke Streets, and in nearby Stephen (Exhibition) Street, were the brothels and disorderly houses of the city's illicit pleasure zone. Melbourne's thousands of immigrant bachelors, far from home and without the immediate prospect of matrimony, were a ready market for prostitution. Visitors to the city in the 1850s were astonished by the effrontery of Melbourne's harlots, who commonly paraded the streets in broad daylight. Two doctors, who carried out an inquiry into the local vice trade in the early 1850s, complained that many eating houses were 'nothing better than brothels in disguise', and two years later complaints were still being made about 'the large number of depraved females who nightly throng the streets of Melbourne'.[19]



The most common disease in gold-rush Melbourne, if not the most serious, was homesickness. The city's isolation and dependence made its immigrant citizens chronically hungry for news from home. Communications both within the colony and with the outside world were still very rudimentary. Melbourne had no railways until 1854, no regular omnibus services, no telegraph communication with the Port Phillip Heads, and no regular mail deliveries. It is no wonder that the arrival of the overseas mail ships, carrying the news, now three months old, from England, was such a celebrated event in the city's social life. When the flag was raised on Flagstaff Hill, thousands of Melburnians made their way to the General Post Office on the corner of Bourke and Elizabeth Streets. A contemporary recalled the scene:

> A small cupola, containing a clock, rose a few feet above the low roof on the south or Bourke Street side of the edifice and along the street frontage was not a very broad verandah, slightly elevated above the footpaths, and standing a little way back . . . It was a place of rendezvous for friends and a lounge for vagrants. Tramps would select it as a dormitory during the summer months and it shared with the old Eastern Market the distinction of being selected as tribune for popular orators . . . In the early days all newcomers had their letters from the old country directed to them at the Melbourne Post Office and when the monthly mail arrived the verandah and its approaches were thronged. No sooner were

**Figure 3.1** *The General Post Office*, in S. T. Gill and Nicholas Chevalier, *Victorian Illustrated in 1857 and 1862.* (La Trobe Picture Collection, State Library of Victoria)

> the slides at the delivery windows lifted than a scene of confusion and excitement ensued. Men's eyes glistened and their hands trembled as they grasped the letters whose superscription was so familiar to them.[20]

Because it was the most regular port of call in the immigrant city, the GPO also became the focal point for all those trades that catered to the city's immense pedestrian traffic. People calling to collect their mail could stop to do their shopping, to have a drink or to browse in a bookshop. The nearby shops, such as Mars Buckley's drapery, Thomas Gaunt's watchmakers, E. W. Cole's bookstore, and the Royal Arcade, owed their success largely to this passing parade of visitors to the Post Office.

Instant cities were communities pushing hard against the natural, technological and political limits to growth. Between 1852 and 1854, the peak years of immigration, Melbourne's social and physical fabric was stretched almost to breaking point. It was difficult to supply the city with even the basic requirements of life – food, water, shelter, and the protection of the law. Food prices soared and the adulteration of food was rife. Fresh water, which had to be carted from the upstream reaches of the Yarra, sold at extortionate prices.

Providing cheaper and more permanent water supplies called for planning, capital and organisation on a scale that a mobile, atomistic society struggled to deliver. In San Francisco, private citizens had been able to create their own supplies by digging artesian wells to tap the natural reservoirs beneath the city.[21] In Melbourne, on the other hand, as the waters of the lower Yarra became progressively more polluted, locals were forced to bend their collective will to the construction of a comprehensive scheme of pure water reticulation. Several plans were advanced to solve the problem in the early 1850s. A water tower filled from a pump on the riverbank might improve reticulation but the supply would be no purer or more abundant. A pipeline to Dight's Falls, about four miles upstream, might temporarily improve the supply, but the pumping station could be put out of action in time of flood. The most satisfactory solution, and the most expensive by far, was a plan to harvest the more distant Plenty River by constructing a new reservoir at Yan Yean, thirty-five miles from the city, and to convey water to the city by pipes and aqueducts. The infrastructure needs of the new city far outran its resources. In 1853 the new Victorian government was persuaded by financier Anthony Gabrielli to float a loan for £1 million, an almost unimaginable sum at the time. The construction of the Yan Yean water supply was an impressive tribute to the optimism that gold instilled among Melburnians.[22]



By the winter of 1852 Melbourne was in the midst of an acute housing crisis. As fast as unsuccessful diggers drifted back from the goldfields, thousands of newcomers flooded in from abroad. 'Where are we to put them?' asked the *Argus* despairingly. 'We are crammed already. Our houses are already too full for either health or decency . . . We cannot build fast enough with bricks or stone, and houses must be had, and that within these few months.'[23] With most of the tradesmen and labourers running away to the diggings, the city's remaining builders were unable to keep up with the demand for new dwellings.

By 1852 the housing crisis had become so serious that the government sanctioned the creation of a Canvas Town along the south bank of the Yarra in South Melbourne. Homeless immigrants were offered timber, canvas and discarded sails and permitted to erect their improvised dwellings on Crown land in return for a weekly rent of five shillings. Richard Horne took up residence there in mid-1852, after returning from the goldfields and before taking up a government job as an Assistant Goldfields Commissioner. As a bachelor living in two tents, a 'sitting room' and a 'bed room', and taking most of his meals with well-to-do friends in town, Horne lived more comfortably than most Canvas Towners.[24] Families with babies and young children would look back on a more distressing time 'when tents were blown away, and the rain poured down, and children [were] born, and young wives and young mothers wept through the night, hungry and cold, deploring the day that they ever left Old England, and comfortable homes'.[25] By 1853, with the housing crisis no nearer solution, the residents of Canvas Town had begun to do what they could to make their flimsy dwellings a little more comfortable and rainproof. The government, fearful that the temporary camp was in danger of becoming permanent, ordered the site to be cleared of all wooden structures. The Canvas Towners, supported by the press, protested that they had only sought to relieve the hardships of innocent women and children. For the time being the government stayed its hand, but a year later the forcible dispersion of the settlement was well under way.

A more enduring legacy of the housing crisis was the decision of the colonial government in October 1852 to exempt the city's nearer suburbs from the Melbourne Building Act.[26] The abolition of controls over the minimum size of subdivisions, and the materials and size of dwellings, enabled working people to buy their own small allotments, to build their own simple wooden or iron houses, and to extend or embellish them as they found the time and means to do so. Land advertisements addressed to 'the Working Classes' extolled the new liberties – 'far beyond any hitherto enjoyed' – to build as they wished.[27] The new suburbs may have been a boon to their working-class residents, but they appeared less attractive to sophisticated town dwellers straight off the boat from London.



One afternoon in 1852 William Howitt took a stroll from his city lodgings across Eastern Hill towards Richmond where he was confronted by 'an immense suburb' of little wooden tenements. 'This', he remarked,

> is one of the first things which has impressed me with the rapidly running torrent of immigration. Here is a new settlement in all its newness. The houses are some of them incomplete, others just erecting. A balder and more unattractive scene cannot meet the eye of man. Every single tree has been levelled to the ground; it is one hard, bare expanse, bare of all nature's attractions, a wilderness of wooden huts of Lilliputian dimensions, and everywhere around and amongst them, timber and rubbish, delightfully interspersed with pigs, goats, and dogs innumerable.[28]

The instant city assumed something of the form and character of the ramshackle cities of the Third World, with their encircling *favelas*, *kampongs* and shanty towns. While the central part of the city was overcrowded with bachelors' lodgings, the perimeter of makeshift suburbs was inhabited by families, often female-headed families, temporarily or permanently abandoned by their gold-seeking menfolk.[29]

In 1854 Joseph Greathead, a missionary employed by the newly founded Melbourne City Mission, made a door-to-door survey of families in Collingwood. As he moved among the tent- and cottage-dwellers of the new working-class suburb, he meticulously noted the economic and moral condition of each neighbourhood. In the northern portion of the district, towards Clifton Hill, the residents were generally more respectable and prosperous – 'neat, clean people' with their own copies of the Scriptures and 'many useful books'. But the poor, largely Irish Catholic inhabitants to the south-west, along Simpson's Road (Victoria Street), impressed him less favourably. 'The men', he noted, 'generally manifest a disinclination to converse upon, or listen to, subjects of a religious nature'. Usually, however, it was not a man, but a weary, and sometimes distressed, woman who answered the missionary's knock at the door. 'Many mothers sadly complain of the depravity and general disobedience in the absence of their Fathers of the children, especially the larger class of boys', Greathead observed after visiting several distressed families in the neighbourhood of the ironically named Easey Street. Knocking at the door he had met one woman in tears. Was there anything wrong? he inquired. 'Wrong – yes – all is wrong and going to hell', the distraught wife replied.[30]

Some observers of the new working-class suburbs, like Howitt, deplored what they conceived as a plot by the colonial government and its allies, the squatters and speculators, to fob the working classes off with tiny town allotments instead of viable agricultural holdings. Rather than peopling the land with a prosperous yeomanry, they had created a bloated 'town of dog-kennels and rabbit-cotes'.[31] But it is unlikely that all the new suburbanites took as dim a view of their circumstances. Compared with the rented stone tenements of the cities from which they had come, the self-built, ramshackle cottages of Collingwood and Richmond represented a kind of liberty.[32]



While relaxing building standards increased the supply of housing, it also increased the dangers to human life. The lack of a reticulated water supply meant that Melbourne was exposed to high risks of epidemic disease and fire. In a city built mainly of wood, with no system of gas lighting or cooking, fire was a perennial hazard. A lamp or candle had only to capsize, or a log fall out of the grate and ignite some straw or bedding, and an entire heavily populated neighbourhood could be consumed in flames. Many New World cities experienced devastating fires. Some, like the great fires of San Francisco (1851) and Chicago (1871), destroyed the greater parts of the city.[33] Between 1851 and 1855 Melbourne experienced fifteen or twenty big fires and, although none compared with the worst American outbreaks, by 1854 Melbourne was also becoming 'notorious for its conflagrations'. It was the cautionary example of San Francisco that prompted Melburnians to maintain the ban on new wooden building in the commercial centre of the city while relaxing controls in the nearer suburbs. 'Do not let us mix the wooden town with the brick and stone town, and have San Francisco fires in six hours levelling the whole city to the ground', pleaded the *Argus*.[34]

The fire danger peaked during the summer of 1853–54 when several large warehouses, crammed with merchandise, mysteriously ignited. On 13 November a fire broke out in Pye and Moxon's store in Flinders Lane and swept through several neighbouring warehouses before igniting the Edinburgh Castle Hotel. As the blaze took hold, enthusiastic bystanders broke through doors and windows to try to salvage some of the landlord's alcoholic treasure. Even after the building had been burned to the ground, 'many of the most degraded of both sexes' were reported to be 'quaffing champagne and eating sardines and cheese out of the mingled heaps of smouldering and drenched rubbish'.[35]

More dangerous than the risk of incineration was the ever-present threat of epidemic disease. The human costs of the gold rush were apparent in recurrent outbreaks of typhus, dysentery and other contagions. Melbourne, despite its young population, suffered a similar death rate to older, slum-ridden British cities. The young, especially babies, were most at risk; more than 80 per cent of Melbourne's dead in 1853 were children under 2 years.

The pressures of instant growth had left Melbourne with a deadly legacy of muddy streets, gimcrack housing and polluted streams. Visitors were nauseated by the stench of nightsoil and dead dogs decomposing in the streets. In 1853 a government committee drew attention to the condition of a section of the city's main retail district:

> In the block bounded by Great and Little Bourke Street, Elizabeth Street and Swanston Street, there is a space of one hundred square yards hitherto occupied by a green putrid semi-liquid mass, partly formed of the outpourings of surrounding privies; and in the blocks north and south of this one, the very rights of way are similarly saturated.[36]



Public health was a notorious problem for frontier cities everywhere – Wyatt Earp and Billy the Kid were more likely to have died from typhoid or dysentery than from gunshot wounds. San Francisco had experienced serious outbreaks of cholera (1851) and smallpox (1854). In Britain, the publication of social reformer Edwin Chadwick's famous sanitary report on English towns (1842) was a threshold in the development of public consciousness of environmental disease.[37] Following the British lead, Melbourne established a Board of Health to investigate sanitary hazards, and later attracted some talented doctors and engineers to tackle the problem. By the time officials could respond to the emergency, however, urban development had run out of control. Not until the end of the century did Melbourne get an underground sewage system or reduce infectious disease rates to satisfactory levels.[38]

Conservative colonists feared, as David Goodman has well shown, that the social dislocation of the gold rush might culminate in anarchy and revolution. The arrival of thousands of avaricious young men, adrift from the restraining influence of home and kin, exposed to the hazards and temptations of frontier life and susceptible to the appeals of radicals and revolutionaries, inspired dread among the governing classes.[39] S. T. Gill gave vivid expression to these fears in his contrasting portraits of the 'improvident digger', bent upon riot and debauchery, and the 'provident digger', contemplating self-advancement and domesticity (Plates 3.3 and 3.4).

Official responses to the crisis centred on the forceful suppression of disorder and the provision of cultural amenities designed to civilise the population.[40] The city centre, where the young men were concentrated, was guarded by no fewer than six police and military establishments, including foot and mounted police stations and infantry and cavalry barracks. Mounted police officers armed with pistols, sabres and carbines patrolled the streets around the clock. This was a significant departure from homeland practice, where military barracks were usually located on the outskirts of the town rather than in the town itself.[41]

No sooner did contemporaries admit their fear of anarchy, however, than they rushed to cite evidence of the continued loyalty and sobriety of the newcomers. The external disciplines of police and cultural agencies were seemingly less powerful than the self-discipline of religion and respectability. Visitors applauded the way in which the new society was almost unconsciously assuming the shapes of the old. While the drunken revelry of Bourke Street aroused fears of social anarchy, the city's burgeoning suburbs inspired more consoling thoughts. It was almost as though the disorder of the rush had produced its own antidote: 'These suburban residences inform the stranger . . . of the quiet yet

firm manner in which the Anglo-Saxon race takes root in these colonies; that there is in them a determination to rear up the homely institutions of their fatherland . . .', observed one contemporary.[42] To freer spirits this Anglo-conformity seemed almost excessive. Richard Horne considered this 'constant tendency to imitate and "reproduce" the old country' so overdone as to amount to a 'morbid prejudice, an absurdity, and an injury'. But on this, as on other issues, Horne was a maverick. If Melbourne eventually became the most suburbanised and privatised of Australian cities, it was in part, perhaps, as a reaction against the very public and rowdy city it once had been.

## Notes

1 For population growth of capitals, see J. W. McCarty. 'Australian Capital Cities in the Nineteenth Century', in J. W. McCarty and C. B. Schedvin (eds), *Australian Capital Cities: Historical Essays*, 1978, Sydney University Press, Sydney, pp. 9–25; for the broad patterns of Australian and North American growth, see L. Frost, *The New Urban Frontier: Urbanisation and City Building in Australasia and the American West*, 1991, University of New South Wales Press, Sydney, and D. Hamer, *New Towns in the New World: Images and Perceptions of the Nineteenth Century Urban Frontier*, 1990, Columbia University Press, New York, esp. pp. 158–9.

2 G. Barth, *Instant Cities: Urbanization and the Rise of San Francisco and Denver*, 1975, Oxford University Press, New York; W. Kelly, *Life in Victoria or Victoria in 1853 and Victoria in 1858*, 1859, Chapman & Hall, London, vol. 1, p. 268; George Train,

23 June 1853, in G. F. Train, *A Yankee Merchant in Goldrush Australia: The Letters of George Francis Train*, 1970, Heinemann, Melbourne, p. 25.

3 *Argus*, 19 July 1854.

4 Kelly, *Life in Victoria*, p. 43.

5 Kelly, *Life in Victoria*, p. 31.

6 G. Serle, *The Golden Age: A History of the Colony of Victoria 1851–1861*, 1963, Melbourne University Press, Melbourne, ch. 2.

7 W. Howitt, *Land, Labour and Gold*, 1855, Longmans, London, facs. edn 1973, Sydney University Press, Sydney, vol. 1, pp. 35–6.

8 Train, *Yankee Merchant*, p. 28.

9 E. Carlton Booth, *Another England*, 1869, Virtue, London, pp. 69–71; S. Mossman and T. Bannister, *Australia Visited and Revisited*, 1853, Addey & Co., London, reprinted 1974, Ure Smith, Sydney, pp. 91–3.

10 Compare R. W. Lotchin, *San Francisco 1846–1956*, 1974, Oxford University Press, New York, ch. 3.

11 Booth, *Another England*, p. 69.

12 On the transformation of club life by the gold rush, see P. de Serville, *Pounds and Pedigrees: The Upper Class in Victoria 1850–1880*, 1991, Oxford University Press, Melbourne, pp. 255–60, 352–60.



13 Howitt, *Land, Labour and Gold*, vol. 2, pp. 300–1.

14 T. Pierson diary, 1 September 1853, La Trobe MS 11646.

15 *Argus*, 23 August 1854.

16 Kelly, *Life in Victoria*, p. 77.

17 A. Brown-May, *Melbourne's Street Life: The Itinerary of Our Days*, 1998, Australian Scholarly Publishing, Melbourne, chs 5, 6.

18 J. Bailey, 'Cleansing the Great Unwashed: Melbourne City Baths' in G. Davison and A. Brown-May (eds), *Melbourne Centre Stage: The Corporation of Melbourne 1842–1992, Victorian Historical Journal*, vol. 6, nos. 2&3, October 1992, pp. 141–53.

19 *Argus*, 14 August 1851, 17 May 1854.

20 A. Garran (ed.), *The Picturesque Atlas of Australasia* (1888), reprinted as *Australia: The First Hundred Years*, 1974, Hamlyn, Sydney, pp. 224–5.

21 Lotchin, *San Francisco*, pp. 181–3.

22 D. Dunstan, *Governing the Metropolis: Melbourne 1850–1891*, 1984, Melbourne University Press, Melbourne, pp. 138–46.

23 *Argus*, 3 May 1852.

24 R. H. Horne, *Australian Facts and Prospects*, Smith Elder, London, pp. 23–4; compare A. Blainey, *The Farthing Poet: A Biography of Richard Hengist Horne, 1802–84*, 1968, Longmans, London, pp. 197–202.

25 Phillip Rayson to 'Watch Friends' in England, 16 September 1859, MS 3895–7, Royal Historical Society of Victoria.

26 On the background to this decision, see B. Barrett, *The Civic Frontier: The Origins of Local Communities and Local Government in Victoria*, 1979, Melbourne University Press, Melbourne, chs 13–15; M. Lewis, 'The First Suburb', in Fitzroy History Society, *Fitzroy: Melbourne's First Suburb*, 1989, Hyland House, Melbourne, pp. 16–22.

27 Stubbs and Son advertisement, *Argus*, 18 October 1852.
28 Howitt, *Land, Labour and Gold*, vol. 1, p. 15; for a description of a typical working man's cottage, see J. Sherer, *The Gold-Finder of Australia*, 1853, Clarke & Beeton, London, Penguin facs. edn, 1973, pp. 339–40.
29 C. Twomey, '"Without Natural Protectors": Responses to Wife Desertion in Goldrush Victoria', *Australian Historical Studies*, vol. 28, no. 108, April 1997, pp. 22–46.
30 Diary of J. Greathead, La Trobe MS 116.
31 Howitt, *Land, Labour and Gold*, vol. 1, p. 276.
32 Compare G. Davison, 'Colonial Origins of the Australian Home', in P. Troy (ed.), *A History of European Housing in Australia*, 2000, Cambridge University Press, Cambridge, pp. 6–25.
33 Lotchin, *San Francisco*, pp. 174–81, and L. E. Frost and E. L. Jones, 'The Fire Gap and the Greater Durability of Nineteenth Century Cities', *Planning Perspectives*, vol. 4, 1989, pp. 333–47.
34 *Argus*, 14 November 1853.
35 *Argus*, 14 November 1853.
36 *Argus*, 17 January 1853.

37 Lotchin, *San Francisco*, pp. 184–8; E. Chadwick, *The Sanitary Condition of the Labouring Population of Great Britain (1842)*, ed. by M. W. Flinn, Edinburgh University Press, Edinburgh, 1965.
38 Dunstan, *Governing the Metropolis*, ch. 5.
39 S. T. Gill, *Improvident Diggers in Melbourne*, H7849, and *Diggers at the Theatre*, H339, Picture Collection, State Library of Victoria; D. Goodman, *Goldseeking: Victoria and California in the 1850s*, 1994, Allen & Unwin, Sydney, ch. 3, examines the colonists' fears.
40 D. Goodman, 'Fear of Circuses: Founding the National Museum of Victoria', *Continuum*, vol. 3, no. 1, 1990, pp. 18–34.
41 R. Haldane, *The People's Force: A History of the Victoria Police*, 1986, Melbourne University Press, Melbourne, ch. 2.
42 Mossman and Bannister, *Australia Visited*, pp. 101–2. See also G. B. Earp, *The Gold Colonies of Australia*, 1853, Routledge, London, p. 196.

PLATE 3.1 Teale and Whittock, *Bird's Eye View of Melbourne*, 1854. (La Trobe Picture Collection, State Library of Victoria)

PLATE 3.2 S. T. Gill, *Queen's Wharf*, in Gill and Nicholas Chevalier, *Victorian Illustrated in 1857 and 1862.*
(La Trobe Picture Collection, State Library of Victoria)

PLATE 3.3 S. T. Gill, *Improvident Diggers in Melbourne.*
(La Trobe Picture Collection, State Library of Victoria)

PLATE 3.4 S. T. Gill, *Diggers at the Theatre*, 1853
(La Trobe Picture Collection, State Library of Victoria)

hapter 4

# Labour and Trade Unionism in Victorian Goldmining

## BENDIGO, 1861–1915

*Charles Fahey*



Since the publication of his history of Australian mining, *The Rush That Never Ended*, Geoffrey Blainey's view of working conditions and industrial relations in the Victorian goldmining industry has been largely unchallenged. For Blainey, the relationship between miners and employers was a cordial one. There was, he wrote, no 'sharp rift between shareholders and wage miners. Thousands of miners were themselves shareholders, for gold mining was the one industry in which the humble employee was in the best position to speculate.' Share speculating was not their only avenue of hope; he argued that on

> Ballarat and many declining fields hundreds worked as tributers in the Cornish manner, and rented goldmines from moribund companies and paid them as a tribute a share of the gold they mined. Some made fortunes from the rich patches of the gold, and bought white villas in Bendigo or chocolate-soil farms near Ballarat. Others retired with the gold they had stolen.

This world of speculating miners had a profound impact on the behaviour of unions – with the hope of always making their fortune from share speculating or a lucky tribute, miners were loath to use their unions as battering rams and so exercised moderation. Where men dreamt of becoming rich, class antagonism found little fertile ground.[1]

This chapter will, however, challenge Blainey's sanguine interpretation of working conditions and industrial relations in the Victorian mining industry through an examination of Bendigo, the foremost quartz mining centre of Victoria, from the formation of capitalist mining companies in the 1860s until the collapse of the industry during World War I.

**Figure 4.1** *New Chum Reef, Bendigo, 1875.* (Photo: Nicholas Caire)

In the years after 1860 shallow alluvial mining declined dramatically in Victoria, and those who remained in mining found themselves employed in two principal forms: deep-lead alluvial mining and quartz reefing. Bendigo's miners worked in the latter type of mines. Initially, the quartz reefers simply bashed away with gads and hammers at the surface outcrops, or mined quartz spurs in primitive quarries. When these surface deposits were exhausted the bold reefers chased the reefs below the surface. As soon as they did this, costs rose dramatically. By the early 1860s miners had penetrated 100 feet below the surface, and extra effort was required to haul the rock. Once hauled, the quartz rock had to be crushed and amalgamated with mercury. Initially, miners worked on small claims, but once substantial depths were reached they had to hire help to haul the rock. Sinking in search of reefs was also expensive and risky. When the water-table was struck, claims had to be pumped free of water, and machinery had to be erected to crush the quartz. The impact of this was that miners simply could not afford to sink claims on their own account.

Outside capital had to be introduced – at first from local traders, and later through the formation of limited liability companies.

Such steps towards capitalist mining, well under way in the early 1860s, were given added impetus by the need to sink to around 500 feet in the late 1860s and early 1870s. The success of a handful of claims at these depths awakened a share-market boom and massive investment and profits in the early 1870s. However, by the late 1870s gold at this level had been virtually exhausted, and there was a need to sink deeper still. This required further infusions of capital to erect more powerful hauling machinery. With hand labour, progress was slow, but in the early 1880s mechanised drilling sped up the process of sinking so that reefs around 1000 feet were explored. Yet, by late in that decade, the cycle of reduced yields had returned and the field once again faced a crisis. Another period of exploration followed, and in the early 1890s mines had penetrated well below 2000 feet. Such deep sinking necessitated huge investment in labour costs and brought about massive changes in surface plant. Larger winding engines and steel head frames replaced the whims and wooden poppet legs. By the early twentieth century, mines were being sunk more than 3000 feet below the surface.[2]

**Figure 4.2** *Old Chum Hill, Bendigo, 1875.* (Photo: Nicholas Caire)

Quartz reefing was an industry, therefore, that required considerable and risky investment to win gold, yet the structures for raising this capital remained unstable throughout the century. In the early 1860s, a number of local businesses joined together to float limited liability companies, while a handful of individuals owned claims that produced sufficient wealth at this time to finance further exploration. Some of these men were eventually to be among the wealthiest in the colony. John Boyd Watson, for example, was perhaps the largest urban landowner in Victoria's capital, Melbourne. His great rival, George Lansell, eschewed investment outside of Bendigo but exercised enormous sway in his home town. Suzanne Hunt's chapter shows us the grandeur of his garden, akin to that of British gentry.

The typical speculator, however, was a man of limited means. A couple of dozen men featured among the colonial elite with fortunes of over £10,000, but many investors held estates of only a few thousand pounds. Most of these men had risen from humble circumstances in Great Britain and Germany. In periods when yields were high, particularly in the early 1870s, these men were prepared to finance exploration. They also demanded high and regular dividends, and companies seldom set aside reserves for lean times. During times of investor excitement many bogus companies were formed with the simple intention of winning capital gains from rising share prices. When returns inevitably declined during the mid-1860s, the late 1870s and late 1880s, companies grew starved of investment capital. Moreover, as mines went deeper, risks rose in proportion and, in turn, investor unwillingness to commit savings mounted. As share capital was forfeited by small investors, stock became more and more concentrated in a narrow group of investors led by George Lansell and Ernest Mueller.[3] In the early twentieth century these two investors held around 40 per cent of the shares in Bendigo mining. Both of them held most of their shares in the central Bendigo mining district. This district declined in importance in the early twentieth century and was replaced by Eaglehawk. Here, the industry was also dominated by a small group of speculators, whose main spokesman was Hay Kirkwood.

The development of the quartz mining industry transformed Bendigo from a simple goldfield into a provincial city of some pretensions. In the 1850s, population was highly unstable, with men constantly on the move from rush to rush. The change to quartz mining helped to build a more stable and diverse community. Most importantly, quartz mining encouraged the migration of individuals and family groups from the mining counties of Great Britain. Unquestionably, the strongest influence in shaping the cultural character of Bendigo was the migration of Cornishmen fleeing declining copper and tin mines. By 1881, Cornish migrants made up almost half of the mining workforce. Along the lines of reefs, these men and their families built distinctive communities. On freely available Crown land – in suburbs with colourful names such as Long Gully, Golden Square and California Gully – miners erected humble weatherboard cottages. Owners often built these houses with no architectural pretensions, and added skillions as their families grew. And grow they did. In the first decades of quartz mining, birth rates were

**Figure 4.3** *Shaft Sinking at 2000 Feet, c. 1898.* Shaft sinking was the most skilled mine work. (Photo: J. B. Edwards)

**Figure 4.4** *Great Northern . . . at 1780 Feet.* Miners at work on a compressed-air rock drill. (Photo: J. B. Edwards)

exceptionally high and family sizes large. To cater for this enormous expansion of the 'native' population, the colonial government erected substantial schools that towered over the humble cottages of miners. The miners themselves added to the distinctive character of their villages by constructing solid brick Wesleyan, Primitive Methodist and Bible Christian chapels. The only other substantial buildings in these communities were the homes and shops of local traders and artisans, and the occasional mansion built by a mine owner amidst mining detritus and plant.[4]

In the 1860s and for most of the 1870s, mines were worked by the traditional Cornish method of tap and hammer mining. Compressed-air rock drills were introduced in the late 1870s. These drills were essential in reducing the cost of deep sinking and exploration as they could remove twice as much rock as tap and hammer mining. However, we can overestimate the impact of machinery on the workplace. Even with rock borers, much of the work remained hard manual labour. Mullock and quartz continued to be shovelled and trucked by hand, and the building of workplaces required the manual packing of stone, and huge hardwood timbers had to be manoeuvred in confined spaces to construct stulls and lagging. In 1878 there were only two compressed-air drills on the field; the number jumped to twenty-six in 1881, and seventy-five were at work in 1889. In effect, this meant that there was less than one borer per claim and that the typical miner worked with hand drills.

To understand the place of mining unions we must also explore the market for mining labour. The most important point to note is that, for the period of quartz mining from the 1860s to the early 1900s, there were simply too many miners for the available profitable work. Furthermore, as mines went deeper and claims were amalgamated, unemployment rose to frightening proportions, although its true level was hidden by the use of tribute labour.

The halcyon days for Bendigo miners were in the early 1860s, when quartz claims were being opened out for the first time and shallow alluvial mines were still large employers. In this period, it was not unusual for miners to be paid 50 shillings or more per week, and skilled tradesmen around mines could command much more. Even humble labourers trucking rock could ask for 45 shillings per week. For men who had migrated from Cornwall and other British districts, where wages rarely exceeded 20 shillings per week, these rates must have seemed high. Yet these prosperous days were short-lived. In 1861 the number of Europeans working in Bendigo quartz claims was only 1052; ten years later it was 4594. Labour was also recruited from other Victorian goldfields. The first crisis for miners came during the summer of 1865–66, when drought struck. Without water the mining of Bendigo's quartz reefs ground to a halt. Between November 1865 and June 1866, unemployment was widespread and employers used the drought to reduce wages. In these months weekly wage rates fell to 45 shillings for ordinary miners. This remained the standard wage for the remainder of the century.



Prosperity returned to the field in the late 1860s and early 1870s with the discovery of a number of extremely rich reefs, yet patterns of investment ensured an unstable labour market for miners. Many of the claims floated during the prosperity of these years were bogus affairs where little mining was undertaken. Even in active mines, when gold was not found, the companies closed down and labour was terminated. Moreover, the wealth of the companies on the field was uneven. In 1871, 1872 and 1873, dividends of £1,745,482 were returned for an investment of £826,787. Only half of the companies on the field paid dividends, however, and the remainder had to survive on the calls of shareholders. Small shareholders were reluctant to pay these, and companies that did pay dividends in these years rarely set aside reserves for future exploration, preferring to prodigally return profits to investors. These factors combined to produce a mining management regime that often left labour stranded.

In seeking a lease from the Crown, all companies covenanted to employ a minimum number of employees. In all mining operations, the largest cost, apart from dividends, was labour. When yields declined, company directors often cut the number of miners they employed to the bare minimum listed in their covenants. Many even tried to cut their complement of miners below these numbers, and some simply failed to work their claims. Idle mines, or those working at a reduced capacity, often watched the yields of their neighbours. If neighbouring claims on the same line reef found gold, then exploration might resume in the hope of finding the same reef. This practice was known throughout Victoria as 'shepherding'. The net result was that many mines operated intermittently, and labour requirements were adjusted accordingly. Unless individual miners worked in one of the dozen or so major companies, this resulted in broken employment and a regular search for work. From January 1871 to December 1877, a period when the Bendigo goldfield was generally prosperous, Richard Pope, a Cornish miner who has left us a detailed diary of his work experiences, worked in no fewer than seventeen mines. His average length of employment was just over three months.[5] Newspaper reports also frequently referred to the need for miners to move around a number of claims each year looking for work, with the result that lost time reduced nominal wage rates from 45 shillings to as low as 30 shillings per week.



If employment was unstable in this period of prosperity, conditions deteriorated thereafter. Although there are no official statistics on the rate of unemployment in the late 1870s, newspapers consistently reported closed and abandoned mines, and considerable distress among mining families. From reports in the contemporary press of the number of unemployed, we can estimate that at least 9 per cent of miners were unemployed in 1879 and 1880. With the introduction of compressed-air rock drills in the early 1880s, prosperity returned to the Bendigo goldfields. Output rose dramatically. Yet this increase in production was achieved in part through the amalgamation of claims, and unemployment rates remained high. Even in 1884, when dividends were £318,048 and calls £154,697, the rate of unemployment was almost 10 per cent. Five years later Bendigo's mines would become unprofitable in the aggregate for the first time. Unemployment reached almost 16 per cent in 1888. While the early 1890s saw a revival in mining fortunes, unemployment continued to rise as claims were amalgamated. By the late 1890s unemployment had reached crisis proportions. In 1900 at least one in five miners experienced some periods of no work. Yet the true rate of unemployment was probably much higher, for mining companies reacted to the decline in dividends by the introduction of the tribute system.

The tribute system was an integral part of the normal operation of Cornish mines. Every six months, the mine owners, or adventurers, would open their mines for inspection and divide them into sections. A 'Dutch auction' would be held, and miners would bid to work each section for an agreed percentage of the ore won.[6] In Victoria, tributing never became the preferred method of working mines and only replaced wage work when gold yields declined. Under the Bendigo system of tributing (the system was similar on other Victorian fields), the miner rented a portion of the mine from the owner or company and, as a rent, paid a proportion of the gold won. The percentage varied from 15 to 30 per cent of the gross yield of gold, but the miner had to pay the company for hauling, carting and crushing the quartz. The miner had also to pay for timber, candles and explosives and for the sharpening of tools. An alternative was the 'halves system', where the miner received only half the gold won, with the company paying for hauling, carting and crushing. The company also provided timber, candles and other mining materials. Tributes were let for specific periods, usually six to twelve months. Yet, for most of the nineteenth century, tributes were not legally binding on the mine management and could be terminated at their discretion. All evidence indicates that tributes were only let when mines could not be profitably run through wage labour, and tributers seldom made half-wages.[7]

Comments by contemporaries as well as one official return suggest that tributing was, from the late 1880s, an extremely common form of employment. A return compiled by the Mines Department in 1889 recorded one in five mine employees as working on tribute. Five years later, at the select committee hearings into tributing, Mr Praed of the Miners Union testified that 500 of his 2000 members were employed as tributers.[8] In July 1898 the editor of the *Bendigo Independent* claimed that 700–800 miners were thrown out of work by the 1897 Mining Act, which attempted to reform the tribute system. This number represented 18–20 per cent of men engaged in mining at Bendigo.[9]



The first attempt at labour organisation on Bendigo was an Eight Hours Association, formed in the winter of 1865. A combination of working miners and sympathetic businessmen, this organisation chose its time poorly; drought in the summer led to massive unemployment and the collapse of the association.[10] A more enduring body was the Bendigo Miners Union, formed in 1872. This was one of the more prosperous years in the history of Bendigo quartz mining, and in February the newly formed union demanded an eight-hour day. Several mine owners responded to this demand by reducing the wage rate from 45 shillings to 42 shillings. Miners quickly joined the ranks of the new union, and mass demonstrations demanded the full wage rate for eight hours. In the boom conditions mine owners quickly accepted the miners' demands and agreed to eight hours at 45 shillings. Leading this agitation was Robert Clark, the son of a Durham pitman who had worked as a goldminer in the early 1860s and then found employment as a journalist soon after.

From this time, the Bendigo mining field remained industrially quiet until August 1874, when a new mining accidents statute was introduced. This Act, based on British precedent, stipulated that in the case of accidents mine management was prima facie responsible, and if its negligence was confirmed by a subsequent inquiry the company could be sued by injured miners or their families. In response, petulant mine managers drew up an agreement that all miners had to sign before receiving employment. Among the terms of this agreement were clauses that stipulated that miners would submit to strip searches for gold, and that they themselves were responsible for the safety of their workplaces. The miners reacted indignantly; mass meetings and the threat of strike action forced the owners to relent. With mining dividends providing an aggregate return of 165 per cent on capital invested, it made little sense for major mine owners to suffer long periods of industrial turmoil.[11] Although the miners won both these battles of the early 1870s, they were essentially limited actions. Winning of the eight-hour day was a major achievement, but there was no attempt to use the buoyant conditions of the early 1870s to win increased wages.

The final major confrontation between miners and mine owners occurred at the end of the 1870s boom. It was precipitated by the attempts of a number of mine owners, who were experiencing a decline in profitability, to cut costs in August and September 1879 by a reduction in the nominal mining wage rate from 45 shillings to 42 shillings. This bid prompted strike action and mass demonstrations by the miners. At the height of the conflict miners flooded into the union and membership peaked at 1850 members. Unlike previous disputes, which tended to be short-lived, this one lasted more than a month. At its peak 200–250 men went on strike and were supported by strike pay from the union.



By contemporary standards this appears to have been a minor dispute, but in the colonial context this was a new form of conflict. Most Victorian industrial disputes at this time were based narrowly on small craft unions, and confined to individual shops. This dispute involved semi-skilled workers and threatened to bring a whole community to a standstill. Throughout the dispute tensions were high. Mine owners saw, in the recent decline in dividends, a threat to the independence they had won in Victoria, while insecurity of employment had also become a part of the working world of miners. The strike itself took place against a background of political crisis and economic recession. For many men of property, Victoria at this time seemed destined to be governed by mob rule or at least the rule of demagogues such as the Premier, Graham Berry. The manner in which the strike was solved only served to confirm this. In November 1879 the colonial government threatened mine owners with a termination of their leases if labour covenants were not followed. Mine owners had little option but to comply, yet it is clear that from this time companies began to use the tribute system more consistently in order to reduce costs.[12]

For the Miners Union, the end of the strike was not without its own problems. Membership quickly fell away and general meetings often attracted as few as ten or fifteen members.[13] Part of the problem lay with the character of the union leadership. Since the early 1870s it had been dominated by Robert Clark. By the mid-1870s his ambitions had transferred to colonial politics and, strikes apart, he had little time for the daily business of running a union. He did, nonetheless, introduce a bill in 1877 which confirmed the existing eight hours as the legal working day in Victorian mines.[14]

In the late 1870s the initiative for union activity shifted to the alluvial mining district of Creswick. A miners' union had been formed at Creswick sometime in the early 1870s but it was generally inactive. In August 1878 mine directors sought to reduce the price of contracting in the Ristori mine. During the crisis the moribund local union was reformed, under the leadership of W. G. Spence and John Sampson, and the Ristori directors gave in to the miners. A further strike occurred on the Creswick field when owners attempted to blackball Spence and Sampson. In the early years of deep-lead alluvial mining at Creswick the return on capital was truly astounding and protracted strikes made little economic sense for investors. Between 1879 and 1884, for example, the Ristori mine paid its shareholders £231,650 for an investment of only £6650. Under these circumstances mine investors came quickly to terms with life with the union.[15]

Safely established in Creswick, Spence turned his attention to uniting all miners in Victoria, and in 1883 the Amalgamated Miners Association (AMA) of Victoria was established. In Bendigo, government of the local miners' union had always been tense. There was a split between working miners and middle-class leaders such as Clark. During the disputes of the 1870s Clark's voice was often one of moderation, warning the miners to avoid violence. As a politician, he drifted further towards a conservative stance in the late 1870s and early 1880s. In March 1880, for example, he lectured his electors that 'nothing should be done . . . to create a breach between the employer and the employed, they must have no conflict between capital and labour'.[16] During a constitutional crisis of 1879–80 he, like the mine owners, supported the opponents of the populist Premier Berry. By the early 1880s he had alienated several committee members of the Miners Union and he was in chronic ill health. Members accused Clark of serving his own political ends and suggested that he had done little of the real work of raising money during the 1879 strike. He had favoured the more 'congenial work of spouting in St James' Hall on Saturday evenings'.[17] Amidst this dissension it was possible for the Creswick leadership to gain control of the Bendigo branch in 1882. Mr J. Davey from Ballarat, and the vice-president of the AMA, promised the Bendigo branch 'a society of miners, managed by miners, and for the mutual and lasting benefit of miners; not a strike association, officered by politicians'.[18] After a series of public meetings Davey convinced sufficient Bendigo members of the value of amalgamating with Creswick, and in May 1882 the Bendigo miners joined Spence's union.[19]

In the early 1880s the AMA quickly established itself as an integral part of the community in Victorian mining towns. The annual miners' sports day, for example, was one of the major public holidays, and the benefit functions of the union were crucial in the lives of countless mining families. In his memoir, *Australia's Awakening*, Spence boasted that this friendly society function was a means of recruiting normally apathetic miners, and that through it they were able to institute the closed shop in Creswick.[20] In an industry characterised by constant danger, accident relief was a major benefit. In the late 1880s there was an annual average of 7 fatal accidents, 23 'serious' accidents and 370 accidents described as 'not serious', among the 2000 members of the Miners Union in Bendigo. For a joining fee of one shilling and a weekly charge of sixpence, injured unionists were entitled to a subsistence wage of £1 per week. In the event of death their next of kin received £110. Permanently incapacitated members could buy themselves out of the fund with the payment of a lump sum. These payments were usually £50 or less and allowed the member to establish a livelihood, albeit precarious, in a small business – driving a cab or raising poultry, for example. Unlike its earlier predecessors, the Bendigo union in the 1880s was controlled by working miners or, in the case of its long-serving secretary, Peter Phillips, miners incapacitated by accident.[21]

Although this accident fund was a critical factor in keeping many families out of the clutches of institutional charity, the union did little to win wage increases in the years after 1880. And while this appears to be a sign of conservatism, or failure, the union did try to guard the interests of its members within the confines of an appalling labour market. Even when the union was prepared to use strike tactics in the 1870s, these were essentially defensive affairs. Not even in the boom of the 1870s, with miners pouring in from other declining goldfields and from overseas, could the union force wage increases. Even the victory of 1879 was not a clear one. Forced to reopen his mines, George Lansell did so in a grudging way and frequently employed only sufficient men to meet the requirements of the labour covenants. A year later he caused major disruption by refusing to pump water from mines adjoining his 180 mine, and threw some 200 men out of work for several months between April and July.[22]



As the miners' watchdog in the 1880s, the union carefully monitored the enforcement of the mining safety regulations and quickly reported any suspected breaches to the Department of Mines. It also policed the observance of the labour covenants by mining companies. Most importantly, the union monitored the honouring of the wage rate. In February 1886 a number of members began to report that they were being replaced by young men employed at 42 shillings, rather than the union rate of 45 shillings.[23] A sub-committee, headed by Secretary Phillips, visited the offending mines and convinced the management to abide by the standard rate. The union also became adept at lobbying local politicians. In a largely working-class electorate, politicians could not ignore a body that at times represented 2000 men, and the union frequently corresponded with local members. Through such men the union protested against infringements to the mining regulations, fought for improved ventilation in the mines and won the concession of a seven-and-a-half-hour day.[24]

In the 1880s the union debated the need for a closed shop in local mines, and in March 1888 decided to ballot members on this issue. More than 2000 ballots were sent out to members, returning an overwhelming vote for compulsory unionism: 1514 in favour and only 188 against. Although this ballot had the impact of inducing 170 wage-earning non-members to join the union, fifty men still refused to sign up. In a climate of extensive unemployment the union made no attempt to force the issue through industrial action. The only pressure they could assert was the moral force of the ballot. Furthermore, in the discussion about inducing miners to join the AMA, no attempt was made to coerce tributers. It made little sense to force men who were battling to feed their families to pay union dues.[25]

By the end of the 1880s local officials had become alarmed at the widespread use of tribute employment and the low rewards won by tributers. The indefatigable Phillips began to collect information from tributers and from published company reports on the amount of gold obtained by tributers. Through the press he published articles on the issue, and in 1890 he successfully lobbied local members to pressure parliament to set up a royal commission into goldmining. This commission heard considerable evidence on the issue but no further action was taken. Continuous union pressure resulted in the formation of a Select Committee on Tributing in 1893. This committee took evidence on most goldfields from both local managers and union officials. Unionists pointed out that miners had no security of tenure and argued that a subsistence wage should be the first call on all gold produced by tributers. They also argued that many companies used the system as a means of avoiding calls, using tributers to undertake exploration work then throwing them aside when gold was found.



Reform, however, took another three years of agitation. When it finally came in 1897, the consequences showed how powerless local miners had become. An amendment to the Mines Act stipulated that all tribute agreements had to be placed in writing and lodged with the local mines inspector. More importantly, the Act provided that all tributers were to be paid half-wages as a subsistence wage, and this was to be the first cost deducted from any gold won. When the Act was introduced mine owners claimed that it created too much bookwork and they simply dismissed most of their tributers, adding further to the prevailing economic distress. Attempts were made to find a compromise, and meetings were held between the union and mine owners, but the latter refused to relent. In the end, the new Act was simply ignored, and desperate miners offered themselves for tributing contracts without bothering to register them with the mines inspector.[26]

The issue irrevocably split the union. Branch meetings held to discuss this matter ended in disorder.[27] By the end of the century the Bendigo branch, the largest AMA one in Victoria, was industrially impotent.[28] Members simply lost interest, meetings were poorly attended and tensions arose between committee men and Secretary Phillips.[29] During the late 1890s, Phillips repeatedly sought permission from the committee to deal with the question of non-members and the maintenance of a standard wage for miners, but his fellow unionists refused to be drawn into a sustained fight. They probably agreed with the local liberal newspaper, the *Bendigo Independent*, that 'the ordinary individual does not pick a quarrel with his bread and butter, unless he has the assurance that by doing so his bread will be returned to him buttered on both sides'.[30] Desperate individuals occasionally resorted to violence – in July 1898 a leg on the head frame of Lansell's Red White and Blue mine was mysteriously dynamited – but such actions were denounced by the union and most miners simply struggled to make ends meet as best they could.[31]

In the first decade of the new century, the Miners Union adopted a more militant stance despite worsening employment conditions. This change in direction was due to a number of factors: the opposition of employers to wages boards; the rise of a strong local Labor Party; and new blood in the management of the union.

In 1896 the Victorian parliament passed a temporary measure to introduce minimum wages in six 'sweated trades'. Early in the new century this measure was made permanent, and the legislation was quickly extended to other industries. By the end of 1906 there were forty-nine boards and almost 50,000 operatives covered by minimum wages. Critically, these boards gave regular wage increases. Between 1901 and 1910 labourers in the clayholes of Bendigo saw their weekly wages increase from 40 shillings to 50 shillings. During the same period the nominal rate for miners remained at 45 shillings per week, a wage rate that had been ruling since the 1870s. Miners first applied for a wages board in 1905. According to their secretary, they registered 'to seek justice in a peaceful manner and not by way of strikes, which were prevalent and mostly unsuccessful in the early days of their organisation'.[32] Attempts to win a wages board were opposed by employers. Negotiations continued through 1906 and 1907 and always met with the same response: the industry could not afford pay rises. This intransigence gave local Labor Party politicians a powerful platform on which to seek support, and by 1907 Labor controlled all Bendigo seats in the Legislative Assembly.[33]



During the long battle to win a wages board, a number of younger men joined the local executive of the Miners Union. One of the more prominent of these was John R. Little. Born in Eaglehawk in 1875, Little entered the mines as a young lad, working on a winch. He worked in several local mines before he migrated to the West Australian goldfields. When he returned to Bendigo in 1904 he became a union steward and in 1908 he was elected general secretary of the Bendigo branch. In 1913 he left there to become general secretary of the Federated Mine Employees Association, a body formed in 1909 out of the old Amalgamated Society of Miners in order to act as a nationwide union. Little combined union activity with local politics and was elected as a Labor candidate to the Eaglehawk Borough Council in 1907. He was one of a number of new executive members who had experience on other fields and an interest in Labor politics. Among a minority of miners this caused some dissension, but mass meetings in December 1912 endorsed trade union support for the Labor Party.[34]

With increasing demands for wages justice, local mining investors inflamed class tension when they insisted that all miners should be searched at the end of their shift. This reflection of distrust was deeply resented by miners, many of whom were staunch chapel attenders. When an attempt was made to enforce the policy at the Virginia mine in December 1907, the miners struck. After a prolonged dispute, the first since 1879, the men returned to work without success. This defeat bred resentment, yet it also fostered a renewed effort to push for a closed shop. Resentment was also heightened by the publication of Dr Walter Summons' reports on miners' phthisis (tuberculosis), which clearly demonstrated that deep quartz reefing was killing Bendigo miners.[35]

In February 1911, after six years of agitation, a wages board was appointed. The structure of these boards did not favour quick reform. Boards were composed of equal numbers of employees and employers, and presided over by a neutral chair. The miners' board quickly became bogged down in argument, and sittings extended over ten months. The final award was at best a compromise. Although wage rises were generally made for all classes of labour, these proved less satisfactory than miners had hoped. Indeed, on some alluvial fields near Ballarat, wages actually decreased. The inevitable result was industrial disturbance: bitter strikes were waged in the Ballarat and Stawell districts. In Bendigo the local union was dismayed, and agitation arose to abandon the wages board system and seek justice through federal arbitration. In April 1913, during a federal election campaign, a federal log of claims was served.[36]



In Bendigo the failure of the wages board led to a hardening of attitudes. Local officials pushed to enforce a closed shop. They argued that the cost of arbitration was expensive and that non-members were simply freeloaders on the union. The consequent move to recruit members galvanised local organisers. With the help of the secretary of the local branch, stewards in each mine approached non-union members and cajoled them into joining the union. Local investors reacted angrily to this action, and recruiters were frequently sacked. In the case of the Red White and Blue mine this led to a brief stoppage in December 1912.[37]

The question of non-unionist labour was brought to a head during the 1913 election campaign. A key issue in the Labor Party platform became preference for trade unionists in arbitration awards, particularly the need for an award for agricultural labourers in rural districts. In the midst of the campaign the Bendigo branch of the Miners Union conducted a ballot for compulsory unionism. The result was a strong endorsement of compulsory unionism, and the threat of a general strike hung over the election campaign. Although Labor lost the election nationally, the miners of Bendigo threw their support behind Labor and after thirteen years in federal parliament the local liberal candidate, Sir John Quick, was defeated. His opponent, Jack Arthur, the son of a Ballarat miner, had represented the agricultural implement makers in the famous Harvester case of 1907, and much of his campaign was centred around the extension of the Arbitration Act. By the end of May only forty-two underground workers out of a total workforce of 2000 had refused to join the union. The union sought permission from employers to gain access to mining leases to discuss membership with recalcitrant miners. When this was refused, miners in two Eaglehawk mines struck in late June. This strike dragged on for almost four months and, although strikers refused to return to work, police protection was provided to 'scabs' and the mines were kept going with heavily reduced manpower.[38]

Defeated in this strike, the Miners Union turned to the Federal Arbitration Court. Once again, employers stonewalled the action and sought legal measures to delay a hearing. When a hearing was finally granted in 1915, miners were determined to win wage rises despite the state of the industry. On the first day of the court's sitting, Justice Powers questioned John Little:

> You first of all get your living wage – every man must get that, according to the practice of the Court, and then the other matter comes in. If it could clearly be proved to the Court that an industry could not pay at a particular time, should you press the claim should be paid Mr Little – you know what the result would be?

Little replied, 'We say, your Honor, that if the industry is dying we have no right to die with it'.[39] The miners pressed on with their case.

The first federal award followed the lead of the Miners Union, choosing the path of significant wage increases over preservation of the mining industry. The base wage of a miner was increased from 54 shillings to 64 shillings and sixpence, and there were significant increases for other grades of labour such as labourers and battery feeders. Through arbitration, miners rejected the low wages that had ruled for most of the decade. The increase in costs did not lead to a flight of capital; in the end the miners were robbed of victory by the simple vagaries of geology. In 1916, 1917 and 1918 both yields and profits declined. Employment in the mines also collapsed in these years.



It is mistaken to see goldmining as an industry free of class tensions and industrial conflict. On Bendigo from the early 1860s a clear cleavage developed between the interests of mine owners and wage-earning miners, and this was little ameliorated by miners' participation in mining shareholding or gold production under tributing. Very few miners became shareholders, while tributing clearly favoured mining companies and mine owners. In the generally prosperous decade of the 1870s, miners and mine owners fought a number of bitter actions over wages and conditions. Throughout the 1880s and 1890s there were no major industrial actions in Bendigo, yet tensions remained high over such matters as tributing. Working in a climate of high unemployment, most union officials feared taking industrial action and worked for reform through local politicians. A transformation in union activity came with the introduction of state labour market regulation. Yet for many years miners were denied access to wages boards. This undoubtedly strengthened the resolve of miners, leading the local Miners Union to become more militant in the years immediately before the war. By 1913 union officials had resolved to win wage increases through the Federal Arbitration Court, even if this meant the end of their industry.

The pre-war years may have witnessed a new determination by miners to win wage justice, but this did not reflect the emergence of divisions between capital and labour, for such divisions had been manifest from the very formation of quartz mining companies. Quartz mining was an inherently unstable industry. A small number of great fortunes were won, yet the costs of production were high and the returns to most investors were uncertain. This mix produced relations between mine owners and miners every bit as hard as the famous quartz of Bendigo.

## Notes



1 G. Blainey, *The Rush That Never Ended: A History of Australian Mining*, 1963, Melbourne University Press, Melbourne, pp. 299–300; For a recent important work that accepts Blainey's interpretation, see R. Markey, *The Making of the Labor Party in New South Wales, 1880–1900*, 1988, New South Wales University Press, Sydney.

2 For the transformation of Bendigo mining, see the Quarterly Reports of the Mining Registrars and the Annual Reports of the Mines Department. The process can also be followed through the local press, particularly the *Bendigo Advertiser* and the *Bendigo Independent*. The yield of gold from the Bendigo field was: 4,519,543 ounces in 1851–60; 2,220,554 ounces in 1861–70; 2,550,566 ounces in 1871–80; 1,847,254 ounces in 1881–90; 1,977,648 ounces in 1891–1900; and 2,092,963 ounces in 1901–10.

3 This picture of the wealth and background of investors has been compiled from probate records and death certificates. At the most about 4 per cent of miners owned shares in the 1870s and 1880s.

4 This picture is compiled from a variety of sources, including birth certificates and municipal rate records. In 1871 the population of Bendigo was 28,577; by 1881 it had risen to 35,584. The city had lost population by 1891 but revival in mining in the early 1890s increased it from 34,089 in 1891 to 39,141 in 1901. In both 1871 and 1881 around 80 per cent of Cornishmen in Bendigo worked as miners. In 1881 the age-specific fertility for women aged 25–29 was 379 births per 1000 women; for the cohort 30–34 the figure was 293 per 1000 women.

5 Calculated from the diary of R. Pope, State Library of Victoria, Manuscripts Collection, MS 11918.

6 For the Cornish system of work, see R. Burt (ed.), *Cornish Mining: Essays on the Organisation of Cornish Mines and the Cornish Mining Economy*, 1969, David & Charles, Newton Abbot.

7 The problems of poor tributers is a recurring theme in the diary of the mine director Isaac Edward Dyason, State Library of Victoria, Manuscript Collection, MS 10813. Dyason, a native of Kent, was a director in many of the field's major

mines and managed the affairs of the so-called 'Quartz King', George Lansell. Dyason frequently noted that tributers made only £1 per week. See his diary 11 November 1879, 4 February 1880, 4 April 1893 and 8 June 1895.

8 *Bendigo Independent*, 9 September 1893.

9 *Bendigo Independent*, 30 July 1898.

10 The progress of the Eight Hours Association can be followed in the *Bendigo Advertiser* from July 1865 to July 1866.

11 For the best account of these disputes, see M. Roper, 'Public Ritual in a Goldmining Community: Bendigo 1865–1885', Monash University thesis, 1985 (see ch. 6 for 1870s strikes). In 1872 investors received £683,140 for an investment of £360,300. In 1874 the return on £188,384 was £500,615.

12 For this strike, see Roper, 'Public Ritual'; for the reactions of one mine owner, see Dyason diary.

13 For the Bendigo Miners Union, see K. Peoples, 'To What Extent Does the Second Phase of Trade Unionism in the Bendigo Miners Association, 1882–1888, Represent a Growth in Trade Unionism and a Development in Both Unionism and Working-Class Identity over the First Phase, 1872–1882', University of Melbourne thesis, 1975.

14 For Clark, see D. Potts, 'Robert Clark', in D. Pike (ed.), *Australian Dictionary of Biography*, 1969, vol. 3, Melbourne University Press, Melbourne, pp. 407–8. 

15 For the Miners Union in Creswick, see M. McAleese, 'Creswick in the 1880s', Monash University thesis, 1968.

16 *Bendigo Independent*, 16 March 1880.

17 *Bendigo Advertiser*, 20 April 1882.

18 *Bendigo Advertiser*, 22 April 1882.

19 See *Bendigo Advertiser*, 9 May 1882, for half-yearly meeting when the crucial vote on amalgamation was taken.

20 W. G. Spence, *Australia's Awakening: Thirty Years in the Life of an Australian Agitator*, 1909, Worker Trustees, Sydney, p. 33.

21 For the Bendigo Union in the 1880s, see Bendigo Miners Association, Minutes 1885–89, Noel Butlin Archives Centre, Australian National University, P 103/4. Statistics of accidents for the same years were given in the Annual Reports of the Mines Department.

22 The dispute over drainage was reported extensively in *Bendigo Independent* from April to July 1880. For a description of idle mines, see *Bendigo Independent*, 4 May 1880.

23 *Bendigo Advertiser*, 5 February 1880.

24 These dealings with local politicians are detailed in the Union Minute Book.

25 For the vote on the closed shop, see *Bendigo Independent*, 17 and 21 March 1888.

26 For the failure of the 1897 amendment, see the report of the Bendigo Miners Association in *Bendigo Independent*, 1 March 1899. See also *Bendigo Independent*, 30 July 1899, and for the attitude of mine owners, see Dyason diary, 24 November 1899.

27 *Bendigo Independent*, 13 July 1898.

28 In all likelihood the AMA at Ballarat was also quiescent in the years after 1870.

Weston Bate does not detail any strikes or other industrial action. W. Bate, *Lucky City, The First Generation at Ballarat, 1851–1901*, 1978, Melbourne University Press, Melbourne.

29 The half-yearly meeting for January 1899 was attended by only about fifty miners. *Bendigo Independent*, 23 January 1899.

30 *Bendigo Independent*, 5 August 1899.

31 For this and a similar incident, see *Bendigo Independent*, 13 July 1898.

32 For the impact of wages boards, see Reports of the Chief Inspector of Factories, Work-rooms and Shops, 1901–10. Evidence of J. R. Little, Secretary of the Federated Mining Employees Association of Australia, Arbitration Transcript, Australian Archives Series 1958/7/1915/30, p. 15.

33 For the Labor Party in Bendigo, see C. Cleary, *Bendigo Labor: The Maintenance of Traditions in a Regional City*, 1998, Colin Cleary, Bendigo.

34 For the career of J. R. Little, see *Bendigo Advertiser*, 23 July 1913.

35 W. Summons, *Miners' Phthisis: Report on the Ventilation of the Bendigo Mines*, 1906, Stilwell & Co., J. Kemp, Acting Government Printer, Melbourne; *Miners' Phthisis, Report of an Investigation at Bendigo into the Prevalence, Nature, Causes and Prevention of Miners' Phthisis*, 1907, Stilwell & Co., Melbourne; for union protests about the poor state of ventilation, see *Argus*, 19 and 29 October 1907.



36 For the reaction to the wages board, see *Argus*, 26 and 29 January 1912, and *Bendigo Advertiser*, 25 and 29 January, 2 February 1912; for the federal log, see *Bendigo Advertiser*, 5 April 1913.

37 *Bendigo Advertiser*, 5 December 1912.

38 This dispute can be followed in *Bendigo Advertiser* from April to end of October 1913.

39 Miners' transcript, p. 15.

# Mullock Heaps and Tailing Mounds

Chapter 5

## ENVIRONMENTAL EFFECTS OF ALLUVIAL GOLDMINING

*Barry McGowan*



Alluvial goldmining was a form of accelerated erosion, and many of the scars that it has left on the landscape are permanent and alarming. Many streams and rivers were worked along almost their entire length, not just once but several times, and often by different techniques. Some alluvial workings appear relatively benign and are characterised by a seemingly inoffensive mass of shallow holes and hummocks of dirt. Elsewhere, the landscape appears to explode before one's eyes. There is no subtlety about these workings. The holes cannot be described as pits but, rather, quarries – thirty, sixty or even a hundred feet deep, and spread over many acres of ground.

The environmental impact of mining, in particular alluvial goldmining, has been commented upon by some historians. For example, Geoffrey Bolton described the impact of the gold rushes as 'almost entirely destructive', and commented unfavourably on the voracious demand of the mining industry for timber and on the effects of waste sludge on streams and rivers.[1] Geoffrey Blainey has also commented in a similar vein, implying that the early pioneers gave scant regard to the environment in their pursuit of wealth and prosperity. However, as shall be discussed, not all pioneers had such disregard.[2] In general, though, historians have tended to pass over as of little consequence the environmental effects of mining, unless they are highly visible.[3]

This chapter includes a discussion of the technologies used in all types of alluvial goldmining, in particular the more capital-intensive forms such as hydraulic sluicing and dredging. The main focus, however, is the environmental effect of these activities, for example siltation, pollution, erosion and deforestation. Examples will be taken from mining fields in south-east New South Wales and north-east Victoria. Some of these fields were among the richest and most heavily worked areas of alluvial goldmining in Australia. Particular attention will be given to the political debate surrounding the issue of dredging in north-east Victoria.

**Figure 5.1** Alluvial Mining Claim, Blind Creek, near Harden, New South Wales. (Source unknown)

At its most basic, alluvial goldmining resulted in heavily scoured creek beds and gullies worked down to bed rock and, adjacent to this, on the shallow leads, numerous small shallow shafts and hummocks of dirt. These have sometimes been referred to as 'small claim' or 'pothole tailings'.[4] Claims were worked by washing river or stream dirt in a gold pan. As an advance on that, a cradle or long tom with a bucket to raise water or a short twelve-foot box sluice was used, the miners working as small cooperative parties. The cradle was an open box with rockers underneath, a movable hopper with a perforated screen on the top, and inside the box a number of slides with riffles. Long toms were a type of sluice box and were used where the bottom of the workings was below the adjacent creek bed and there was insufficient fall, or where the water supply was inadequate. Box sluices were also used where the fall was not sufficient. They were set on trestles, the incline depending upon the nature of the wash dirt, which had to be elevated or carried to the sluice. The more difficult the wash, the steeper the grade required. Where gullies were worked to a shallow depth, the process was known as 'gully raking'.

Individual claims were often very small and not worked to any great depth. The environmental effect was more significant where a large number were established in the same area. On many claims, almost from the outset, difficulties arose from a lack of water, which required the construction of water races and catchment dams. In addition, where the wash dirt contained a large amount of clay, puddlers were constructed. Puddlers consisted of a large circular hole in which the wash dirt and water were mixed. The dirt was broken down by triangular harrows attached by a chain to a shaft located in the centre of the hole, the shaft being yoked to a horse.

On the larger claims the network of dams and water races was complex and extensive. Some of the races could be many miles in length, and their remains can often be seen on the side of hills or roads. Where the hills were particularly steep, stone embankments reinforced the races which were often flumed across gullies, sometimes on large wooden trestles, or by inverted siphon. On the Shoalhaven River in New South Wales, the Limekilns race was 24 miles long, with branch races of between 8 and 10 miles in length, and the Spa race was 16 miles long.[5]

**Figure 5.2** *Chinese Sluicing near Beechworth*, engraving after drawing by Nicholas Chevalier, in *Illustrated Australian News*, 28 January 1867. (National Library of Australia)

Elsewhere, the workings resembled rectangular quarries, of 130 × 130 feet or more, and ten or more feet in depth. The method of working was referred to as paddocking and involved, first, the removal of the overburden by barrows (stripping) to tips and, second, the removal of the wash dirt to the sluice boxes for washing. The boxes were set on trestles, which were moved back into the cutting as work progressed. On most fields, horses and drays eventually replaced the barrows. Excess water on the creek-bed claims at the Araluen and Jembaicumbene fields in south-east New South Wales led many proprietors to introduce steam engines and construct extensive tail races for drainage purposes. At Araluen, the size of some of the claims warranted the use of horse-drawn skips or locomotives to assist in transporting the stripping.[6]

Another form of working was ground sluicing. This was only used where the bottom was sufficiently high to provide the necessary fall. It was the main method of working elevated ground. Often the ground sluices were only channels cut down to bed rock, but they were also built of sawn planks and boards. The wash dirt was dug or broken out of the claim and conveyed to the sluice by a stream of water run over the face of the claim from water races. These claims could be very large and often resembled large, shallow quarries. For example, the Pipeclay 1 diggings on the Shoalhaven River were around 1000 feet across at their widest point and 1200 feet at their longest, the face of the diggings varying in height by up to 16 feet.[7]



Mercury pollution may also have been a consequence of alluvial gold-mining. It is difficult to ascertain the prevalence of mercury usage during the gold-rush era. Its use was, however, advocated in contemporary accounts of common sluicing procedures. For example, the practice was to sprinkle some mercury into the upper sluice boxes, from which it gradually progressed to the lower ones. Mercury was also placed in the lower boxes. The quantity used depended on the amount of fine gold in the dirt.[8]

Hydraulic sluicing was used on elevated ground, where the drift and overburden were too deep and often too poor to be worked by any other method. In essence it involved the removal of the auriferous drift by the use of water conveyed under pressure to a hose or hoses, which would then be turned against the face of the workings. After the mine was opened up the gravel banks were undercut by the hoses. The gravel was reduced by the fall and swept by the hoses into the sluice boxes, until such time as the claim had been washed to bed rock. By this method larger areas could be processed in much less time than would be the case with ground sluicing, which relied on manual labour. A distinction can perhaps be made between the use of pipes or canvas hoses with nozzles and work performed by a giant nozzle. The latter was only introduced into Victoria from California in the mid-1880s and is distinguished from the former by the much larger quantity of material that could be removed.[9]

Centrifugal sluicing or jet elevating was similar to hydraulic sluicing, with the difference that the dirt was hosed into a sump hole at the base of the workings, and sucked or pumped up by a gravel pump into a vertical elevator pipe. The pipe was mounted on wooden scaffolding and was sometimes

**Figure 5.3** Hydraulic Sluicing with Monitor at the Pioneer and Union Claim, Mitta Mitta, 1888. (Minerals and Petroleum Victoria)

referred to as a jet elevator. From there the wash was conveyed into sluice boxes, which sat on the scaffolding above the face of the workings. With centrifugal dredging the gravel pump and boilers were mounted on a barge, which was stationary until it was necessary to move it nearer to the working face, at which time the site was dammed and flooded and the barge floated in. This form of mining was developed in the Yackandandah area of Victoria in the 1880s and soon became the prime means whereby the river flats were worked.

In appearance, the remains of the hydraulically sluiced claims resemble massive quarries. For example, the Spa South site on the Shoalhaven River in New South Wales is about 1000 feet long and 165 feet wide, with a face of 60–100 feet. At this claim it was expected that one nozzle alone would hose 2000 cubic yards of wash per day. The Limekilns 2 site on the Shoalhaven River comprises two main areas, one measuring 660 × 60–100 feet, with a face of up to 85 feet in height. It adjoins another site of similar size. The face of the centrifugal sites could be of a similar dimension but, as they were on lower and more poorly drained ground, the floor of the workings often resembles a small lake or swamp.[10]

**Figure 5.4** Centrifugal Dredging, Staghorn Flat, Yackandandah, Victoria. (Yackandandah and District Historical Society)

Bucket dredges were not introduced on most fields until the early 1900s, by which time all other forms of alluvial goldmining operations in the vicinity had ceased. The dredges were akin to large floating factories, which excavated the bottom of streams and creeks by using a long chain of buckets mounted on the pontoon along with much of the other equipment. They moved under their own power by using the buckets for cutting a passage through the wash, breaking down the gravel in front and stacking it behind after processing. In the earliest dredges the tailings were emptied back into the excavation made by the buckets. Subsequently, however, tailings elevators were used to ensure that the tailings were deposited away from the excavation.

Alluvial goldmining, even in its more primitive forms, affected the environment from the outset and was not as benign in its effects as present-day landscapes might suggest. Bolton quoted an English visitor, Louise Meredith, who thought that alluvial mining left a landscape 'more irredeemably hideous than the bleakest mining village in any English coal or iron district'.[11] In 1853, William Howitt, the radical journalist, wrote of the amazing propensity of the diggers to fell trees, for no sooner had they done their day's work than they began felling. Of the creek-bed (or wet) diggings on Spring Creek, he wrote:

> Many of these wet diggings are from ten to twenty feet deep; and not only are they thus flooded with fetid water, but the sides continually tumble in, and require to be cased with slabs or sheets of stringy-bark . . . These deep and unshapely abysses, are black with mud, in which lie beams and poles, and masses of stringy-bark.

Of the dry diggings:

> Every yard of ground is there dug up . . . The whole surface is honey combed with holes, from ten to forty feet and more deep . . . In some place large square masses have been dug out solidly; and the excavations resemble the foundations of some ancient city, like Nineveh, laid open to the day.

On visiting nearby Reid's Creek he was compelled to write:

> For nearly two miles a wide valley is completely covered by tents and the soil turned upside down by diggers. A more rowdy and uninviting scene I never saw . . . All the trees were cut down; the ground, where it was not actually dug up, was eaten perfectly bare by lean horses . . . more shabbiness and apparent wretchedness it would be difficult to conceive.[12]



Streams and rivers were immediately affected by the work on wet diggings. The waste water, sludge and debris from dry diggings also eventually ended up in the waterways through the tail races.

In Victoria, an inquiry into the sludge question was held in 1886. Its report stated that, from the earliest days, 'the creeks and rivers were looked upon as the natural outlets for the debris from the mines' and that 'no thought had been given to the ultimate effects in filling up water courses'. The inquiry concluded that the damage caused by mining sludge of all descriptions was far more serious than generally supposed and that this injury was likely to greatly increase in the future. Some of the damage could not be cured, for example the silting-up of large clear waterholes in the creeks and rivers, which caused the sludge to overflow on to adjacent agricultural lands. This had led to the destruction of vegetation and fruit trees and to harmful effects on cattle and sheep.[13]

Some of the inquiry's observations on the hydraulic sluicing claims provide an interesting insight into the dimensions of this problem. For example, at the Union Company's claim at Mitta Mitta, 50,000 cubic yards of dirt had already been washed directly into the Mitta Mitta River, for only seventy-six ounces of gold. At the time, the company was washing 5000 cubic yards per week, and at the adjacent Pioneer claim 1333 cubic yards were being washed each day directly into the Mitta Mitta River by means of a giant nozzle. Combined, both companies had about sixty-one acres left for working, with an average face of 100 feet. This would take about twelve years to work out and would involve the removal of over eight million cubic yards of dirt. Unfortunately, there was no possible means of stacking the sludge from either company.[14] At Omeo the largest company was the Oriental, the tailings from which were delivered down the gully into Livingstone Creek, which had been blocked up, thus causing the submersion of much cultivated land. It had taken eight months to work out one acre, which contained about 58,000 cubic yards, and about thirteen years work remained. This claim was worked until 1904, when the Sludge Abatement Board prohibited the washing of gravel into the creek.[15]

At Beechworth, the Rocky Mountain and Fletcher's claims both had many years of prospective work ahead of them and had already discharged a large quantity of water containing sludge into Spring Creek, from where it flowed into Reid's Creek. This creek had already been destroyed and further injury to agricultural land on the creek and the Ovens River was anticipated without the addition of further sludge from the larger contemplated new works. Reid's Creek was so much filled that, about four miles from Wangaratta, the old watercourse had been completely obliterated by the deposits of sludge and a new channel had been formed. At the Rocky Mountain claim there were about fifty-six acres left to work, with an average face of about thirty-five feet.[16]



By the early 1900s, bucket dredging was widely used, and by 1913 the environmental consequences had become a subject of serious political debate. Both the Melbourne *Age* and the Melbourne *Argus*, but particularly the former, were relentless in their pursuance of this issue. For example, in January of that year, in an article titled 'The desolating dredges', the *Age* commented on the conversion of fertile land by 'these engines of destruction' into a 'permanent waste of shingle'.[17] The following month, the Premier of Queensland visited the Ovens valley and expressed his amazement that the Victorian government would allow a few dredges to devastate the area. The *Argus* reminded its readers that a few years previously an officer of the Agriculture Department had made similar comments, only to be censored for trespassing on the jurisdiction of the Department of Mines. It was further commented that since that time the dredges had been encouraged to carry on their work of destruction. Even the Sludge Abatement Board had, by its own admission, been ineffective, and in its report for 1912 it had complained of the 'natural tendency on the part of a good many mine-owners to avoid all possible outlay on the disposal of their sludge and tailings'.[18]

The *Age* was particularly critical of the attitude of the Minister for Mines and his department and called for dredging operations then in progress to be 'peremptorily forbidden'.[19] One issue to engage the ire of the *Age* was the minister's unfavourable and unhelpful, if not misleading, response to a deputation from the Australian Natives Association (ANA) who were protesting against the pollution caused by the dredges. Other ministers, such as the Minister for Water Supply, did not share this complacency. Nor did local residents and organisations such as the Ovens River Anti-Sludge and Pollution

**Figure 5.5** Rocky Mountain Extended, Beechworth, Victoria. (Robert O'Hara Burke Memorial Museum, Beechworth)

Association.[20] For instance, a petition had been sent by Ovens valley residents to the State Rivers and Water Commission, calling for an investigation into the effect of dredging.[21] The involvement of the ANA indicated that the issue was more than just regional. For example, other organisations to enter the fray were the Chamber of Agriculture and the Waterworks Trust Association.

A consequence of these concerns was that a further inquiry was held in 1913. The *Age* was far from satisfied, however, referring to the in-house investigation by departmental officers as a 'sham inquiry'.[22] On a more optimistic note, though, it noted the diligence with which the officers conducted the inquiry. In some districts the issue had, by this time, reached crisis point.[23] For example, the *Argus* commented in May that, at Bright, the residents were divided into two hostile camps, one in favour of dredging and the other, not.[24] In the Harrietville area the crops and orchards had vanished and 'all was a scene of desolation, complete and depressing'. On a subsequent visit, the reporter stated that 'nothing remained but huge boulders and large quantities of gravel and pebbles with here and there a deposit of sand'. The beach of

St Kilda was 'fertile in comparison'. Some of this land had been under cultivation for fifty years, and now it was a 'waste of rubble'.[25]

The inquiry found, as its predecessor had done twenty-six years earlier, that there were numerous instances in which current stream levels were several feet higher than the original surfaces. For example, the Sandy Creek valley and Staghorn Flat area of Yackandandah Creek had been covered with deposits ranging in depth by up to four feet. This siltation had been attributable to the early sluicing and more recent hydraulic sluicing rather than to dredging, which was the main concern of the inquiry. At the Rocky Mountain claim at Beechworth, an area of about eighty acres had been worked out and more than three million cubic yards of dirt removed, of which all but a fraction remained in the old workings or in Reid's Creek into which the tail water passed. The mass had passed into the Ovens River and accounted for much of the surface elevation through the Tarrawingee and Wangaratta districts down to the Murray River. At the ground sluicing operations at Punds on Three Mile Creek, the workings had contributed to millions of cubic yards of material filling up the creeks and alluvial flats downstream.[26]



Siltation was exacerbated by the practice of constructing and deepening the tail races. The larger the claim, the deeper the race. On the Beechworth and Yackandandah diggings, a number of tail races were constructed or deepened by cutting or blasting through the bed rock of existing stream beds. Obviously, such works hastened the deposit of sludge and waste water downstream and left a permanent mark on the watercourses. Some of the best examples are at Woolshed Falls on the Reid's Creek diggings, Newtown Falls on the Spring Creek diggings of the Rocky Mountain Company, Osborne's Flat and at what is now known as Yackandandah Gorge. At Newtown Falls the tail race proved insufficient and the Rocky Mountain Extended Company was forced to construct a tunnel under the town, creating an entirely separate outlet for the sludge. On its completion in January 1880 the tunnel was regarded as one of the greatest engineering feats in Australia. In addition, streams were often diverted so that the bed could be worked. One of the most impressive examples of this is at Woolshed Falls, where Reid's Creek has been diverted through a channel cut into solid rock and overburden to a depth of 130–170 feet in places. It was common practice to cut channels across river flats or stream bends for the same purpose.

The 1913 inquiry was particularly unimpressed with the effects of dredging on the environment. Covenants protecting streams from siltation from alluvial goldmining activities, particularly dredging, were not embodied in leases until 1906, well after the commencement of dredging. Prior to that, restrictions placed on dredgers and sluicers were nominal, and nowhere in the leases were conditions inserted as to the amount of earth or other substances which might be discharged into streams from the workings. 'In no case was any effort made to restrict the amount of polluting matter that might pass into and down the district main water-courses or to restore in any degree the surfaces of dredged areas.' The current system of stacking tailings prevented the escape of earth or debris in large quantities during normal weather conditions, but not in heavy floods.[27] Some idea of the quantity of earth removed can be gleaned from dredging statistics for the Araluen field in the Braidwood district of New South Wales. For example, the Araluen Valley, Araluen Central, Tewksbury, Victorian Araluen and Redbank dredges, to name but five, excavated a combined total of 2.1 million cubic yards of earth in 1912 and 2 million in 1913.[28]

There had, however, been some measure of success in curtailing the sludge arising from centrifugal sluicing and dredging, despite the massive size of some of these operations. Provision had to be made for the disposal of the debris into lower-lying land, the local creek or through 'impounding' by means of a dam or an embankment across a depression. This latter practice was now demanded by the Sludge Abatement Board. With impounding, the tailings were passed on to this area until it was filled, when the barge site and other excavations were flooded and the barge floated into a new position for future operations. This left the old site and previously sluiced areas available for the deposit of future streams of tailings. The sluiced water was retained within the excavations, pumped up and repeatedly used. No debris was therefore allowed to pass beyond the limits of the claim.[29]

Notwithstanding the riches to be won from dredging, and the resoiling  conditions in the leases, the inquiry considered that dredging was an ephemeral activity and should be excluded from all lands of agricultural value. It also considered that under no circumstances should permission be given for dredging or sluicing of the bed or bank of any river or principal stream, and it recommended against the construction of dams on streams which were primary waterways for the purposes of flotation, siltation or precipitation. Perhaps the most important recommendation was that no lease should be granted on land valued at over £3 an acre.[30] The inquiry also reflected on the effects that dredging had upon the look of the countryside. It admitted that bucket dredging produced an appearance of devastation and desolation but that the passage of time might ameliorate this effect if some species of vegetation, of tree or shrub growth, could be promoted on the dredged areas.[31]

The inquiry also quoted from the Conservation Commissioners of California, who in 1912 stated that they had been unfavourably impressed with the destruction of land produced by these operations. Hydraulic mining had been regulated because it had been destroying large areas of agricultural land, and the regulations were of such a nature as to stop all such operations. Dredging 'turned the land upside down, putting the top soil at the bottom and the bottom cobbles on top of the spoil heaps. The result was that the productivity of the land was practically destroyed.' The commissioners questioned whether it was in the state's interest to 'permit the spoilation [*sic*] of large areas of land and the turning of them into hills of unproductive scenery of cobblestones'.[32]

Another important environmental effect of dredging not mentioned by the 1913 inquiry was deforestation. As can be seen from the comments of William Howitt in 1853, timber felling was enthusiastically pursued from the earliest days. However, with the advent of dredging, deforestation reached new heights. There was an increased demand for timber to construct the dredges, a process which was in reality a form of shipbuilding. For example, each of the Tewksbury Dredging Company dredges on the Shoalhaven River required 100,000 feet of timber for their construction.[33] At Araluen and Jembaicumbene the demand would have been much greater as both fields were home to many more dredges, almost all of which were successful over a number of years. More important than this, however, was the prodigious demand for fuel for the boilers on both the bucket and centrifugal dredges. In 1913, for instance, the difficulties in securing timber supplies for the Araluen dredges had led to a 20 per cent increase in the cost of firewood, but even at that price there were insufficient hauling teams. In the Yackandandah and Sandy Creek areas the demand for firewood for the centrifugal and bucket dredges would have been of a similar dimension. The treeless aspect of these various fields bears witness to these effects.[34]

**Figure 5.6** Jembaicumbene Gold Dredging Company, Jembaicumbene, New South Wales. (Source unknown)

There is some merit in reflecting on the parting comment of the Californian Commissioners to the 1913 inquiry, concerning the healing effects of time on the landscape. The key concern was that some measure of revegetation occur on the sites. Where this has happened, even the most heavily worked areas have shown some regeneration, though in some instances it has been a process spread over 100 years. On the bucket-dredging fields, the effect of dredging on the streams was immediate and permanent, the streams now resembling little more than a series of interconnected dredge ponds along much of their length. The channels cut by the dredges are clearly visible, and the tailing mounds can often be seventy feet long and twenty feet high. In addition, the ground is strewn with water-worn pebbles and discarded or broken equipment, such as cables, buckets, ladders, screens, winding gear and boilers, and occasionally dredge pontoons. Fortunately, because the fall and slope are small, erosion has been minimal.[35]

Many sluiced landscapes, particularly those worked by hydraulic methods, have been less fortunate. In southern New South Wales and north-east Victoria, almost without exception, there has been no effort to reclaim these landscapes. In many instances natural regeneration has occurred, and trees and shrubs have grown on the eroded slopes and floors of the claims. However, the accentuated slope and fall of the ground and the very thin or skeletal nature of the topsoil on many of these claims usually ensure the continuation of some measure of erosion.



For natural regeneration to occur there needs to be a conjunction of favourable factors, the main one of which is the absence of further disturbance for a lengthy period. Where, for example, there has been extensive grazing, track construction or further mining, erosion has continued unabated and has eventually encompassed a much larger area. The constant wear and tear and grazing of the cattle, in particular, does not allow shrubs and trees to grow and exposes the land to further degradation through the invasion of noxious weeds such as broom and blackberry. Some particularly bad examples of these processes can be seen on the Araluen and Major's Creek fields in southern New South Wales.[36]

Notwithstanding the environmental damage caused by these forms of mining, many of these landscapes have a different and more valuable dimension. A number of them are of heritage value, in some instances of national significance, as they represent evidence of past technologies and life ways, particularly where there are mining camps located nearby. In north-east Victoria, the Oriental claim at Omeo, the Pioneer and Union claims at Mitta Mitta, and Yackandandah Gorge, Woolshed and Newtown Falls at Beechworth are not only of national heritage significance but are also valuable tourist attractions. The Rocky Mountain claim at Beechworth has also been incorporated into Lake Sambell, which is the main recreation area in the town.[37] In New South Wales all hydraulically sluiced landscapes along the Shoalhaven River – for example, the workings at the Spa, Limekilns, Bombay, Spring Creek, Oallen and Bungonia – have been nominated for inclusion on the Register of the National Estate.[38]

Another form of alluvial mining was deep-lead mining. In the Chiltern and Rutherglen areas, alluvial mining was dominated by an extensive system of deep leads, which were auriferous deposits or gutters buried by the eruption of basalt or lava or the deposition of new layers of sand and gravel. Exploitation of these leads required shaft sinking, similar in method to that employed in reef mining. In this system of mining, the two main environmental impacts were deforestation and pollution. The latter was basically controlled by the construction of sludge dams, which ensured that the waste material did not find its way directly into streams and rivers. Although cyanide was used extensively on the tailings to assist in gold recovery, and mercury used in the crushing processes, these activities appear in most instances to have been conducted well clear of waterways. Possibly the most important effect was the use of prodigious quantities of timber for the boilers which powered the pumping and hauling equipment, and the treeless aspect of these districts may be attributed at least partly to that.[39]

Regardless of the perspective, alluvial goldmining wrought considerable changes to the Australian landscape. Many of these changes, such as altered watercourses or raised stream beds, are now permanently incorporated into the rural landscape. Some landscapes have to a degree been reclaimed by the bush, but elsewhere they continue to erode and cause loss of productive land and occasional siltation. Where the deforested areas have been resumed for agriculture and grazing they are likely to remain treeless, whereas in other cases secondary tree growth has occurred. It is clear that many of these processes have been largely arrested in Australia. Further, a number of the landscapes are now of national heritage significance and even tourist value. However, in other countries where alluvial goldmining continues unabated, it causes the same devastation and environmental havoc that was such a familiar part of the Australian scene some 100–150 years ago, and the benefits of heritage significance and tourism are a long way off.



## Notes

1 G. Bolton, *Spoils and Spoilers*, 1981, George Allen & Unwin, Sydney, pp. 69–70.
2 G. Blainey, *The Rush That Never Ended*, 1981, Melbourne University Press, Melbourne, pp. 363–4.
3 Heathcote is another who has also commented upon the impact of mining on flora and fauna, and stream systems, and the slowness with which such scars healed. R. L. Heathcote, *Australia*, 2nd edn, 1994, Longman Scientific & Technical, Harlow, pp. 156–8.

4 N. Ritchie, 'Archaeological Interpretation of Alluvial Gold Tailing Sites, Central Otago, New Zealand', *New Zealand Journal of Archaeology*, vol. 3, 1981, p. 62.
5 B. McGowan, *Braidwood to Bungonia*, 1996, Barry McGowan, Canberra, pp. 67, 275.
6 McGowan, *Braidwood to Bungonia*, p. 139; B. McGowan, 'Conservation and Heritage Overview of the Araluen Catchment Area, Incorporating the Araluen, Bell's Creek and Major's Creek Goldfields', 1998, prepared for the Lower South Coast Catchment Management Committee, Canberra, pp. 25–44, 73–6.
7 McGowan, *Braidwood to Bungonia*, p. 139; McGowan, 'Conservation and Heritage Overview', pp. 25–44, 73–6.
8 *Town and Country Journal*, 28 March 1874, p. 498; *New South Wales Agriculturist and Grazier*, 9 July 1881, pp. 19–20.
9 Board of Inquiry, *Report into the Sludge Question*, 1887, Government Printer, Melbourne, p. xii.
10 McGowan, *Bungonia to Braidwood*, pp. 57–61, 229, 259–62.
11 Bolton, *Spoils and Spoilers*, p. 69.
12 W. Howitt, *Land, Labour and Gold*, 1972, Lowden Publishing Co., Kilmore, pp. 99–101.
13 Board of Inquiry, *Report into Sludge Question*, pp. vii–viii.
14 Board of Inquiry, *Report into Sludge Question*, pp. xii, xxvi, 16–17. 
15 Board of Inquiry, *Report into Sludge Question*, p. xxvii; Parks Victoria, information sheet.
16 Board of Inquiry, *Report into Sludge Question*, pp. xxiv–xxv, 3–11.
17 *Age*, 9 January 1913, p. 9.
18 *Argus*, 7 February 1913, p. 6.
19 *Age*, 10 January 1913, p. 10, and 17 February 1913, p. 8.
20 *Argus*, 27 February 1913, p. 6; *Age*, 27 February 1913, p. 7, and 28 February 1913, p. 7.
21 *Age*, 4 March 1913, p. 6.
22 *Age*, 12 March 1913, p. 8.
23 *Age*, 16 May 1913, p. 11.
24 *Argus*, 14 May 1913, p. 13.
25 *Argus*, 15 May 1913, p. 13.
26 Dredging and Sluicing Inquiry Board, 'Report upon Complaint of Injury by Dredging and Sluicing', paper presented to both Houses of Parliament, Melbourne, 1914, pp. 3–4, 14.
27 Dredging and Sluicing Inquiry Board, 'Report', pp. 5–6.
28 McGowan, 'Conservation and Heritage Overview', pp. 11–12.
29 Dredging and Sluicing Inquiry Board, 'Report', p. 15.
30 Dredging and Sluicing Inquiry Board, 'Report', pp. 15–16.
31 Dredging and Sluicing Inquiry Board, 'Report', p. 18.
32 Dredging and Sluicing Inquiry Board, 'Report', p. 12.
33 *Braidwood Dispatch*, 11 December 1909.
34 McGowan, 'Conservation and Heritage Overview', p. 79.
35 McGowan, 'Conservation and Heritage Overview', p. 95.

36 McGowan, 'Conservation and Heritage Overview', pp. 85–7.

37 In more recent times a large number of sluiced landscapes have been added to the inventory of heritage sites in this region. P. Freeman and Associates, 'Indigo Shire Heritage Study', draft only, February 2000.

38 B. McGowan, 'Historic Mining Sites Survey of the Shoalhaven and South West Slopes Districts of New South Wales', report to the New South Wales Department of Planning and the Australian Heritage Commission, 1995, pp. 67, 80, 132, 155, 176–7.

39 S. Hunter, 'The Chiltern Goldfields', *Memoirs of the Geological Survey of Victoria*, no. 1, 1903, Department of Mines, Melbourne, p. 9.

# Part II Immigrants and Ethnic Relations

# 'Men of All Nations, except Chinamen'

hapter 6

## EUROPEANS AND CHINESE ON THE GOLDFIELDS OF NEW SOUTH WALES

*Ann Curthoys*

The Chinese on the goldfields . . . the very phrase evokes contradictory images:  long lines of Chinese men walking to the Victorian goldfields from Sydney, Melbourne and Robe; marked by their pigtails, their clothes, their carrying poles, their skin, their hair, their eyes; portents of a large population and an ancient culture to the north, harbingers of a much larger peaceful invasion perhaps. We imagine a few wealthy men, with many poor men under their control; Chinese secret societies, men gambling and smoking opium, perhaps consorting with European women; men digging, washing, cradling, sluicing, staying behind on worn-out goldfields when many others had gone; and trading, buying and selling. We see them looking after their own, living in separate parts of the diggings, sometimes by order of the Gold Commissioner, sometimes by choice; and fleeing from angry European diggers who wanted them off the goldfields and out of the country. There were men trying to adapt and men trying not to adapt, to be who they knew how to be.

Our treatment of these Chinese men shames many of us, others perhaps find it understandable; in any event their treatment remains a matter of uneasiness. The Europeans thought these men barbarians, and the Chinese, in their turn, thought of the Europeans in the same way. In later decades they would prove adept at managing the European society they encountered. Then we conjure a later image, of lonely Chinese men, after the rushes were over, working market gardens, growing and selling vegetables, often the first to do so in dry inland towns, the butt of children's pranks and jokes, thought harmless and indeed useful though perhaps a little mad. Perhaps they were.

These Chinese men perch at the edge of our historical consciousness, figures of fun and shame, a marker of our colonial origins and the colonial vestiges in our present culture. They truly were a harbinger, not of an invasion of millions of people as some feared, but of later dilemmas for policy makers and for citizens, dilemmas about immigration policy and citizenship, and of modern Australia's ambivalence about a place we call Asia. As a nation, our fears and our hatreds, as well as our interest in the exotic and the Other, live with us still. In imagining those who observed, met, liked, hated, loved and traded with those Chinese gold seekers we see, ultimately, ourselves as we were and in some respects continue to be. The historical narratives we continue to weave around them are one pathway to deciding what kind of society and polity we would ultimately like to create.

Here I explore the interactions between Chinese and Europeans on the goldfields of New South Wales. As we see when comparing my chapter with Patrick Bertola's, interaction in New South Wales had some distinctive features: the gap between the arrival of the gold rush and the dispossession of indigenous peoples tended to be much longer there than in the other Australian colonies – twenty or thirty years, as against five or less. The result was a somewhat more settled and less frontier-like society. Disagreement over how to respond to the Chinese presence was much sharper than elsewhere, with the result that legislation to restrict their entry took longer to be enacted. This delay gave time for the emergence of a particularly bitter and violent situation at Lambing Flat, which saw repeated and violent anti-Chinese riots in 1861. Those riots now exemplify and represent anti-Chinese racism in nineteenth-century Australia.



Chinese men came to the goldfields of New South Wales for several reasons. In the middle decades of the nineteenth century, men left the southern Chinese provinces of Fujian and Guangdong for places in the Pacific and South-East Asia, hoping through migration to solve economic difficulties at home. Suffering from the effects of economic dislocation in their region, these men sought new sources of wealth to sustain their lineage and to buy land, while women stayed behind to keep the family together. In the 1840s, a contract system of emigration to various parts of the world – the West Indies, Cuba, Singapore, Panama, Martinique, California, Sumatra, Tahiti – developed.[1] The Australian colonies participated in this system from 1848 to 1852, when several thousand Chinese emigrants were brought here under indenture by pastoralists seeking additional and cheaper sources of labour.[2] This scheme, however, was stopped by the British government in 1852, after riots and demonstrations against it in Amoy, the exit port in China. The pastoralists in Australia had not found the Chinese indentured labourers as useful as they had hoped; many absconded from service on the pastoral stations, and some of these absconders moved on, like everyone else, to try their luck on the goldfields of New South Wales and Victoria.[3]

With the end of the indenture scheme, a much more lucrative option emerged – gold seeking. Men from Fujian and Guangdong began emigrating in search of gold to California from early 1850 and to Victoria from January 1853.[4] In place of the earlier contract system of emigration from Amoy, these

**Figure 6.1** *Two Portraitists from Melbourne*, in *Punch*, 22 January 1857. (National Library of Australia)

men embarked from Hong Kong on a 'credit-ticket' system, under which they borrowed money for their fare from brokers in China, to be repaid later from earnings on the goldfields. They were organised into groups of between 30 and 100 men, connected to one another by kinship and locality, under the supervision of a leader of some wealth. Their kinship connections meant they could develop strong group cohesiveness and purpose, attributes which allowed them to survive difficult times much better than many other groups of diggers. They came to Victoria in their thousands; by June 1855 there were 17,000 on the goldfields.[5]

These gold-seeking Chinese immigrants arrived to find the goldfields in a state of rebellion over questions of high licence fees and much else, culminating in the Eureka uprising of 1854. The ensuing royal commission investigated complaints against the Chinese, noted the vast increase in Chinese immigration in the first half of 1855, and recommended that their entry be restricted. In June 1855 an Act – with the evasive title 'An Act To Make Provision for Certain Immigrants' – imposed an entry tax of £10 and a restriction on the entry of Chinese to one per ten tons of a ship's register.[6] In response, traders in Hong Kong decided to stop the ships at Sydney, with most Chinese passengers disembarking and walking overland to the Victorian goldfields, while a token number would be taken on to Melbourne.

A little under half of the Chinese gold seekers who arrived in Sydney under this plan did indeed walk overland to Victoria. Many others, learning on arrival that there were goldfields much closer to hand in New South Wales itself, went no further. Of the 898 men who arrived in Sydney in mid-1856, more than 400 went north to the Rocky River goldfield. In the nearby town of Armidale, they were met with curiosity, one newspaper reporting that 'their singular appearance was the subject of admiration to a tribe of youngsters'.[7]

Other Chinese gold seekers went west; by August, many were on the old, declining fields around Bathurst, contributing to their revival. Still others went south, to Braidwood or on to Victoria.[8] Wherever they went, the Chinese men caused initial interest in the towns. When a group of about 140 camped at Bathurst in early July, the local paper recorded that their camp had been visited 'by almost every person in the township'.[9] They were seen as interesting and picturesque and, when a second group camped there a month later, many townsfolk went to watch them cooking and eating, to witness, as the *Sydney Morning Herald* put it, 'the novelty of the proceedings'.[10] Later, when these same two groups arrived at the Stoney and Louisa Creek diggings further west, just south of the present town of Mudgee, their unusual appearance and their practice of walking in long single file, carrying their possessions on bamboo poles, caused laughter.[11] Mirth and astonishment seem to have been common themes of first contact.[12]



Fewer Chinese gold seekers came to New South Wales the following year. During the first half of 1857, shipping interests in Hong Kong tried to avoid the Victorian tax by landing 14,000 Chinese gold seekers at the tiny township of Robe, in South Australia, to walk overland to the rich Victorian fields.[13] When the South Australian government followed the Victorian one by imposing restrictions, Sydney again became the chief point of entry. The number of Chinese immigrants rose dramatically, from 327 in 1857 to 10,691 in 1858, mainly in the first nine months of the year. A further 12,000 arrived in Sydney in the following three years, about half going on to Victoria or returning home. The rest stayed in New South Wales, so that by April 1861 there were almost 13,000 Chinese living and working in New South Wales.[14] By this time, they formed about a quarter of all miners in the colony.

Like other miners, Chinese joined the big rushes. They also worked on the small residual goldfields, becoming important at Rocky River to the north, Tambaroora, Sofala and on the Meroo to the west, and at Braidwood and Adelong to the south. They lived in their own camps and villages, often had their own merchants who ordered goods direct from Sydney and China, and rarely worked for Europeans.

Wherever they went on the goldfields, the Chinese encountered opposition and hostility. This came largely from European miners, not from government officials. The government appointed Gold Commissioners to allocate claims, regulate labour relations and maintain law and order, and its attitude was that the Chinese should be protected and collisions between European and Chinese miners prevented.[15] When attacked at Rocky River in August 1856, Chinese miners ran to the commissioner's camp for protection: three European

**Figure 6.2** *Chinese Gold Miners*, in Fr. Christman, *Australian*, Leipzig, 1870. (National Library of Australia)

miners were charged with assault, and one was eventually found guilty.[16] The Lands Department, which oversaw goldfields administration, saw the Chinese as useful contributors to the goldmining industry. The Secretary for Lands told one assistant commissioner that the Chinese, if 'judiciously managed', could lead to 'a fuller development of our mining resources than heretofore has taken place'.[17] In this he had the support of the business community, who hoped for more from the goldfields. The *Sydney Morning Herald*, in a leader on the sudden influx of Chinese to New South Wales in 1856, agreed.[18] The hope that the Chinese would assist the revival of the New South Wales goldfields meant that calls for anti-Chinese legislation to copy that of Victoria and South Australia were not popular in parliament, leading to the withdrawal in November 1857 of a Private Member's Bill on the matter.[19]

Most European miners saw it differently – they wanted the Chinese kept out. Why did they object so much to the Chinese? After all, many of them were recent immigrants too, and about 10 per cent were continental Europeans. We find a clue in the common complaint that the Chinese were 'taking the gold'. To the European miners the Chinese were not revivers of the goldfields but, rather, competitors for an increasingly scarce resource. As one petition put it, the Chinese 'are now stripping these goldfields of their wealth, to the injury of the Colonists'.[20] The definition of 'colonists', it seems, could be stretched beyond its original 'British' meaning to include continental Europeans and Americans, but it could never include Chinese. Ultimately, their objection was a racial one: they regarded Chinese as outsiders and interlopers who had no right to share in the wealth of the colonies. This racial identity had not yet developed the nation-building function it was to fulfil later, but it was crucial to the expression of belonging and exclusion within each of the Australian colonies.



The hatred of Chinese does not seem to have been the product of long or bitter experience. It emerged very quickly on the goldfields, and in any case the Chinese miners appear to have done little to offend. There were some particular problems, such as the frequently expressed complaint that they wasted water, a scarce and always valuable resource. Perhaps their methods were more water-intensive; more likely, their mere presence on a water-scarce goldfield intensified competition for water. They were invariably described as barbarians, as lacking in morality or a British sense of justice, and sometimes attacked for bringing disease. Yet a reading of the newspaper reports of miners' complaints against the Chinese leads me to the conclusion that the animus against them was based less on any particular matter than on the feeling that they had no right to be present at all.

The miners were not alone in their calls to exclude the Chinese from the goldfields and the colony. Local newspapers such as the outspoken *Bathurst Free Press*, the *Yass Courier* and the *Empire* began to join the anti-Chinese cause.[21] In Sydney there were signs of concern as well; in March 1858 one ship with Chinese immigrants was greeted by about 100 youths with hostility and stone throwing.[22] In newspaper commentary, those who supported the European miners in wanting to exclude the Chinese spoke of the Chinese as immoral, filthy, idolatrous and unchristian. They would contaminate the British character of the colony and, if sexual contact and mixed-race children should result, would threaten the purity of the Anglo-Saxon race. They were an inferior people who could not assimilate and would inevitably form an inferior group in society.[23]

Such racial antagonism and desire for exclusion meant that all the elements were present in 1858 for New South Wales to follow Victorian and South Australian precedents and exclude Chinese immigrants from the colony. It almost did. Chinese were arriving in very large numbers, almost 1000 per month. By March, about 5000 were concentrated on the western goldfields, especially at Tambaroora. On 11 March the European miners there drove the Chinese off the field. Two days later they held a meeting, after which a petition was sent to the government expressing their opposition to Chinese miners – they wasted water, would exhaust the diggings and might bring disease.[24] The department instructed the Gold Commissioners to keep the peace by keeping Chinese and European living areas separate and by reserving separate domestic water.[25] Further conflict was averted when the Chinese left the field. Similar bad feeling was reported at Louisa Creek, Tuena Creek and Adelong, and petitions calling for the prohibition of Chinese immigration were sent from the Meroo, Turon and Rocky River diggings.[26]



In response to the large number of arrivals and miner unrest, public concern increased.[27] The *Bathurst Free Press* was an outspoken supporter of the miners, saying the diggers had clearly indicated they would always oppose 'a race with whom they have little more in common than with a race of baboons or a tribe of ourang-outangs'. It went on to argue for self-protection and a rejection of the 'cosmopolitan dogma' of those who objected to the anti-Chinese position.[28] The government decided to take action. Over three months in 1858 a Chinese Immigration Bill similar to the Victorian Act was debated and then passed by the Legislative Assembly.[29] Where New South Wales differed from the southern colonies, however, was in the political complexion of its Upper House. The Legislative Council was dominated by pastoralists who were enthusiastic advocates for increasing the supply of labour, and especially of what they perceived as cheap Chinese labour. In debates on the Chinese Question, as it came to be called, the pastoralists and their representatives in the Council consistently saw the Chinese as economically valuable, fairly civilised, and harmless. By the time the Council came to consider the bill in late October, the flow of immigration had almost stopped and agitation subsided. In this relatively calm atmosphere, the Council rejected the bill on the grounds that the Chinese were useful sources of labour who could help develop the colony.[30]

In 1860 the pattern of goldmining in New South Wales changed, bringing a revival of both Chinese immigration and anti-Chinese feeling. As the residual fields were worked out, good discoveries attracted a very large number of migratory diggers, anxious to try new fields rather than take up the more laborious and increasingly company-dominated systems of mining on the old ones. The first of these big rushes was to Kiandra in the Snowy Mountains, where 1900 miners gathered in August 1860.[31] Snow, however, imprisoned the field, making it unworkable, and within a month it was clear that the high hopes that had been held for the goldfield would not be realised.[32] Here anti-Chinese feeling was high, exacerbated by the original high hopes for the goldfield and its subsequent failure. Miners complained of Chinese wastage of water and of their 'dirty habits'.[33] To avert a collision, Gold Commissioner Cloete moved the Chinese camp a mile away from the European camp, a degree of official segregation not commonly practised in the colony.[34]

A much more successful field was discovered in June 1860 at Lambing Flat, 250 miles south-west of Sydney, near the present township of Young. It was, many said, a 'poor man's goldfield', able to sustain a large number of diggers, though not to make fortunes. By September, several hundred diggers were working there and, as the failure of Kiandra became apparent, the exodus to Lambing Flat from other goldfields grew until there were over 1300 miners working there in November.[35] Many Chinese came north from Victoria to escape the Victorian licence laws, which were discriminatory against Chinese and which, from January 1859, were rigorously enforced.[36] The Lambing Flat goldfield had several camps – Little Spring Creek and Blackguard Gully, where the Chinese predominated, and Stoney Creek, Spring Creek and the Wombat, mainly for the Europeans.



A letter to the *Empire* newspaper on 24 February 1860 had predicted that the first 'large paying goldfield discovered in this colony will be the battlefield and the grave of the Chinese question'. How prescient this was. The Lambing Flat goldfield, the first major field for some years and perhaps the largest ever in New South Wales, witnessed the largest anti-Chinese riots and directly stimulated the passing of restrictive legislation on the Victorian model. It is the Lambing Flat riots which are most remembered today.[37]

When Chinese miners arrived at Lambing Flat, they entered a volatile situation. By the time the field was proclaimed on 27 November 1860, and David Dickson appointed Resident Commissioner, strong anti-Chinese feeling was already apparent – as the *Yass Courier* had put it, the Chinese 'are exceedingly unpopular on this goldfield'.[38] Violent clashes between Europeans and Chinese began immediately. The exodus from other fields continued, so that the mining population swelled to at least 9000 by the end of January.[39] Chinese were sometimes driven away, but they always returned. Rumours spread among the European diggers that the Chinese intended to arm themselves against any European attack and, indeed, to claim Lambing Flat as a Chinese goldfield. There were also rumours of thousands more Chinese on their way to the Flat.[40]

A large gathering of about 1500 miners and traders, many armed with sticks, assembled for an anti-Chinese meeting at Lambing Flat on 27 January 1861. It was the first of many well-attended and angry meetings that year. The issue for the meeting was defined by the chair, a prominent storekeeper named Charles Allen, as one of deciding whether Burrangong (the correct but little-used name for the diggings which included Lambing Flat) was to be 'a European goldfield or a Chinese territory'. One speaker, John Stewart, asked the meeting: 'Shall we as men and British subjects stand tamely and allow the bread to be plucked from the mouths of ourselves, our wives, and children by those pig-tailed moonfaced barbarians?'[41] The meeting agreed they could not and that the Chinese had to be driven away. A band led the way as a party of diggers set off for the Chinese encampment at Little Spring Creek. The Chinese, having heard of the meeting, had begun packing and, carrying their belongings, they hurried before the attacking party. Europeans lined the creek bed along which they fled, and hurled abuse, the band playing the whole time. They were driven from Blackguard Gully in the same manner. The 1500 or so Chinese miners fled to other goldfields and to various other places, including James Roberts' sheep station at Currawang, more than twelve miles from the Flat. The commissioner and police, vastly outnumbered, were powerless to prevent the expulsion.[42]

Flushed with success, the European miners formed a Miners' Protective League which forwarded to John Robertson, Secretary for Lands, a petition with 3394 signatures demanding government protection for European miners from the Chinese.[43] Meanwhile, Commissioner Dickson went to Yass to get reinforcements. Within days the Chief Commissioner of the Southern Goldfields, P. L. Cloete, arrived to take over, and the commissioners were soon  joined by a detachment of fifteen police of the Southern Mounted Patrol, led by Captain Henry Zouch.[44] The government's instructions were clear: 'The law must be upheld – the Chinese must have the same justice and protection extended to them as other people'.[45] Under police protection, Chinese began gradually to return, more anti-Chinese meetings were held, and they were again driven away.[46] Many fled for a second time to Roberts' station, panic-stricken by the continuing meetings and skirmishes. It was clear that more police or military would be needed if the Chinese were to be protected.

The Executive Council of the colony agreed to send all available forces to the Flat, together with Premier Charles Cowper, as Special Commissioner, to investigate the situation. The Premier's decision to go to the Flat himself is a clear indication of the seriousness of the situation.[47] The same day, 165 troopers, 9 officers and 42 police were sent from Sydney, and 25 more police from Kiandra.[48] When Cowper arrived, he found thousands of Chinese miners waiting just outside the Flat and a Miners' Protective League manifesto 'stuck to every gum tree', listing reasons why the Chinese should be driven away.[49] The league defined its membership as 'men of all nations, except Chinamen', and called for 'equality, fraternity and glorious liberty' and the advancement of the race.[50] Cowper addressed the European miners at two meetings, one at Stoney Creek on 5 March and a second at Spring Creek on 9 March. The latter was attended by more than 2000 men.[51] He implied much agreement with their concerns about the Chinese and urged them to seek a change in the law rather than taking it into their own hands. He also insisted that the Chinese would be reinstated at Blackguard Gully. With additional military aid, bringing the total number of soldiers and police to 290, the Chinese were finally returned to Blackguard Gully, though not anywhere else on the Flat.[52] It proved a Pyrrhic victory for the

Chinese: without water and with the area now yielding little gold, they left within weeks for other goldfields. Many European miners found that Stoney Creek and Spring Creek were giving out, and they too began to depart.[53] All was quiet.

Meanwhile, a failed Private Member's Bill of the previous year which had sought to restrict Chinese immigration was revived, debated and passed in March in the Assembly. Many members spoke in its favour, including Henry Parkes, who stressed the Chinese inability to become assimilated and their threat to 'our very existence as a nation'.[54] Cowper returned to Sydney and reported to the Assembly on his visit to the Flat.[55] The Legislative Council, however, rejected the bill on 2 May. There was little debate. Most speakers argued, as they had done three years earlier, that the Chinese were economically valuable and, as one member put it, 'no lower than French or Germans'.[56] Eleven days later the Council was dissolved at the end of its five-year term.

The failure of the bill was to be expected. The Council's opposition to excluding such a valuable source of labour was well known and, at the end of its life in any case, the Council had little to lose. The leaders in the Assembly, Charles Cowper and John Robertson, concentrated their efforts on another front, a Goldfields Act Amendment Bill to empower them to keep Chinese miners away from troublesome goldfields.[57] Although passed in principle by both Houses, the bill had not become law when the parliamentary session ended on 10 May.[58] Gold Commissioners were instructed, nevertheless, to demarcate those fields where Chinese should be prohibited and to issue Chinese with special mining tickets.[59] The military was withdrawn from Lambing Flat, the senior commissioners returned to their usual duties and two inexperienced commissioners, George Clark and J. Lynch, were left with about twenty police to maintain order.[60]



Almost as soon as the military had left, tensions revived on the Flat. The Europeans made new finds at Tipperary Gully. The Chinese started to flow back into the areas now abandoned by Europeans and began also to work a new area at Back Creek, where they did well. Thousands of Europeans began to flood back to the Flat, until all the old gullies, and the new, were filled with miners – the Europeans at Tipperary Gully, the Chinese at Stoney, Spring and Back Creeks, and the Wombat. News arrived at Lambing Flat of the arrival of almost 1500 new Chinese immigrants by sea in Sydney on 16 and 17 June. Once again the Europeans decided to drive the Chinese away.[61]

The riot which resulted was larger and more violent than that of 27 January. Our knowledge of it comes mainly from four sources – the *Sydney Morning Herald*, which had a special correspondent present, the *Miner*, the *Empire* and the *Bathurst Free Press*.[62] It began at about 10 a.m. on Sunday 30 June when 2000–3000 European miners and traders, armed with pick handles and some guns, assembled at Tipperary Gully with the purpose of driving off the Chinese once and for all. English, Irish and American flags were flown and a large banner carried, with a Eureka Cross painted in the centre and with the words 'Roll Up, Roll Up' down the sides and 'No Chinese' across the top and bottom. This flag can still be seen, in the museum of the Young Historical Society

(Plate 6.2). Chinese in the area were attacked and their pigtails cut off. The crowd then marched on to a Chinese camp, six miles from the Flat – 900 Chinese took flight, forty tents were burnt and tools, clothing, furniture and food were destroyed. The same thing happened at Back Creek, where, with six police and the commissioner watching, a group of 1200 Chinese were caught as they fled and forced to drop their swags, which were then burnt in six or seven bonfires. They were rounded up, struck with bludgeons and whips, pulled up against the saddles of horses and their pigtails cut off. The procession then returned to Lambing Flat, now flying a Chinese silken sign appended to the corners of which were these cut-off pigtails.

Hundreds of Chinese sought refuge, yet again, at Roberts' station at Currawang. Many Europeans also left, fearing arrest.[63] That night all was still, the place lately occupied by the Chinese now dark and desolate, with the exception of the fires still burning with Chinese property.[64]

A week later, Zouch returned to the Flat with extra police, bringing the total number to fifty.[65] Twelve days after the riot, on 11 July, the more senior commissioner, Griffin, arrived.[66] On 14 July three men were arrested.[67] Up to 1000 miners attacked the lock-up and attempted to have the men released. The troopers charged and police fired. Many were wounded by the troopers' swords, some were trampled by horses, and one miner, William Lupton, was shot dead.[68] The next day the three men were released, and the police, the Gold Commissioners, the bank managers, the unpopular *Sydney Morning Herald* correspondent, the editor of the *Miner* and several storekeepers hurriedly left the area.[69] On 17 July martial law was declared at the Flat and a detachment of troops was sent from Sydney.[70] While they were on their way, Lupton's funeral was attended by 4000 diggers.[71] Once the military arrived, four more men were arrested and about 250 Chinese were escorted from Roberts' station and reinstated at Back Creek.[72] In September, juries acquitted nine of the ten men arrested for rioting, and the judge had to quell the applause which broke out when the verdict was given. One man, Claremont Owen, was sentenced to two years gaol for his part in the riot at the police camp on 14 July. Three leading figures in the riot, who were arrested subsequently, had their charges dropped, while one of the leaders, Charles Spicer, finally arrested in November, was tried in March the next year and sentenced to two years imprisonment. He was released before his sentence was completed.[73]

Most newspapers now agreed that Chinese immigration must be restricted. It was generally felt that the presence of Chinese miners had brought about a disintegration of law and order within a peaceful community. For the first time, anti-Chinese meetings were held in Sydney, starting with a large meeting on 31 July. Chaired by the Mayor, it had eight MPs, four aldermen, a minister of religion, two representatives from Lambing Flat and 'four other well known gentlemen' on the platform. The aim of the meeting was to dissociate the anti-Chinese cause from the violence at the Flat, and to assert that it was essentially a lawful, and indeed nationally imperative, cause.[74] Meetings were held around the colony, and petitions flowed in.[75]

was essentially a lawful, and indeed nationally imperative, cause.[74] Meetings were held around the colony, and petitions flowed in.[75]

When parliament reconvened on 3 September, the Legislative Assembly was virtually unchanged but there was now a new Council. After a political crisis over the Land Acts which the old Council had rejected, a list of twenty-one nominees for the new Council was drawn up jointly by Charles Cowper and the new Governor, Sir John Young, selected so as to give the ministry a working majority.[76] The Goldfields Act Amendment Bill from the previous year was passed, to enable the segregation of Chinese and Europeans onto different fields, a method already tried and found wanting at Lambing Flat.[77] A Bill To Restrict Chinese Immigration was passed by both Houses of Parlia-

**Figure 6.3** *Riots at Lambing Flat: 'Roll Up Roll Up – No Chinese'*, 1861, in *Sydney Mail*, 10 July 1935. (National Library of Australia)

PLATE 6.1 *Chinese Procession at Ballarat Charities Fete, 1875*, in *Illustrated Sydney News*, 11 December 1875. (National Library of Australia)

PLATE 6.2 *Roll Up Roll Up – No Chinese*, flag. (Young Historical Society Inc.)

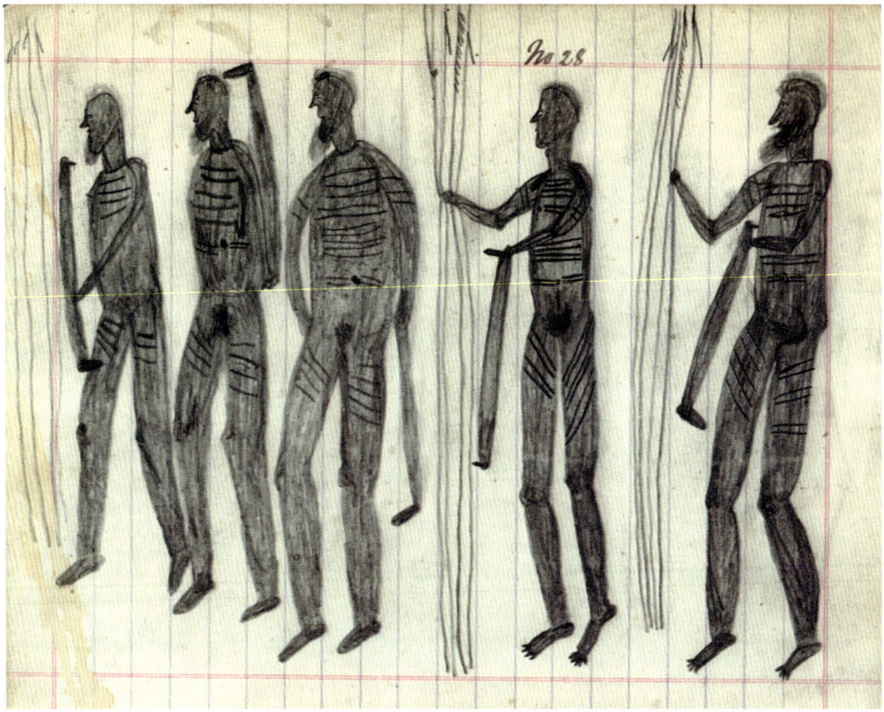


Plate 9.1 Oscar, *Palmer River Fighting Men (Warriors)*, 1890s.
(National Museum of Australia)

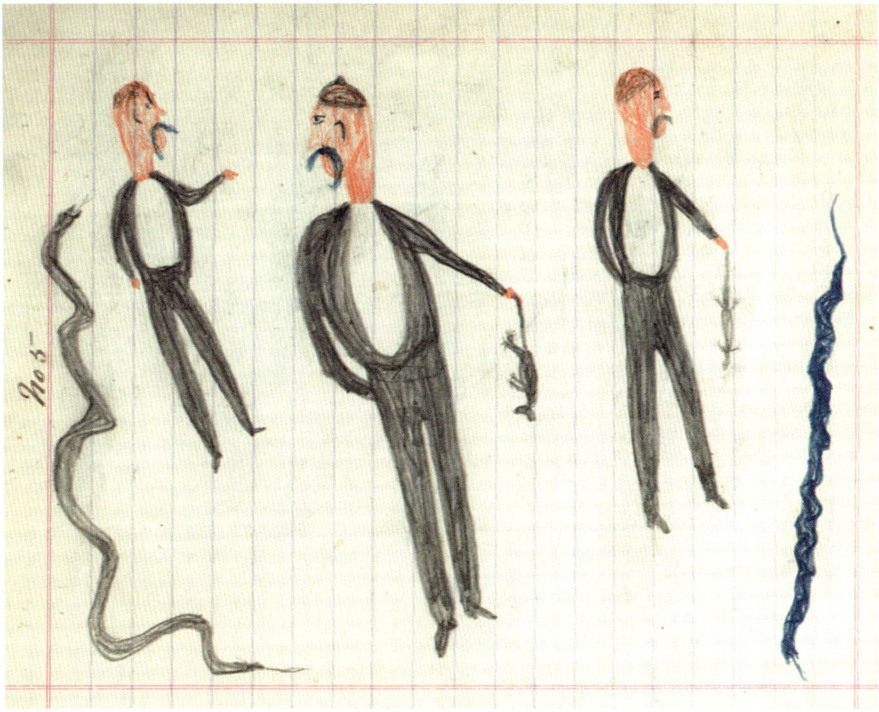


Plate 9.2 Oscar, *Some Cooktown Swells*, 1890s.
(National Museum of Australia)

Those who wanted to exclude the Chinese from colonial society on the grounds of their supposed difference, inability to assimilate, immorality, disease, but above all their simple lack of a right to share in the wealth of the colony, had prevailed.

In the years following the passing of the Chinese Immigration Restriction Act 1861, the strength of anti-Chinese feeling among the European population of New South Wales declined. For a period of about sixteen years, Chinese and Europeans coexisted in an atmosphere of comparative calm and goodwill. The effect of the Act, together with the decline of the goldfields, meant that Chinese immigration virtually ceased. The last major gold rush was on the Lachlan, north west of Lambing Flat, in late 1861 and early 1862, and the last ship bringing Chinese gold seekers to New South Wales arrived on 15 March 1862.[79] By September of that year, Chinese gold seekers were leaving New South Wales; of those who left by sea in the five years of the operation of the Act, 4034 went to Hong Kong, 770 to Queensland and 115 to New Zealand, presumably for the Otago gold rush. With many returning to China, their numbers in the colony halved between 1861 and 1871, from more than 14,000 to a little over 7000.[80] 

The decline of the goldfields resulted in a lessening of hostility between those Europeans and Chinese who remained. Chinese often worked on goldfields no longer coveted by Europeans or entered occupations such as market-gardening. They were skilled in earning a living from virtually worked-out goldfields and during the 1860s persisted with the unmechanised alluvial methods, although some did turn to deep sinking. As one goldfield after another was practically deserted by Europeans, the Chinese found they were welcomed by local traders and businessmen.

They ceased to live and work in large clusters and now worked in small groups or as isolated individuals, becoming so scattered that there were few identifiable centres of Chinese communal life. Chinese generally became individually known, less anonymous and more familiar. They still faced some difficulties, appearing frequently in the courts on minor charges such as stealing, selling spurious gold and mining without a licence. Many of these appearances concerned crimes committed by one Chinese man upon another, and many more disputes were solved within the community.[81]

In general, although they were never fully accepted, the Chinese in these years were usually tolerated and, in a limited way, valued for their services. The belief in British racial superiority remained, but weakened in the face of growing Chinese conformity to British cultural expectations. Isolated conflicts with Europeans persisted: at Stoney Creek in January and at Jembaicumbene in April 1862; in a public bar at Long Creek near Mudgee and at Lambing Flat in 1864.[82] The main clash in these years was at the Napoleon Reef diggings, Glanmire, on 22 March 1866, when Europeans drove Chinese away from a portion of the diggings and set fire to their huts.[83]

Chinese could now be excluded by law from any goldfield where officials feared European miners might attack them. This meant exclusion from the better fields, which were effectively reserved for Europeans. Chinese could only be allowed on those goldfields specifically declared open to them. Unsurprisingly, as each field declined, it would be declared open to Chinese. Nine months after the collapse of the Lachlan diggings and township in November 1862, for example, both Chinese and Europeans in the district petitioned the Colonial Secretary to allow Chinese miners to enter the field.[84] The Gold Commissioner agreed on 26 February 1864, declaring that they should be kept from fresh leads but should be allowed onto existing workings. Later, they were allowed onto the entire goldfield.

Only at Burrangong, of which Lambing Flat was a part, were the conflicts bitter and lasting.[85] European-only areas were marked by a ploughed furrow which Chinese were not allowed to cross. Several attempts by them (and by local traders on their behalf) to gain access to the goldfield met with little success. In the next three years, Chinese continued to petition the Colonial Secretary for an extension of their area, and European traders and sympathisers sent similar petitions.[86] When Chinese transgressed their boundaries, many were fined or gaoled, and more than 1500 Europeans petitioned for a stricter enforcement of the law in 1864.[87] A public meeting on 31 March 1866 opposed  all attempts to allow the Chinese on Burrangong, and a petition signed by 816 miners was sent.[88] Yet even here, opposition gradually declined. When a new Goldfields Act Amendment Bill, removing the sections of the Act enabling the confining of Chinese to particular areas, was passed by both Houses in August 1866, a meeting of only fifty miners at Burrangong called for a retention of the status quo.[89] The Act became law on 1 January 1867 and, although the Chinese miners appear to have been somewhat apprehensive of violent conflict when they finally entered the Burrangong goldfield, no collision occurred. This field was now in fact a poor one and, of the 2000 Chinese who went there as soon as the new Act was proclaimed, there were only 700 left six months later, and the exodus continued thereafter.[90] A pattern, so much more easily accomplished in other areas, had finally been established at Burrangong: as the goldfield declined, Chinese and Europeans worked the remaining field in harmony.

In the gold towns, Chinese began entering other occupations, with little objection from the townsfolk. Although about 80 per cent of Chinese in 1871 still lived in the goldfields areas, many of them had left mining to become storekeepers, market gardeners, and hawkers. Market gardens were established in towns like Orange, Deniliquin, Hay, Young, Forbes, Bathurst and Carcoar.[91] Local newspapers around the colony reported with interest the development of this useful branch of small-scale agriculture. Late in 1864 one of the largest and most enduring market gardens was established at Deniliquin, probably by Chinese from Victorian goldfields. The garden, a closely fenced paddock within a township of nearly four acres, with one acre then under crop, worked by four men, was described in detail by the *Deniliquin Chronicle*.[92] The Chinese gardeners differed from the English in their close planting, frequent watering (these four men brought a thousand buckets of water a day from the lagoon), and in their placing of manure on top of, rather than below, the soil. By 1865, people were coming to Burrangong from Yass and Goulburn for their vegetables and, in September 1865, Chinese began taking their produce from Burrangong and selling it at Wagga Wagga.[93] A few years later, Chinese also began to enter other rural occupations, both as independent farmers and as workers on pastoral stations – as cooks, domestic servants, station hands, sheep washers, seasonal labourers and shearers.

By 1866, fear of the Chinese had fallen to a low level. In such an atmosphere, some began to argue for the repeal of the 1861 Act on several grounds: the economic usefulness of Chinese immigrants; the role of the Act in hindering the 'free intercourse between nations' in general, and in contravening the spirit of the Anglo-Chinese treaty of 1860 in particular; the groundlessness of fears of their danger to the social fabric of the community; and, finally, the irrelevance of such an Act where Chinese no longer sought to enter. Two petitions for repeal of the Act were presented early in 1866, one signed by 691 clergymen, magistrates, merchants and others, and the other from 125 Chinese residents.[94] The first described the Chinese as honest, sober and industrious, and said some had proved to be good merchants and employees. In addition, the Act interfered with trade between China and New South Wales. The Chinese petition emphasised the law-abiding nature of the Chinese, their desire 'to live in peace and harmony with all classes', and their economic value as consumers, revivers of declining goldfields, domestic servants, employers and storekeepers. J. D. Lang presented the petitions, and the *Sydney Morning Herald* spoke in favour of repeal, in particularly interesting terms, on 1 August 1866. One should distinguish between peoples, it said, on the basis of the standard of 'civilisation' attained and not on the basis of 'varieties of colour and shade'. Moreover, while one could judge the degree of superiority of a civilisation for any one particular time, one could not make such a judgement for all time: 'The fortunes of people changed, some being in the forefront of civilisation at one time, and others at other times'. Thus, India and China had been great 'when our forefathers were painted savages', and the Egyptians, 'even while they had the lineaments of a negro', had once been an intelligent and powerful people. Even though the Chinese had now for a long time been arrested by a stereotyped civilisation, they were now likely, as a result of European penetration of China itself, to become again 'high in the ranks of civilised nations'.



In the debate on the repeal bill, the usual arguments were advanced by both sides.[95] The bill was lost but, a year later, Lang tried again.[96] In the intervening year, the goldfields had further declined and the Goldfields Act Amendment Act had allowed Chinese onto all fields. In commenting on the second reading debate in September 1867, the *Empire* newspaper expressed the feeling of many when it said that, since the danger of Chinese immigration had passed in reality, they could 'indulge in brotherhood' in law.[97] The Act was repealed, and a celebration held in the streets of Sydney.[98]

In the 1870s there was a slight revival of the goldfields, especially at Gulgong, and some Chinese remaining from the earlier gold-seeking emigration went to the new fields, usually forming about 10 per cent of their population. This time there were few clashes and little administrative concern. Some Chinese became successful and highly respected storekeepers. One of the Gulgong stores belonged to Sun Tong Lee and Co. and was photographed by Beaufoy Merlin for Otto Holtermann. In his short story, 'The Ghosts of Many Christmases', Henry Lawson evoked Sun Tong Lee's, where one could get 'strange, delicious sweets that melted in our mouths, and rum toys and Chinese dolls for the children'.[99]

How well did Chinese and Europeans interact in these later years? Many Chinese could not speak English, a clear indication that mixing was relatively slight. At Braidwood, for example, there were still large numbers of Chinese miners who kept to themselves in the 1870s. A permanent court interpreter was sorely needed but never appointed. The best interpreter in the district was Quong Tart, who had come to Braidwood as a 9-year-old boy and acquired European foster-parents. Despite his ability, and despite pleas from the Braidwood court, the government repeatedly refused to appoint him permanently.[100]

On the other hand, some Chinese did make positive overtures of friendship to Europeans, quite often inviting them to celebrate the Chinese New Year. In Hay in 1870 one Chinese man, probably a market gardener or perhaps a storekeeper, invited nearly all the townspeople to partake of a feast to celebrate the occasion. The local paper described the event with approval, detailing the feast, the decorations and the fireworks.[101] Jane Lydon has shown for a later period how important the use of food and feasting was in Chinese overtures to Europeans.[102] Chinese were again hospitable to Europeans, at Reedy Creek in 1875 when they treated them to bottles of spirits on the occasion of a Chinese funeral.[103] The *Town and Country Journal* said of Sofala in 1870 that 'terrestrials and celestials appear to hob-nob together with that degree of intimacy which naturally comes of long acquaintance'.[104]



People *could* get along, but only when the Chinese were seen as a declining force, only when the colonial Australians' sense of European and particularly British political, social and economic dominance was unthreatened. Power and racial knowledge were truly entwined. When the numbers of Chinese rose, as they did in New South Wales from 1877, and more dramatically on the Palmer River and other goldfields in northern Queensland, conflict and racism re-appeared with renewed virulence. But that is another story.

## Notes

1 E. Rolls, *Sojourners: The Epic Story of China's Centuries-Old Relationship with Australia*, 1992, University of Queensland Press, Brisbane, pp. 49–50; P. Crawford Campbell, *Chinese Coolie Emigration to Countries within the British Empire*, 1923, P. S. King, London, esp. pp. 90–100.

2 'Report from the Select Committee on Asiatic Labour', 27 November 1854; *Votes and Proceedings of the NSW Legislative Council*, 1854, vol. 2; Rolls, *Sojourners*, 1992, pp. 45–72.
3 Rolls, *Sojourners*, p. 63.
4 G. Serle, *The Golden Age: A History of the Colony of Victoria, 1851– 1861*, 1963, Melbourne University Press, Melbourne, pp. 320–1.
5 Serle, *Golden Age*, pp. 321, 323.
6 Serle, *Golden Age*, p. 324; Rolls, *Sojourners*, p. 126; K. Cronin, *Colonial Casualties: Chinese in Early Victoria*, 1982, Melbourne University Press, Melbourne, p. 81.
7 *Armidale Express*, 16 August 1856; *Maitland Mercury*, 21 August 1856.
8 One group reached Albury in early October 1856. See *Border Post*, 4 October 1856.
9 *Bathurst Free Press*, 12 July 1856.
10 *Sydney Morning Herald*, 2 August 1856.
11 *Sydney Morning Herald*, 11 August 1856.
12 *Maitland Mercury*, 24 July 1856.
13 Serle, *Golden Age*, p. 325.
14 Arrival figures for 1856 and 1857 and for 1858 till August are in the Report of the Collector of Customs, *Journal of the NSW Legislative Council*, 1858, vol. 3, pp. 313, 315. Departure figures for September–December 1858 are in the Shipping Gazette and Sydney General Trade List, September–December 1858. 1858 arrival figures and figures for 1859–61 are in the Reports of the Immigration Agent, *Votes and Proceedings of the NSW Legislative Assembly*, 1858–59, vol. 2, p. 356; 1861, vol. 2, p. 667, and 1863–64, vol. 2, pp. 1134–5, 1142. The census figures for 1861 are in *Votes and Proceedings of the NSW Legislative Assembly*, 1862, vol. 3, p. 31.
15 The machinery of government control was financed by the imposition of a licence fee, dropping as a result of miner dissatisfaction from the very steep 30 shillings per month in May 1851 to 10 shillings per month in September 1853, and replaced by the much cheaper miner's right of 10 shillings per year in March 1857.
16 Anti-Chinese feeling remained high at Rocky River for several years, with petitions being sent to the Legislative Assembly during the next two years. See *Votes and Proceedings of the NSW Legislative Assembly*, 1857, vol. 1, p. 603, and 1858, vol. 2, pp. 947–8.
17 Secretary for Lands to Gold Commissioner Lynch, 5 August 1857 (copy), Letters from Lands to Gold Commissioners, 1857–60, re. no. 57/2686, NSW Archives Office, 4/6854.
18 *Sydney Morning Herald*, 9 July 1856.
19 *Sydney Morning Herald*, 11 November 1857.
20 Petition from the Meroo, *Votes and Proceedings of the NSW Legislative Assembly*, 1858, vol. 2, p. 947.
21 *Bathurst Free Press*, 24 June, 1 July 1857, 27 February, 20 March 1858; *Empire*, 26 May, 23 July 1857, 4 March 1858; *Yass Courier*, 17 October 1857.
22 *Sydney Morning Herald*, 13 February 1858.
23 See A. Curthoys, 'Racism and Class in Nineteenth-Century Immigration Debate', in A. Markus and M. C. Ricklefs (eds), *Surrender Australia?: Essays in the Study and Uses of History*, 1985, George Allen & Unwin, Sydney, pp. 94–100.

24 *Empire*, 19 March 1858; *Sydney Morning Herald*, 20 March 1858; Letter from Gold Commissioner to Secretary for Lands, 22 March 1858, Lands In Letters, no. 58/930, encl. with 58/1028, NSW Archives Office, 3586.
25 Gold Commissioner to Secretary for Lands, 16 March 1858, Lands In Letters, no. 58/841, encl. with no. 58/1028, NSW Archives Office, 3586; and letter of 22 March cited in Note 24.
26 Gold Commissioner at Louisa Creek to McLean, Gold Commissioner, 3 May 1858. Lands In Letters, unnumbered, encl. with 58/1449, NSW Archives Office, 3587; report of Gold Commissioner Griffin, 23 August 1858, printed in *Journal of the NSW Legislative Council*, 1858, vol. 3, p. 320; *Albury Telegraph and Federation Journal*, 5 June 1858; Rocky River and Meroo petitions, *Votes and Proceedings of the NSW Legislative Assembly*, 1858, vol. 2, p. 947; Turon petition, *Journal of the NSW Legislative Council*, 1858, vol. 3, p. 303.
27 Sydney petition (2469 signatures), *Votes and Proceedings of the NSW Legislative Assembly*, 1858, vol. 2, p. 943; Sydney petition (327 signatures), *Journal of the NSW Legislative Council*, 1858, vol. 3, p. 299; *Bathurst Free Press*, 26 June 1858; *Empire*, 16, 17, 19, 22, 23, 26 and 29 June 1858.
28 *Bathurst Free Press*, 20 March 1858.

29 *Sydney Morning Herald*, 9 and 10 April, 21 May, 19 and 30 June 1858.
30 Debates reported in *Sydney Morning Herald*, 22 and 29 July 1858; bill rejected 29 October, reported *Sydney Morning Herald*, 30 October 1858.
31 *Sydney Morning Herald*, 23 August 1860.
32 *Empire*, 19 September 1860.
33 *Alpine Pioneer*, 31 August 1860; *Sydney Morning Herald*, 21 August 1860.
34 *Braidwood Observer*, 7 July 1860; *Alpine Pioneer*, 31 August 1860; *Sydney Morning Herald*, 7 July 1860.
35 *Empire*, 19 September 1860; *Sydney Morning Herald*, 4 August, 8 October 1860; Gold Commissioner Griffin to Secretary for Lands, 1 November 1860, Lands In Letters, no. 60/5781, NSW Archives Office, 3618.
36 Serle, *Golden Age*, pp. 330–1.
37 Its importance means that the story of the riot has been told a number of times. Previous accounts include T. A. Coghlan, *Labour and Industry in Australia*, vol. II, 1918, Oxford University Press, London, pp. 777–81; M. Willard, *A History of the White Australia Policy*, 1923, Melbourne University Press, Melbourne, pp. 32–3; M. Clark, *A Short History of Australia*, 1963, Heinemann, Sydney; C. Connolly, 'The Lambing Flat Riots', in A. Curthoys and A. Markus (eds), *Racism and the Working Class in Australia*, 1978, Hale & Iremonger, Sydney; W. A. Bayley, *Rich Earth: History of Young and the Shire of Burrangong*, 1956, Young Municipal Council, Young; D.L. Carrington, 'Riots at Lambing Flat, 1860–1861', *Royal Australian Historical Society Journal and Proceedings*, vol. 46, pt. 4, October 1960, pp. 223–43; R. Walker, 'Another Look at the Lambing Flat Riots, 1860–1861', *Royal Australian Historical Society Journal and Proceedings*, vol. 56, pt. 3, September 1970, pp. 193–205; C. Price, *The Great White Walls Are Built: Restrictive Immigration to North America and Australasia 1836–1888*, 1974, Melbourne University Press, Melbourne, pp. 36–9; A. Markus, *Fear and Hatred: Chinese in Australia and California*, 1978, Hale & Iremonger, Sydney; Rolls, *Sojourners*, pp. 165–72.

38 *Yass Courier*, 21 November 1860.
39 *Braidwood Observer*, 2 February 1861; *Bathurst Free Press*, 26 January 1861.
40 *Yass Courier*, 23 January 1861; speeches made at meeting reported by *Yass Courier*, 2 February 1861.
41 *Bathurst Free Press*, 2 February 1861.
42 *Empire*, 4 February 1861; *Bathurst Free Press*, 2 and 6 February 1861; *Sydney Morning Herald*, 30 January 1861; *Braidwood Observer*, 9 February 1861; *Yass Courier*, 13 February 1861.
43 *Miner*, 2 February 1861.
44 *Bathurst Free Press*, 2 February 1861; memo attached to telegram from Gold Commissioner Cloete to Secretary for Lands Robertson, 30 January 1861, Lands In Letters, unnumbered, encl. with no. 61/1964, NSW Archives Office, 3626.
45 Memo attached to telegram from Gold Commissioner Cloete to Secretary for Lands Robertson, 30 January 1861, Lands In Letters, unnumbered, encl. with no. 61/1964, NSW Archives Office, 3626.
46 *Miner*, 20 February 1861.
47 Minutes of meeting of Executive Council, 25 February 1861, Minute Book, minute no. 61/9H, NSW Archives Office, 4/1540; *Sydney Morning Herald*, 28 February 1861.
48 Announced in speech by Cowper in Legislative Assembly, reported in *Sydney Morning Herald*, 27 February 1861.
49 *Miner*, 6 March 1861.
50 *Miner*, 6 March 1861.
51 *Miner*, 6 March 1861; see also Colonial Secretary to Secretary for Lands, 7 and 11 March 1861, Lands In Letters, unnumbered telegrams encl. with no. 61/1964, NSW Archives Office, 3626.
52 *Miner*, 27 March 1861; speeches by Cowper to Legislative Assembly, reported in *Sydney Morning Herald*, 27 March 1861.
53 *Yass Courier*, 13 April 1861; *Bathurst Free Press*, 27 April, 1 May 1861.
54 *Sydney Morning Herald*, 9 March 1861; adjourned debate reported on 13 March 1861.
55 *Sydney Morning Herald*, 21 March 1861.
56 *Sydney Morning Herald*, 3 May 1861.
57 *Sydney Morning Herald*, 17 and 18 April 1861.
58 *Sydney Morning Herald*, 11 May 1861.
59 Meeting of the Executive Council, 10 May 1861, Minute Book, minute no. 61/20B, NSW Archives Office, 4/1540.
60 *Sydney Morning Herald*, 21 June, 2 July 1861.
61 *Braidwood Observer*, 12 June 1861; *Empire*, 8 and 29 June 1861; *Miner*, 22 June 1861; *Sydney Morning Herald*, 21 June 1861; *Yass Courier*, 26 June, 3 July 1861.
62 *Sydney Morning Herald*, 2 July 1861; *Miner*, 3 July 1861; *Empire*, 5 July 1861; *Bathurst Free Press*, 6 July 1861.
63 *Sydney Morning Herald*, 4 July 1861.
64 *Empire*, 5 July 1861.
65 *Sydney Morning Herald*, 8 July 1861.

66 Gold Commissioner Griffin to Secretary for Lands, 12 July 1861, forwarded by Cloete to Secretary for Lands, 16 July 1861, Lands In Letters, no. 61/2390, NSW Archives Office, 3626.
67 Gold Commissioner Griffin to Secretary for Lands, 16 July 1861, Lands In Letters, no. 61/2399, encl. with no. 61/2597, NSW Archives Office, 3627.
68 *Bathurst Free Press*, *Yass Courier*, *Sydney Morning Herald*, all 17 July 1861; Gold Commissioner Griffin to Secretary for Lands, 16 July 1861, no. 61/2399.
69 *Miner*, 17 and 20 July 1861; *Sydney Morning Herald*, 17 July 1861; *Yass Courier*, 17 July 1861; Gold Commissioner Griffin to Secretary for Lands, 16 July 1861, Lands In Letters, no. 61/2597, NSW Archives Office, 3627.
70 Meeting of Executive Council, 17 July 1861, Minute Book, minute no. 61/32, NSW Archives Office 4/1540; *Sydney Morning Herald*, 18 July 1861.
71 *Yass Courier*, 20 July 1861; *Empire*, 18 July 1861.
72 *Sydney Morning Herald*, 5 August 1861; *Braidwood Observer*, 10 August 1861; Col. Kempt to Secretary for Lands, 3 August 1861, and Supt Cloete to Secretary for Lands, 3 and 4 August 1861, Lands In Letters, unnumbered telegrams, encl. with 61/1964, NSW Archives Office, 3626; *Goulburn Herald*, 10 August 1861; *Empire*, 14 August 1861.

73 *Goulburn Herald*, 18 and 19 September 1861; *Yass Courier*, 9 and 16 September, 6 November 1861; *Empire*, 13 November 1861; *Sydney Morning Herald*, 23 November 1861; *Miner*, 4 December 1861; Supt Cloete to Secretary for Lands, 19 August 1861, Lands In Letters, unnumbered, encl. with no. 61/1964, NSW Archives, 3626.
74 *Sydney Morning Herald*, 1 August 1861.
75 *Sydney Morning Herald*, 6 August, 19 September 1861; *Empire*, 12 and 16 September 1861; Major's Creek petition enclosed in Colonial Secretary to Secretary for Lands, 12 September 1861, Lands In Letters, no. 61/3318, NSW Archives Office, 3629.
76 P. Loveday, 'The Legislative Council in New South Wales, 1856–1870', *Historical Studies, Australia and New Zealand*, vol. 11, no. 44, April 1965, pp. 489–94.
77 *Sydney Morning Herald*, 12 September 1861.
78 *Sydney Morning Herald*, 10 October 1861
79 Report of the Immigration Agent, *Votes and Proceedings of the NSW Legislative Assembly*, 1863–64, vol. 2, p. 1151.
80 Reports of the Immigration Agent, *Votes and Proceedings of the NSW Legislative Assembly*, 1863–64, vol. 2, p. 1151; 1864, vol. 1, p. 627; 1865, vol. 2, p. 232; 1866, vol. 1, p. 207; 1867–8, vol. 2, pp. 435, 451; 1872, p. 1141.
81 *Wynyard Times*, 8 April 1862; *Lachlan Observer*, 11 June 1862; *Sydney Morning Herald*, 15 May 1866, 13 June 1871; *Braidwood News*, 8 January 1864.
82 *Goulburn Herald*, 22 January 1862; *Bathurst Free Press*, 14 May 1862; *Mudgee Liberal*, 14 January 1864; *Deniliquin Chronicle*, 17 December 1864; *Sydney Morning Herald*, 10 March 1866.
83 Official in Department of Public Works to Henry Parkes, 23 March 1866, Parkes Correspondence, vol. 32, p. 286, Mitchell Library, A902.
84 Williamson to McLean, 24 August 1863, encl. with Gold Commissioner McLean to Secretary for Lands, 4 September 1863, Lands In Letters, no. 63/5231, encl. with

no. 65/2369, NSW Archives Office, 3685; Gold Commissioner McLean to Secretary for Lands, 19 September 1863, Lands In Letters, unnumbered, encl. with no. 65/2369, NSW Archives Office, 3685.

85 *Burrangong Courier*, 2 August 1862.

86 Petitions sent February, July 1863, May 1864, in Lands Department In Letters, no. 65/2369, NSW Archives Office, 3685.

87 *Burrangong Argus*, 16 September 1865; Petition from Inhabitants of Burrangong to Secretary for Lands, 10 August 1864, Lands In Letters, no. 64/4183, encl. with no. 65/2369, NSW Archives Office, 3685.

88 *Burrangong Argus*, 7 April 1866; Petition from Miners to Secretary for Lands, 4 April 1866, Lands In Letters, 66/2649, NSW Archives Office, 3701.

89 *Sydney Morning Herald*, 10 September 1866.

90 *The Mining Record and Grenfell General Advertiser*, 15 June, 31 August 1867.

91 Extract from *Western Examiner*, 9 April in *Sydney Morning Herald*, 13 April 1864; *Burrangong Argus*, 9 September 1865, 1 September 1866; *Deniliquin Chronicle*, 31 December 1864; speech by Forlonge in Legislative Assembly, 20 September 1867, reported in *Sydney Morning Herald*, 21 September 1867; *Sydney Morning Herald*, 8 October 1867, 9 June 1868.

92 *Deniliquin Chronicle*, 31 December 1864. 

93 *Burrangong Argus*, 9 September 1865.

94 *Votes and Proceedings of the NSW Legislative Assembly*, 1865–66, vol. 2, pp. 233–6.

95 *Sydney Morning Herald*, 15 August 1866.

96 Reports of the debates are in *Sydney Morning Herald*, 4 and 21 September, 5 October 1867.

97 *Empire*, 23 September 1867.

98 Entry for 21 November 1867, *Votes and Proceedings of the NSW Legislative Assembly*, 1867–68, vol. 1.

99 H. Lawson, *Prose Works*, 1956, Angus & Robertson, Sydney, p. 553.

100 Quong Tart to Braidwood Bench, 19 August 1869, Colonial Secretary's In Letters, no. 69/6736, NSW Archives Office, 4/668; Quong Tart to E. Greville, MP, 17 February 1876, encl. with Greville to Colonial Secretary, 29 February 1876, Colonial Secretary's In Letters, no. 76/1565, and other letters on the matter enclosed with Greville to Colonial Secretary, 18 March 1876, Colonial Secretary's In Letters, no. 76/2049, NSW Archives Office, 1/2325.

101 *Deniliquin Chronicle*, 3 February 1870.

102 J. Lydon, *Many Inventions: Chinese in the Rocks, 1890–1930*, 1999, Monash Publications, Melbourne.

103 *Evening News*, 29 June 1875.

104 *Town and Country Journal*, 30 July 1870.

# Undesirable Persons

## RACE AND WEST AUSTRALIAN MINING LEGISLATION

**Chapter**

*Patrick Bertola*



Early in 1973 the Parliament of Western Australia deleted from the West Australian Mining Act several provisions which imposed restrictions on 'Asiatics' and 'Africans' – whether aliens or British subjects. These provisions had hitherto prevented such people from holding a miner's right and from being employed in any capacity associated with mining and mineral leases in Western Australia. The Labor government of the time sought to repeal the provisions in order to remove obstructions to Australia becoming a party to the International Convention on the Elimination of All Forms of Racial Discrimination.[1] In so doing, members reversed the Labor Party's longstanding support for the restrictions, including its opposition to an attempt in 1963 to repeal discriminatory elements of the Act.

The history of these legal provisions began prior to the enactment of the Mining Act in 1904. They developed in the context of a longstanding debate over race and migration in the colony and had their origins in the determination of the colonial administration from the 1880s to prevent the arrival of Chinese immigrants on the West Australian goldfields.

Such attempts at restrictions in mining legislation and regulations might be viewed in terms of a number of strands, the first of which was directed towards 'non-white' people from Asia and Africa. Over time, West Australians developed regulations and legislation that sought to exclude alien nationals absolutely from holding mining property and engaging in mining. They also sought to place severe discretionary limits on people from Africa and Asia who claimed to be British subjects. The scale of penalties for breach of the laws was high and extensive punitive powers were granted to authorities under the legislation.

Another racially exclusive provision not removed until 1963 acted against the employment of Aborigines in mining. The Mining Act required leaseholders to employ a certain quantum of labour, proportionate to the size of a lease.[2] Yet the labour of Aboriginal Australians could not count towards the quantum, a provision that effectively made such labour unattractive to employers and disadvantaged Aboriginal prospectors who might seek to work small leases on their own account.

A third strand was directed against a broad group of 'foreigners' from southern Europe. While little specific legislation impinged upon southern Europeans, section 42 (2) of the Mines Regulation Act 1906 required (underground) mining employees to be able to speak the English language readily and intelligibly.[3] On a number of occasions in the first three decades of the twentieth century, sections of the 'British' labour force sought to exclude 'foreign' workers from the industry through the application of a so-called 'language' test. While language proficiency could plausibly be linked to safety, the introduction and application of the provisions were inextricably bound up with Anglo-centrism and with relations between and among employers and employees in the mining workplace. Debate on this issue, especially in the early 1900s and during the late 1920s and early 1930s, suggests that arguments for exclusion were frequently based on the claim that southern Europeans, like Chinese coolies before them, were arriving as groups under a formal or informal contract, and were being used to depress wages and undermine working conditions. Further legislation circumscribed the role that non-British labour could play in oversight of the workplace: in 1915, a Labor government amended the Mines Regulation Act to prevent any person who was not a native or naturalised British subject from voting in any election for a workmen's inspector of mines.[4] These developments suggest that, while labour might act collectively in respect of 'its own', it could be chauvinistically nationalistic in seeing capacity and civility as synonymous with 'British' or 'European', while Latins, Slavs, or persons from the south of Europe were caricatured as having undesirable qualities that made them incapable of 'real' contributions to labour and its world.[5]

In respect of southern Europeans, the conflation of race, ethnicity and inferiority is perhaps understandable, given that race was then the central organising and classifying concept, a position now held by culture. Arguing that social and racial evolution mirrored physical evolution, nineteenth-century theorists attempted to link social theory, the origins of social groups, and comparative anatomy. In so doing, they concluded that cultural differences, as well as apparent differences in intelligence and 'civilisation', could be explained in terms of permanent and stationary hereditary capacities. As a corollary, theorists posited a qualitative hierarchy of abilities among races, which was manifest in national achievements. Arguably, it was race theory and associated factors that informed and consolidated the solidarity of the 'Britisher', even while carrying a paradoxical sense of racelessness. Specifically, in Western Australia, southern Europeans could be, and were, seen as a race separate from and inferior to the British.

In the debate on immigration restriction in 1897, much of the agitation came from the goldfields. For example, F. C. Vosper, member for the goldfields seat of North-East Coolgardie, noted that

> at some future time it might be considered necessary, in the interests of the Colony, to exclude Greeks, Hungarians, Italians, and other persons who were not coloured . . . [for] [p]ersons of the nationalities mentioned had become a greater pest in the United States than the coloured races.[6]

Similarly, C. J. Moran, member for East Coolgardie, another goldfields seat in the Legislative Assembly, referred to the necessity of using the legislation not only 'to prohibit aliens or Asiatics, but also . . . [to] prohibit certain classes of Europeans from coming into this country'.[7]

Common to all attempts at ethnic discrimination was the objective of establishing the legal bases upon which the presence of others could be limited, or circumscribed, within goldfields economy and society. Moreover, the language of collective race or ethnicity rarely differentiated between regional groups. While 'Asiatic' was primarily synonymous with Chinese,[8] it could also refer to groups as diverse as north-west Indians (in turn, collectively misknown as Afghans) or Japanese pearl divers. Likewise, 'Italian' often included anyone of southern European origin as well as people from the Balkan regions. Further, while 'alien' might be 'Asian', as in the legislation, in goldfields society it was frequently used interchangeably with a word like 'foreigner' to denote someone of Italian, Slavonic (Dalmatian, Croatian, Macedonian, and so on) or Greek origin, whether or not they were a naturalised British subject. Of course, there was another, less respectable language of difference found in words like 'ding', 'dago', 'nigger', 'chow', or 'chink'.



Underpinning the restrictions in West Australian mining legislation was the same sort of restrictive sentiment embodied in colonial legislation in general, and in the Commonwealth Immigration Restriction Act 1901 in particular.[9] Such restrictions drew upon a history of attempts to exclude or control those considered undesirable or perceived as a threat to labour, or to prevent the arrival of a great influx of immigrants. In 1893 during debates on the Chinese Immigration Amendment Bill, S. H. Parker, the Colonial Secretary, argued that, while the Chinese might be described as law-abiding, thrifty, industrious and frugal, 'they do not belong to the European race, and the question is, do we desire Western Australia to be overrun with Chinese, to the exclusion of the Anglo-Saxon race?'[10]

Yet there was conflict within the colonial legislature on the issue of immigration. While maintaining the racial superiority of the Anglo-Saxon, many leading colonists still wanted to make use of 'Asiatic' labour. In the debate on the 1893 bill, G. W. Leake argued: 'It will be a great pity if the colony is deprived of the Chinese, who are assiduous and skilful'.[11] J. Morrison contended that such immigrants would even be 'a means of elevating the working classes, who should be in the position of superintendents over Chinese or employers of Chinese'.[12]

Such demands for access to Asiatic or coolie labour were not new. The terms under which the colony was founded meant that it had had a longstanding shortage of labour and low population growth.[13] Colonists sought to resolve this shortage by attempting to attract assisted British immigrants to the southern areas, by convict transportation and the longstanding practice of using prison labour, by employing Aborigines under a variety of terms in pearling, on pastoral holdings and in agriculture, and by using 'Asiatic'/Chinese/coolie labour under indenture or as 'free' labour in the same industries. C. Harper, pastoralist and MLC for the northern district, noted:

> [t]hese men were imported from Singapore generally, at a very low rate compared with the pay of Europeans, though quite as good, if not better – in many cases, certainly better – men . . . their employers had much more control over them . . . [and they] were, as a rule, more steady in their work than the Europeans.[14]

In short, employers held the coolie who could be obtained 'at about twenty-two shillings a month' to be cheap, docile and constantly productive. Although T. C. Carey 

> had seen the evils attendant upon the introduction of Chinese in the other colonies . . . he had no hesitation in saying that the evils were more than counterbalanced by the good. [For] these men would work at a reasonable rate of wages, and work six days in the week, whereas a great many European laborers in those colonies . . . expected, as a rule, to get paid for six days and only work for half that period.[15]

Moreover, according to Harper, the need for coolies was exacerbated by the alleged perversity of 'British' workers who, on the one hand, would not 'go out of town to perform heavy labouring work in the country' and, on the other, opposed the proposal to supply 'the immediate [labour] requirements of outlying districts'.[16]

Such an appeal to economic efficacy was not, however, universally endorsed. Critics aligned with non-rural factions of the local employing class claimed that the cost of local European labour had not proved an impost and that the colony's payment for importing and repatriating coolies constituted a subsidy for one portion of local capital. Noting that the Perth City Council had 'encountered no trouble in obtaining an adequate supply of labour . . . at wages varying from 4s.6d. to 5s.6d. a day', Parker argued that the rural shortage was largely a function of unwillingness to pay the price of labour as determined in the market.[17]

Critics also argued that importing coolie labour would deter permanent settlement from Europe and undermine the object of establishing a 'British' colony. W. E. Marmion noted that, by introducing coolies who would compete with European labour, the colonial administration would 'prevent the encouragement of that class of immigrants which the country chiefly required for the purposes of settlement, and for increasing the population, and thereby increasing the prosperity of the Colony'.[18] While equating racial purity with economic and social advancement, Marmion and others recognised that their position would not be shared by the majority of the legislature, that is, those whose interests predominantly lay with the rural and pastoral areas. The best they could demand was that employers be responsible for the cost of repatriation.[19]

In 1884, the colony legislated so as to accommodate these differing positions. The Imported Labour Registry Act 1884 regulated the entry, employment and geographic location of 'Asiatic' labour in the north of the colony. In this way, regulation satisfied the demands of the pearling and pastoral industries in the north. The 'leakage' to the south was limited and, following the discovery of gold, the Chinese Immigration Restriction Act 1889 further limited Chinese immigration to the colony.[20]



Because goldmining involved the ownership of property and the prospect of immediate access to wealth, the legislators' attitude to 'Asiatics' generally and to Chinese in particular was restrictive from the outset. Specific measures were modelled on laws introduced elsewhere, particularly those developed in the aftermath of gold discoveries in Queensland.[21] Early in 1886, the Legislative Council addressed the Governor with the opinion that, 'in anticipation of the discovery of payable goldfields in Western Australia, it is desirable to take steps to prevent Chinamen or Mongolians from acquiring licenses to dig upon the said fields'.[22] Governor F. N. Broome advised London:

> Should there be any considerable influx of Chinese, as there certainly would be on declaration of a payable goldfield in this Colony, and should the people and legislature desire to adopt the safeguards which exist in other colonies . . . I am of the opinion that my Government should be authorised to take the necessary action, being of course careful to act with due reflection.[23]

Parliament rapidly enacted legislation embodying such concerns, refining it over almost two decades into a form that remained relatively untouched until the 1960s and 1970s. Effectively, the law banned the issue of miner's rights or business licences to Asiatic and African aliens. It further banned, except by ministerial authority, the issue of miner's rights or business licences to any Asiatic or African claiming to be a British subject; it prevented the grant of leases to Asiatic or African aliens; it granted mining wardens the right to remove Asiatic or African aliens from gold or mineral fields; and it prevented any employer from employing an Asiatic or African 'alien' in or about any mine, claim or authorised holding.[24]

Sections 3 and 8 of the Goldfields Act 1886 contained the original legislative expression of this exclusion. They embodied the first mention of exclusion in mining law and ensured that control over certain people on the goldfields was in place from the time of the earliest gold discoveries in Western Australia.[25] Introducing the bill, S. Burt, the Attorney-General, noted that the legislation represented the desire of the Legislative Council 'that Chinamen and other Asiatic aliens should be excluded from working upon any goldfields that might be discovered in the Colony, at least for a certain period after the goldfield was proclaimed'.[26] In this regard, the Council followed Queensland legislation by imposing an exclusion for five years after proclamation of a goldfield.

In the year of the great gold rushes to Coolgardie, the new self-governing legislature went further, enacting an unlimited bar on 'alien' persons of Asiatic or African background under the provision of its Mineral Lands Act 1892. In 1895, a new Goldfields Act provided that Asiatics and Africans who were British subjects would also be excluded but that the minister could provide an exemption. This Act, W. H. James noted, would embody 'the principle . . . [of exclusion to which] the House, on previous occasions had committed itself beyond recall, and that was with the declaration that Asiatics and Africans who were aliens should not be permitted to carry on mining operations'.[27]



Not content, some members sought to limit Asiatics and Africans further by removing the ministerial discretion on those who were British subjects. During debate on the bill, Moran asserted:

> It is only raising a bogie to say that the British people will be offended if we refuse to allow coloured people, born under the British flag, to come into the colony, and as their advent into the country is objectionable to the white population of the goldfields, I think that this provision should be discarded.[28]

The legislature's rejection of this proposal was purely pragmatic, since the Governor would have reserved the bill to London for Royal assent.

The 1892 Act did not bar aliens from all work on the goldfields. While preventing Asiatic and African aliens from holding general business licences, the legislation's definition of a business activity specifically excluded 'the hawking of farm or garden produce'. In this way, Chinese and Afghans could sell such produce on the goldfields.

Parliament moved further in 1898, barring Asiatics and Africans from holding any interest in a goldmine by virtue of a consolidated miner's right and from employment as a miner. At first, the government sought to replace the Goldfields Act with a new Gold Mines Act. The bill contained the same restrictive provisions as the 1895 Act but sought to add others: to ban Asiatic and African people – both alien and Empire subjects – from goldmining and commercial activity, to impose penalties on Asiatic and African aliens found mining on Crown land, and, in an amendment, to limit the presence of Aboriginal workers on mines. In debate on the bill, Vosper claimed that 'a custom has grown up on the goldfields of employing Asiatics or Africans or Afghans in doing odd jobs about a mine, such as carrying wood or water'; such a custom, he claimed, would lead to serious trouble.[29] Asserting that 'there are plenty of white men to do the work', he argued that the bill should ban the employment of all Asiatics or Afghans, including those who were British subjects, in 'any capacity whatever'.[30] As to Africans, Vosper supported an amendment moved by H. Gregory, Mines Minister, stating that,

> although black in skin, most [Africans] were jolly fellows, and had none of the servile, degrading traits of the Asiatic; but if the American or African Negro were admitted, the result would be that every man with the slightest trace of the tar-brush in his complexion would claim to be an African or American nigger.[31]

In addition, W. Kingsmill successfully moved for an amendment in the committee stages, whereby the labour of Aboriginal Australians would not count as bona fide work in fulfilling the labour conditions on leases or claims. Such a provision was necessary, he argued, to provide the 'legal means of stopping the employment of native labour',[32] something that G. Randell, the

Colonial Secretary, later held was 'also in the interest of European or white workers'.[33] As in 1892, the amendments did not attempt to totally exclude Empire subjects from Africa and Asia, because such measures might lead the Governor to reserve the bill. However, the bill also sought to remove from the Act the right to sell farm and garden produce. This move had the support of Premier John Forrest, who noted that 'no one wished for the continuance of Asiatic hawking'.[34]

When forced to withdraw the bill in September, after it became apparent that parliament would not reach agreement on questions relating to alluvial deposits, Forrest's government then introduced restrictive amendments to the existing Goldfields Act and, significantly, the relevant clauses passed through the committee stages without debate. These restrictive provisions of 1898 expanded earlier legislation and embodied a more comprehensive set of exclusions applicable to Asiatics or Africans, as well as provisions to limit or discourage the employment of Aborigines.

While Vosper and others suspected that ministerial discretion would lead to increases in the numbers of Asiatic and African subjects of Empire associated with goldmining, the provisions did little to affect the general principle. Indeed, the official antipathy was evident in cases of Asiatics applying for exemption from the provisions of the Act. Transmitting a petition from Bishin Sing and others in January 1898, Governor Smith reported that he had been aware

> for some time of the dissatisfaction prevailing in consequence of no distinction being drawn between the various Eastern Races resident in this Colony, especially those who have been among our bravest soldiers in India, fighting against those very Afghans with whom they are now classed.[35]

Later, in 1899, when forwarding a petition from one Ahmed Khan seeking the grant of a miner's right, Smith remarked

> that it is but one of many instances in which Her Majesty's Indian subjects are debarred in this Colony from the exercises of the rights and privileges enjoyed by other of Her Majesty's Subjects. The Hon. Minister for Mines is apparently not satisfied upon the papers submitted to him and the report of the Commandant of the man's identity. The Law appears to give plenary discretion to the Minister . . . who insists upon proof of identity by 'marks' the nature of which he does not specify or even suggest, but I am of the opinion having regard to what the Minister states as to the 'feeling which exists' that no proof of identity in the shape of 'marks' that the wit of man could devise would secure the issue of a Miner's Right in this Colony to any of Her Majesty's Indian Subjects who rightly or wrongly can be classed as an Asiatic.[36]

When the Mining Act 1904 became the principal Act governing all mining in the state, there was little debate concerning the transfer of exclusionary provisions from the Goldfields Act. As before, parliament refined the restrictive clauses and extended them by adding a new section to prevent any person who held a miner's homestead lease from executing a mortgage over it in favour of Asiatic or African aliens, or any person of Asiatic or African race claiming to be a British subject without first obtaining the authority of the minister. The Act also preserved the old clause preventing Aboriginal labour from counting towards the labour requirements on leases. All these provisions remained unchanged for almost sixty years. New sections relating to mineral oil leases and foreign capital that had exclusionary effects were enacted in 1920 and 1926, but were repealed in 1932.[37]



Sections of the Mines Regulation Act 1906 could also limit the presence of non-Britishers in goldmining. The Act prevented the employment in certain departments or positions within mines of those unable to communicate in English. A significant restriction also existed in section 6 (c), limiting the franchise for positions of workmen's inspectors to natural-born Australians or to naturalised British subjects. From time to time, such provisions formed part of the intersection of race and class relations in the mining sector.

Racially restrictive legislation sought to retain a 'British' character and foster 'racial purity' among the population at large and within the goldmining industry. The restrictions rested upon the continuing, though not unchallenged, dominance of ideas about racial hierarchies and the dream of a 'White Australia'. Importantly, as noted briefly above, around the turn of the century such restrictions became more generalised as the notion of racial undesirability was extended to others, particularly southern Europeans. In the context of goldmining, this extension firmly intertwined questions of race with issues of social and economic relations. More generally, legislative structures such as the Mining Act played an important part in the development of racist sentiment and social tensions. The legislation regarding language was justified in part as an attempt to meet the demands of organised labour for some controls, although in practice it did not pose any sustained impediment to the employment of southern Europeans.[38] The state took care to retain sufficient powers of exemption or oversight so that such provisions did not pose a threat to the interests of mining capital.

For a number of reasons, the period up to the 1960s marks something of a hiatus in respect of 'undesirable' migrants within the mining industry. First, there had long been limits on non-European migrants when, from the late 1920s and early 1930s, the Commonwealth had severely limited southern European migration until after the war. Second, the general prosperity of the goldmining industry in the 1930s eased pressures for reductions in the cost of labour. Third, after World War II, goldmining companies found that they could not meet their shortfall of miners from traditional sources in southern Europe or even from Britain. Finally, much of the expansion of the post-war economy was based on attracting and accepting migrant labour, including large numbers from southern Europe.[39] The situation was somewhat different with regard to Asia. Here, antipathy remained, in part as a consequence of the war with Japan,[40] the cold war, and anti-communism. More generally, it reflected a broader Australian tendency to see Asia as an undifferentiated, threatening mass. Thus, while hostility towards southern Europeans lessened, the debate surrounding the attempts in 1963 to repeal section 291 of the Mining Act indicates that this was not true of 'Asiatics'.



During the 1960s goldmining languished and there was a marked shift towards Asia as the destination of mineral exports and as the source of investment in Western Australia for mining, especially for iron ore mining. By 1963 the goldmining industry, which had earlier been the single most important part of the mining sector, was experiencing recession, causing relatively high levels of unemployment and an exodus from the goldfields to other areas of the state. As yet, the large nickel deposits that were to form the backbone of an expansive phase of base metal mining in the late 1960s remained undiscovered, and both the bauxite and iron ore mining industries were in the very early stages of development. Consequently, there is evidence that craft workers from mines simply chose to go to the coastal areas where their skills were in greater demand and conditions were better.

In 1963 the possibility of an 'alien' presence was thrust into this context. On the one hand, labour interests, particularly those in goldmining, perceived a direct threat. Despite its decline, goldmining still remained a significant part of the state economy and, therefore, any changes to mining legislation were bound to be seen as financially menacing and a potential threat to the predominantly 'British' character of its workforce. On the other hand, the lure of mineral expansion encouraged the state to pursue Japanese capital and a Japanese presence.

On 31 December 1962, a shareholder in a copper mining venture at Whim Creek in the north of Western Australia lodged a complaint with the Pilbara warden at Marble Bar, alleging that, during the preceding six months, five Japanese employees of the Dowa Mining Company of Japan had breached the provisions of section 291 of the Mining Act. The complainant claimed that the men were in Australia on an immigration permit, that the Minister for Mines had granted written permission for them to work on the mining lease for the purposes of supervising the installation of new machinery from Japan, but that they had worked outside the terms allowed under the ministerial permit and in so doing had contravened the second provision of section 291. This stated that 'no Asiatic or African alien shall be employed as a miner or in any capacity whatever in or about any mine, claim, or authorised holding'. The warden heard the complaint on 19 March 1963, convicted the Japanese employees and imposed a nominal penalty on each. On appeal in the Supreme Court, Justice Jackson dismissed the convictions, finding that the wording 'shall be employed' in section 291 was intended to be interpreted as being employed by someone who would be liable for breaches under the section.[41]

Even before the case went to the Supreme Court, A. F. Griffith, Minister for Mines in the Brand Liberal/Country Party government, introduced an amendment to repeal section 291. The rationale was entirely consistent with the thinking of the 'development'-minded government, particularly its vision for the development of iron ore mining which was emerging as a major prospect for the Pilbara region. In the context of a developing economy and a mining sector requiring large-scale finance, Griffith's amendment sought to resolve a contradiction that had arisen between the interests of Japanese capital and the state government, on the one hand, and West Australian legislation, on the other. Griffith noted:



> Approaches that have been made to Japan have come generally from the Western Australian deposit holders and, quite naturally, the Japanese desire their own geologists and engineers to examine likely deposits and obtain firsthand knowledge of them . . . Japan today is a major market for Australia's minerals and products. It buys asbestos, tin, manganese, gypsum, copper, ilmenite, and other Western Australian minerals. We hope she will buy iron ore from us in very large quantities.[42]

The attempt encountered opposition from the Australian Workers Union (AWU) and the Returned Services League (RSL), both of which resuscitated the arguments of the past: linking Asian migration to the spectre of an Asian influx; asserting that the state was acting in a highly partisan way against the interests of local mining labour; and using the opportunity to score political points at the expense of the minister and the government. F. J. Wise drew the attention of the minister to a 1903 debate on the Mining Bill and the section that Griffith sought to repeal. He noted that, at the time, R. Hastie had said that 'the people of Great Britain were not in favour of the idea of Australia being overrun with Asiatics'. Indeed, 'the only section at home who would approve of this were those who desired to make money out of the presence of Asiatics in Australia'.[43] Wise added:

> I think they are very pungent words. Later in the Committee stage when this particular section was considered it became obvious that it was put there by design – it was something to protect the very policy and way of life within Australia. I suggest that those circumstances remain today, and we should not weaken the situation one scrap by deleting from our mining laws the words which do not permit aliens to work in the industry . . . [The] minimum effect [of the repeal would be] . . . of throwing wide open to persons who have shares in a copper mine in the north-west, who are not all Australians, the opportunity to employ aliens on which section 291 at present imposes a stricture.[44]

Wise's speech revealed several strands of the racist argument that had long prevailed: its exclusive concentration on Asian migrants and no others; the assumption of a 'way of life' in need of protection; the belief that a mass of Asian labour was waiting to enter Australia; and the alleged connection between industry and the state that existed to drive down wages and conditions, or to replace 'white' labour entirely. To these was added the claim that, while Commonwealth migration legislation imposed a dictation test and, thus, could be used to exclude undesirables, the repeal of section 291 would allow a state government sympathetic to capital to grant permits for Asiatic labour to come into the mining industry, circumventing the Commonwealth legislation.[45]



A second important element in the opposition was the allegedly partisan position of the state in seeking to overturn the principles of the old restrictions when these acted as an impediment to new phases of mining development. H. C. Strickland alleged that the conservative Liberal/Country Party government and Griffith, its Mines Minister, were acting on behalf of large-scale Japanese mining capital. He claimed that 'only since negotiations for the sale of iron ore to the Japanese has the Minister desired to alter the Act'.[46] Responding, R. C. Mattiske contended that it was imperative there be every opportunity for foreign investment and employment to go ahead, for

> it necessarily follows that if we are going to put restrictions on these people to preclude them from coming here and looking after their investments, then naturally they will not be interested in dealing with us . . . All that is intended with this piece of legislation is for those who are prepared to invest their money in purchasing iron ore from this state to have the opportunity to come here to supervise the winning of the ore from the ground and its transport to Japan.[47]

Labor members held that such concerns should not override longstanding principles and apprehensions. For instance, Strickland offered the stock argument for exclusion on the basis that even the smallest concession might lead to a 'flood':

> Our workmen can do all the work necessary around the iron ore deposits to export iron ore. We have the know-how and every thing else, so there is no necessity to bring in Asiatics. The spadework in production has been done in a

> remarkable manner by Europeans, and let us hope that it remains that way. We should not open the crack in the door which will allow even a trickle of Asiatic labour into Australia.[48]

He later enjoined the preservation of an 'essentially British' character in Australia, on the grounds that 'we have the ability . . . to do the work ourselves. If we have not, the Government knows where to get the required men, and it can get plenty of them from Great Britain.'[49]

Others suggested that the main unions both opposed the repeal and derived considerable support for this position from the community. J. J. Garrigan noted that 'the AWU is definitely opposed to it because the white Australia policy is part of its constitution. The RSL is opposed to it.'[50] Indeed, the opposition drew further support from a general suspicion that the changes represented the 'thin edge of a wedge' to break union influence in mining and the debate on migration.

The minister also became a target of the Opposition, who contended that the debate was being rushed in the closing stages of the parliamentary session before the resolution of legal action by aggrieved prospectors. These, members implied, had lost their leases through underhand company actions.[51] The reference to prospectors, with their romantic connotations, was intended to inspire nationalist sympathy for those who had taken the warden's court action against the Japanese nationals. Members accused Griffith, too, of being complicit in issuing permits which gave the appearance of restricted access, when the intention all along was for the Japanese to work the mines. He had supposedly allowed Japanese men to continue work after his department had been notified of their working on or around the mine in non-technical capabilities. When his department expressed concern, his only action was to introduce legislation that would circumvent the apparent breach of the law.[52] Members also alleged that H. W. Woodfield, the principal of Warman Mining and Depuch Shipping and Mining Co. Ltd, and his Japanese associates intended to use Whim Creek as a base from which to investigate nearby iron ore deposits. The implication was that the Whim Creek mine was nothing but a shell for alien interests seeking to gain control of Australian resources. This served to confirm the 'wisdom' that the principals had no serious interest in developing the mine and in providing employment for Australian miners.

Faced with hostility and the accusation of ministerial mishandling of the matter, the government capitulated. Instead of insisting on the repeal of the section, it amended it by adding the words 'without the authority, in writing, of the Minister first obtained'. In effect, the amendment still provided for exclusion but could be overridden with ministerial authority.

Only in 1973 were this and other racially exclusive provisions removed entirely from mining legislation. Arguably, the excision was made possible by the formal abandonment of 'White Australia', by a more liberal social climate in which it was possible for Labor members to 'upgrad[e] our thinking with the times',[53] and by the greater experience of Japanese investment in mining and other industries over a long period. Significantly, it was the Labor Party which sought to enact the change.

This history of racially and ethnically exclusive legislation reveals a longstanding and deeply entrenched bigotry in Western Australia that was evident from a very early period of colonial history and was formalised in its mining laws. The economic and cultural importance of the goldmining sector was reflected in its history as a central site for debate concerning the ethnic composition of the populace – even extending, as Bill Bunbury's chapter shows us, to European immigrants. The issues of race and ethnicity became key expressions of conflict between labour, the state and the mining companies. Racism became inextricably enmeshed with the social relations of production. In the 1963 debate, all of these arguments were reworked in the attempt to prevent the repeal of one restrictive section of the Mining Act. Arguably, the capitulation of the government suggests that these issues retained a populist appeal and that racism directed against 'Asiatics' was still a potent force in the society. The debate also suggests that an understanding of racism is incomplete without some historical analysis of its bases, particularly its expression in the relations of production and the role of the state in West Australian mining history.



## Notes

1 Specifically, the amendment of the Mining Act would help enable Australia to ratify ILO Convention 104.
2 Where a group of leases was held, a leaseholder could consolidate the total labour requirements and employ it on one or more leases within the grouping.
3 Further, section 42 (1) of the 1906 Act provided that 'no person shall be employed in any mine as manager, under-manager, platman, shift boss, or engine-driver unless he is able to speak the English language readily and intelligibly, and to read it whether printed or written'.
4 While the provision did not exclude unnaturalised immigrants or those born outside Australia or Britain from standing as workmen's inspectors, the context of antipathy towards foreigners certainly did. This section remained in place until 1956.
5 See Bill Bunbury's chapter for further discussion.
6 *Western Australian Parliamentary Debates* (hereafter *WAPD*), 11, 23 November 1897, p. 498. Vosper, a journalist, had come to the Murchison goldfields in 1892. In

1894, he had founded the main newspaper in Coolgardie and later, in 1897, the *Sunday Times* in Perth.

7 *WAPD*, 11, 22 November 1897, p. 495. Originally the MLA (member of the Legislative Assembly) for Yilgarn, in 1897 Moran, a contractor's agent and later a farmer, became an independent member for East Coolgardie.

8 The reference to and concern with Chinese is clear in the Governor's despatches to the Secretary of State for the Colonies. See, for instance, 'Policy to be pursued respecting Chinese in the Colony' in which Governor Broome canvassed the need for machinery to prevent an 'influx' of Chinese and the necessity of being able to limit their presence on goldfields in the colony; Broome to Stanley, 24, 16 January 1886, Governor's Despatches to Secretary of State, SRO of WA, WAA 617, Acc.390, WAS 1166, item 15 (1883–86) [hereafter, WAS 1166/15].

9 For example, the colony's 1897 Immigration Restriction Act contained provision for a dictation test.

10 *WAPD*, 2, 4 October 1893, p. 1031. Parker, a lawyer, was born into a pastoral and farming family in York (Western Australia).

11 *WAPD*, 2, 4 October 1893, p. 1031. Leake had practised law in various capacities in the colony since 1850. At the time he was a nominated member of the Legislative Council and chair of the Aborigines Protection Board.



12 *WAPD*, 2, 4 October 1893, p. 1032. Morrison, a nominated member of the Legislative Council, was a stock and station agent who held land at York and who developed and sold properties in the Perth area. For further comments on migration from Asia from the representatives of pastoral and pearling interests in the Legislative Council, see the debate on the Imported Labour Registry Act (1884) in *WAPD*, 8 (old series), 1883. The Act applied to natives of India, China, Africa, the islands of the Indian and Pacific Oceans, and the Malayan archipelago.

13 See, for example, comments and citations of the critique of E. G. Wakefield in Karl Marx, *Capital*, vol. 1, introduced by Ernest Mandel and translated by Ben Fowkes, 1976, Penguin, Harmondsworth, in association with *New Left Review*, pp. 932, 934, 936.

14 *WAPD*, 3, 27 June 1878, p. 132.

15 *WAPD*, 5, 2 September 1880, p. 315. Carey, a surveyor and former civil servant, was MLC (Member of the Legislative Council) for the farming seat of Vasse in the south-west of the colony.

16 *WAPD*, 4, 6 August 1879, p. 40.

17 *WAPD*, 4, 6 August 1879, p. 41.

18 *WAPD*, 3, 27 June 1878, p. 133. Marmion, a prominent merchant in Fremantle, also developed substantial pastoral, pearling and maritime interests. Underlying Marmion's argument were claims that the Chinese contributed little to the local economy, repatriated savings, and would be a charge on the state for the cost of their return to their homeland.

19 *WAPD*, 3, 27 June 1878, pp. 133–4. Such pastoral/rural interests enjoyed official support. Commenting on objections to Western Australia's proposals in a communication to London, Governor Robinson noted that '[it] seems unreasonable that the Eastern Colonies should force on this distant and

exceptionally situated Colony their policy on the Chinese question, and to prevent us from importing an occasional batch of Chinese to supply the unmet wants of this labor market' (Robinson to Kimberley, 26, 25 January 1881, WAS 1166/14).

20 For more detailed discussion of these questions, see Ian van den Driesen, Essay IV, 'Immigration and Immigration Policy in the Post-Convict Period', *Essays on Immigration Policy and Population in Western Australia*, 1986, University of Western Australia Press, Nedlands, pp. 96–150.

21 This question of property appears as a central issue in another industry and suggests that sections of local society supported entry of Chinese and 'Asiatics' as labourers but were not prepared to see them as owners of significant property. In the case of the Shark Bay pearl fishery: 'The white pearlers . . . have represented to me that their livelihood is being destroyed by the Chinese, who . . . acquire boats of their own, and, in fact are rapidly monopolising the pearling grounds . . . [The pearlers proposed to lease the whole of the grounds and claimed that] [i]f this be not done, the European pearlers . . . say that they will be driven off the banks and ruined by the Chinese.' Broome proposed letting tenders, noting that 'it would be against public policy to accept a tender made by Chinese' (Broome to Stanley, 24, 16 January 1886, WAS 1166/14). Broome later reported that the government had 'protected' European interests by buying the boats and plant of the Chinese pearlers who 'have left, or are leaving the Bay'. He noted that at one point it had appeared that the Chinese would forcibly resist ejection, but that the despatch of eight additional armed police and the offer of a liberal price to the 'Chinamen' had 'smoothed away all difficulties' (Broome to Stanhope, 305, 3 December 1886, WAS 1166/16).



22 Broome to Stanley, 24, 16 January 1886, WAS 1166/15.

23 Broome to Stanley, 24, 16 January 1886, WAS 1166/15.

24 Labour agitation contributed significantly to the accretion of these provisions.

25 There also existed legislation in respect of imported labour generally, and Chinese immigrants in particular. One of the first regulatory mechanisms in respect of goldmining was an 1854 ordinance under Act 17 Vic.17 to provide for the preservation of order in cases of any gold discovery. See Broome to Derby, 125, 16 November 1883, WAS 1166/15, for comments on how the ordinance might apply to the anticipated discoveries in the north of the colony.

26 *WAPD*, 11, 29 July 1886, p. 307. Burt, a lawyer and grazier holding interests in his own right and with Premier John Forrest and Alexander Forrest, was at the time a nominee member of the Council.

27 *WAPD*, 8, 15 August 1895, p. 595. James was a barrister and liberal MLA for East Perth opposed to the Forrest Party.

28 *WAPD*, 8, 13 August 1895, p. 575.

29 *WAPD*, 12, 9 August 1898, p. 845.

30 *WAPD*, 12, 9 August 1898, p. 845. See also the amendment moved in committee by Gregory, *WAPD*, 13, 15 September 1898, pp. 1757–9.

31 *WAPD*, 13, 15 September 1898, p. 1758. For discussion on the significance of descriptions of the Asiatic as 'servile', see L. Layman, ' "To Keep Up the

Australian Standard": Regulating Contract Labour Migration 1901–50', *Labour History*, 70, May 1996, esp. pp. 27–8.

32 *WAPD*, 13, 16 September 1898, p. 1762. Kingsmill was a geologist who had managed mines in the Pilbara goldfields. He represented that area in the Legislative Assembly and supported the Forrest Party at the time.

33 *WAPD*, 13, 13 October 1898, p. 2382.

34 *WAPD*, 13, 15 September 1898, p. 1750. Later, in debate on the Goldfields Act Amendment Bill, Randell noted: 'This . . . has been almost universally asked for, especially by the workers on the goldfields of the colony'; *WAPD*, 13, 13 October 1898, p. 2382.

35 Smith to Chamberlain, 4, 27 January 1898, WAS 1166/18 (pts 1–5).

36 Smith to Chamberlain, 38, 26 August 1899, WAS 1166/18 (pts 1–5).

37 Under the provisions, 'foreign' capital was prevented from acquiring or holding a lease in relation to mineral oil. Arguably, the effect of the sections of the Act was to protect local capital specifically and British capital generally. At the time, the latter was the main source of large-scale finance capital for the mining industry in Western Australia.

38 In 1916, however, the circumstances of World War I and the overriding nature of war-time security regulations allowed authorities, under popular pressure, to interrogate, remove from Kalgoorlie and intern 'hostile' 'Austrians' – Slavic people from the Austro-Hungarian Empire.



39 Significantly, few of these went to the goldfields of Western Australia.

40 In debate on the bill to repeal section 291, G. Bennetts, Labor MLC, reminded members that 'we should not forget that Australians were badly treated by the Japanese during the last World War'; *WAPD*, 164, 11 September 1963, p. 939.

41 The original leaseholder appealed Jackson's decision to the Full Court of the Supreme Court. In November 1963, Justices Hale, Negus and Wolff (CJ) upheld Jackson's finding; see *Western Australian Law Reports*, 1964, pp. 112–23.

42 29 August 1963, p. 720. Griffith was a member of the Legislative Council.

43 Cited in *WAPD*, 164, 3 September 1963, p. 756. Wise, a former Labor premier and MLC for the North Province, had been Minister for Industrial Development in the Labor government that had unsuccessfully sought to develop an iron ore industry in the late 1950s. Hastie was a member of the Legislative Assembly.

44 *WAPD*, 164, 3 September 1963, pp. 756, 757.

45 See, for example, the comments of J. Dolan in the debate, *WAPD*, 164, 11 September 1963, pp. 933–5.

46 *WAPD*, 164, 17 September 1963, p. 1057. Strickland, the leader of the Opposition in the Council until 1962, was the member for the North Province.

47 *WAPD*, 164, 11 September 1963, pp. 941–2. Mattiske was Liberal MLC for the Metropolitan Province. Similar arguments can be found in a major dispute that occurred in relation to the employment of non-British labour on dredges in Port Hedland in 1965.

48 *WAPD*, 164, 11 September 1963, p. 944.

49 *WAPD*, 164, 11 September 1963, p. 945.

50 *WAPD*, 164, 11 September 1963, p. 941. Garrigan, a goldminer, had been a member of the AWU and the RSL, and was MLC for the South-East Province.

51 These are to be found in the very long speech of A. M. Moir, MLA for Boulder–Eyre; see *WAPD*, 166, 6 December 1963, pp. 3844–67.

52 Among others, Moir claimed that the Japanese had breached the Mines Regulation Act in a number of respects; *WAPD*, 166, 6 December 1963, pp. 3853–5.

53 A. R. Tonkin, *WAPD*, 198, 10 April 1973, p. 806; L. D. Elliott, MLC, *WAPD*, 198, 17 April 1973, p. 938.

# Golden Opportunities?

hapter 8

## IMMIGRANT WORKERS ON WESTERN AUSTRALIA'S EASTERN GOLDFIELDS, 1900–1965

*Bill Bunbury*



Kalgoorlie's Golden Mile, once dotted with the pitheads, sheds and machinery of underground mining, is now a mile-long hole in the ground – the Super Pit. It is an open-cut mine, and the third and final phase of goldmining in Western Australia's eastern goldfields. As the Super Pit was excavated in the 1980s, it cut through underground mines which date back to the early 1900s. The long, winding descent to the floor of the pit takes the modern miner past socket-like holes in the rock wall – horizontal cross-sections of old mining shafts.

The Super Pit is the last chance to reach sources of gold ore. The old underground mines are now one big quarry and anything down there can be re-treated for traces of gold. At the summit, piles of mining debris, rock, and dusty wooden pit-props, trucked up from below, await reworking. The pit-props suggest a historical, if not a chemical, alchemy. Their presence in the old shafts links the two original goldfields industries: timber cutting and underground mining.

From their outset at the beginning of the twentieth century, both of these industries depended to a large extent on immigrant labour, much of it from Italy and Yugoslavia. These groups, fleeing poverty and in some cases persecution in Europe, provided the timber cutters and carters of the woodline and many of the miners who went underground below the Golden Mile.

Initially, gold discoveries were alluvial. Goldmining centres grew around sites of lucky strikes, like Coolgardie in 1892 and the even luckier Kalgoorlie after 1893. Prospectors roamed the bush in search of promising veins, sometimes smelting the gold ore on the spot with the aid of the fiercely combustible hardwoods of what was then a well-forested region. Much of the current deforestation around these centres is due to the early prospectors' need for instant fuel, often obtained by axing live trees and, as fuel became more scarce, using gelignite to blow out the stumps.

Forest clearing developed on an even larger scale when, at the turn of the century, large companies opened up the huge underground resources of the Golden Mile and needed fuel for gold processing, rigid timber for pit-props, and firewood for a growing city. A syndicate was formed in 1899 to bring in huge quantities of timber from the extensive woodlands surrounding Kalgoorlie, via an extensive network of temporary bush railways known collectively as the woodline. The woodlines ran for hundreds of miles in all directions from Kalgoorlie.

In the mines themselves most underground workers were Anglo-Australian, ex-alluvial diggers. Many were 't'othersiders' from Victoria's goldmining centres. Here too, however, some came from Italy and the Balkans, the latter commonly though not always correctly described as Slavs. Equally inaccurately, they were officially referred to as Austrians, a term which affected the treatment they were to receive during World War I. 'Austrian' was a badge that a few wore with pride as members, albeit subjects, of the Austro-Hungarian Empire. In reality, they were more likely to be Dalmatians, Croats, Montenegrins and Serbs, men who had left some of the poorer subject-states of that empire in the hope of a better deal in 'the working man's paradise' of the new Commonwealth of Australia.



It is noticeable that official reports of this time generalise sweepingly about the characteristics of 'aliens' – a term widely used during World War I when the xenophobia of 'British' Australia shaped both official and personal perceptions of foreigners. In Kalgoorlie itself, the mining companies drew up lists of 'aliens' working on the goldfields. Inevitably, these lists focused on Austro-Hungarian subjects then working on the Kurrawang woodline. Suspicion of foreigners was not confined to the bosses. The Boulder branch of the Mining Union passed a motion stating that their workers refused to work with enemy subjects.

'Austrian' subjects were now under intense scrutiny. They had to register as aliens and were measured, photographed and fingerprinted. In August 1915, without warning, the authorities interned non-naturalised 'Austrian' woodliners and goldminers. They sent them first to Rottnest Island and then to Holsworthy Camp, near Liverpool in New South Wales. Even after the war the authorities were slow to release them, and more than thirty died there from Spanish influenza when it swept Australia in 1919.

One would expect Italians to have fared better than 'Austrians' on the goldfields during World War I, given Italy's support for the Allies. Yet local factors ensured that they too were unpopular. At war's end returning soldiers expressed resentment at what they saw as half-hearted participation by Italy for the Allied side and made no secret of this opinion to their immigrant compatriots on the goldfields. Many soldiers returned to find 'their jobs' taken by 'foreigners'. As they saw it, Italians had taken the place in the mines of enlisted diggers who had chanced their lives on the Western Front.

**Figure 8.1** Jack Sonsee, *A Group of Loaders with Their Tally Clerk*, date unknown.

A letter dated 18 November 1919, from a captain on the General Staff to the Director of Military Intelligence, Victoria Barracks, gives an unambiguous sense of official attitudes towards Italian workers:

> The Italian during the war prospered. He was not called to take an active part in it – either by way of being a soldier as our own people were, or by being placed under restrictions as was his rival on the goldfields – the Jugoslav. The Italians were able to get plenty of work and save money for a trip to Italy when the war ended.
>
> The Italian has never been liked, but the prosperous Italian, free of the claims or depressions of the war was actively disliked. In many instances he did not stand prosperity well. He affected a flashiness of behaviour which was irritating especially to the Returned Soldier.[1]

In effect, the Italians had become scapegoats for a range of complex inter-union and political disputes which affected both miners and returned soldiers. The Returned Services Association (later the Returned Services League) was responsible for much of the post-war tension. Returned soldiers held well-organised meetings, demanding the immediate expulsion of all Italians from the goldfields, a resolution later modified to specify only single Italians.

The situation was serious enough for local police to warn many Italians to leave the town, at least temporarily. Several immigrants who ran hotels and shops went for good. Others were encouraged to find work in the farming districts west of Southern Cross, at the limit of the wheat belt. More than 700 Italian workers left the woodlines at this time, causing a shortage of labour and consequent loss of production.[2] A letter to the Military Department in Kalgoorlie from solicitor R. D. Lane illustrates his clients' concern for their future:

> Some of the Goldfields Italians have consulted me with reference to their position in the country. It has no doubt come to your knowledge that about two months ago the business premises of these Italian people were raided by the mob and their goods and chattels broken and stolen. All were ordered to leave the place, numbers were chased from their work. Many have since been unable to regain employment. Some of these people being destitute and unable to get work and at the same time terrorised by threats from a certain quarter, are most anxious to leave the country, if possible.[3]



The authorities hoped to clear Kalgoorlie of the irritant 'foreigner' factor. Mines further out in the bush, remote farms and the woodline were all convenient dispersal points for the unwanted 'alien'.

Despite military pressure at both an unofficial and more subtly an official level, there was some sympathy for the 'aliens' situation. The Social Democratic League took up the case of an alleged trouble-maker, Louis Francis, an Italian, said to be an agitator and incidentally later found to be Russian. The league wrote to the *West Australian* on 18 September 1919, reporting its own carrying of a resolution:

> That an emphatic protest be recorded against the illegal deportation of a goldfields resident by the RSA and that the league call on the government to have the man returned at once, in order that he may be dealt with by the civil authorities if necessary and that in the event of the Government refusing to accede to this request, the district council of the ALP [Australian Labor Party] be requested to cause the cessation of all work on the goldfields, until such time as the Government extends to goldfields residents that measure of justice and freedom which is so much talked about.

The league found some support among the miners. The Miners Union had already protested at a meeting on 18 August at the spineless manner in which the government had acted in not providing protection for all citizens of the community. The *West Australian* had reported on 19 August 1919 that

> the speakers made it clear that they were not actuated by any feeling of friendship for the Italians, for the presence of an undue proportion of foreigners had always been a matter of irritation to Britishers, but they stood by the principle of liberty

> of the subject, which in their opinion was being violated. They could see no justice in visiting the crime of one man upon the whole Italian community.

Fifteen years later, it could truly be said that the (alleged) crime of one man was visited on much of the community. The occasion – the notorious anti-Italian and Yugoslav riots of the 1934 Australia Day weekend.

I have recounted this story more fully elsewhere,[4] but what happened that weekend can be summarised as follows. The spark was the unfortunate and quite accidental death of Edward Jordan, on Saturday 27 January, outside a Kalgoorlie hotel. Jordan, a popular footballer, had been refused service in the Home From Home Hotel by the barman Claudio Mattaboni, on the grounds that he had drunk too much. The following day Jordan returned to the pub, apparently to settle scores with Mattaboni. A scuffle followed and Mattaboni pushed him out into the street. Jordan fell back on his head. Unfortunately, his skull was unusually thin and he died within hours in the local hospital. The *Kalgoorlie Miner* printed the story the next day. The actual cause of Jordan's death had still not been determined and, in the absence of an informed explanation, rumour quickly spread that an Italian had stabbed him.

The following day, Monday 29 January, was the official Australia Day holiday. Ill-feeling built up all day against 'foreigners' and by evening a small group of men had begun wrecking the Home From Home Hotel, dragging kegs of beer into the street, consuming them, and looting and burning shops and cafes run by Yugoslavs and Greeks:



> We were appalled to see the breaking of bottles, furniture and furnishings being thrown over the balcony, and then of course the flames. There were cars, which were not very numerous then, flying down the street, with people singing and shouting, which amidst the trouble of those whose homes had been wrecked, was rather appalling to the rest of us.[5]

The rioting, by this time beyond the control of the outnumbered police, went on for a further two days and culminated in an assault on Dingbat Flat, in the adjoining city of Boulder, where many Italians and Yugoslavs lived. Determined to defend their homes with jam-tin bombs and any other weapons they could find, the residents put up a desperate resistance, but by sheer weight of numbers they were driven from their defences.

Two men lost their lives in this assault: one Australian, Charlie Stokes, and a Montenegrin miner, Joseph Katich. The rioters then burnt down immigrant homes on Dingbat Flat, driving families into the bush where many remained for several days, short of water and food but too terrified to come back to what had been their home. Many left the goldfields altogether, to join relatives or friends in Perth, and never returned:

> Dad was very worried at the time. He said 'Well, we'll leave!' The neighbours were very good. Actually they were all Australians, apart from one Italian family across the road. They shifted any furniture of value out of our home, put it in

> different homes. They filled the car and we set off. And right along the road from Kalgoorlie, almost all the way to Southern Cross (150km distant) there were people walking like so many refugees, their bundles on their backs, little children walking along with their mothers, holding their hands. It was pretty grim![6]

The Kalgoorlie riots of 1934 rank as Australia's worst ethnic conflict. They are still spoken of by locals as Kalgoorlie's Day of Shame. It is of course important to acknowledge, as evident from the above, that many Anglo-Australians did what they could to prevent the worst of the violence. Yet such a serious incident cannot be easily written off, in the manner of many locals, as 'down to a few young fellers inflamed by grog'.

Rolf Gerritsen, in his detailed account of the riots, draws our attention to prevailing economic conditions.[7] Australia was still in the grip of the Great Depression and, while Kalgoorlie as a gold producer was one of the few places where men could find work, miners struggled to retain employment in a cheap and highly competitive labour market. Coupled with the strains of competition, there were allegations of ethnic bribery. It was claimed that Italians paid 'sling-back' to ensure shifts underground. It is not clear how widespread this practice was on the goldfields – it was certainly common in Italy, where miners' labour was easily exploited – but where it existed in Western Australia it was also exploited by the Anglo-Saxon shift-bosses who accepted the bribes.



Add to this the nature of the work itself. Mining was and still is an inherently dangerous and potentially unhealthy occupation. All miners risked 'dusting' – the dreaded silicosis. Even more immediate was the physical danger of working underground, with the risk of death by drowning or roof collapse. These risks were thought to increase when fellow workers were not proficient in English and did not understand or heed safety precautions.

Preoccupation with language extended beyond its functional aspect. English-speaking Australians resented people speaking another language, an attitude common enough both before and after World War II. The resentment was sometimes based on a vain suspicion that the foreigners were talking about them. In reality, of course, most immigrants were more concerned with their own affairs.

Cultural differences also played a role. Many Australians resented the habit, common to both Yugoslav and Italian men, of talking in large groups on street corners and blocking the pavement as they did so. For newcomers, the street corner may have replaced the rural piazza, a place to meet and talk. Nevertheless, it proved an irritant, as did their insistence on playing bocce rather than cricket or football – habits which can be seen, as Gerritsen argues, as the natural inclination of people not fully at home in their new environment and needing to retain valued parts of their old life. In inter-war 'British' Australia, these harmless rituals were not always tolerated.

There was also a widespread fear that the goldfields would be overrun by immigrants, although in reality 'foreigners' constituted a small proportion of the 1930s goldfields population. Finally, 'foreigners' were resented for not

Plate 9.3 Oscar, *Dispersal Usual Way: Some Good Shooting*, 1890s. (National Museum of Australia)

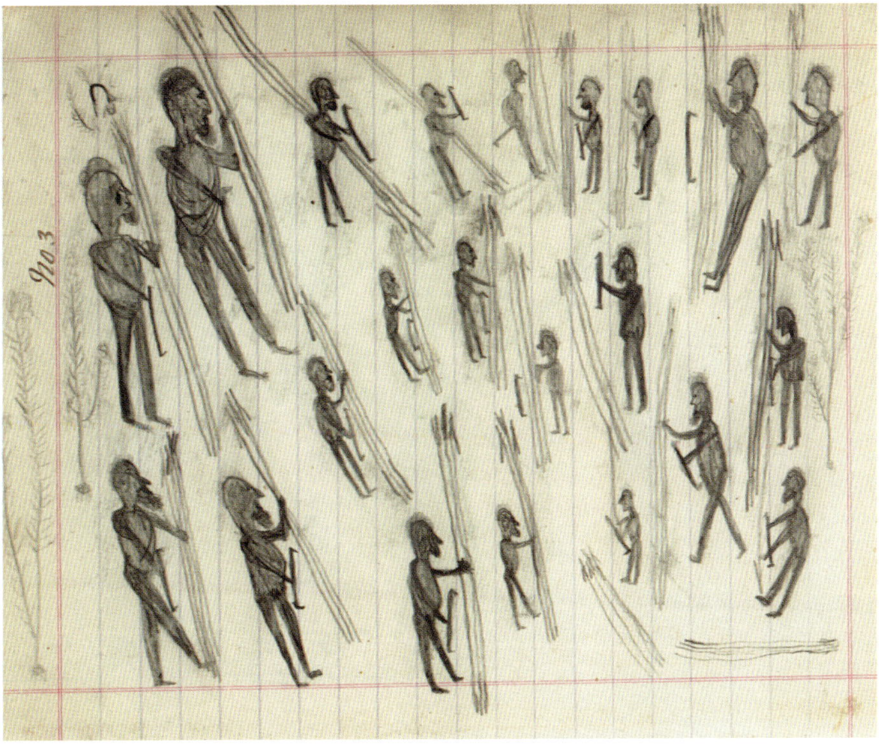


Plate 9.4 Oscar, *Palmer Blacks Preparing for Battle*, 1890s. (National Museum of Australia)

PLATE 9.5 Oscar, *Camooweal White Ladies*, 1890s.
(National Museum of Australia)

PLATE 9.6 Oscar, *Police Boys Doing Duty (Lynch Law)*, 1890s.
(National Museum of Australia)

PLATE 10.1 W. Hodgson, *Some Real Australian Cricket: An Incident in the Goldfields*, in A. F. Calvert, *My Fourth Tour in Western Australia*, Heinemann, London, 1897. (National Library of Australia)

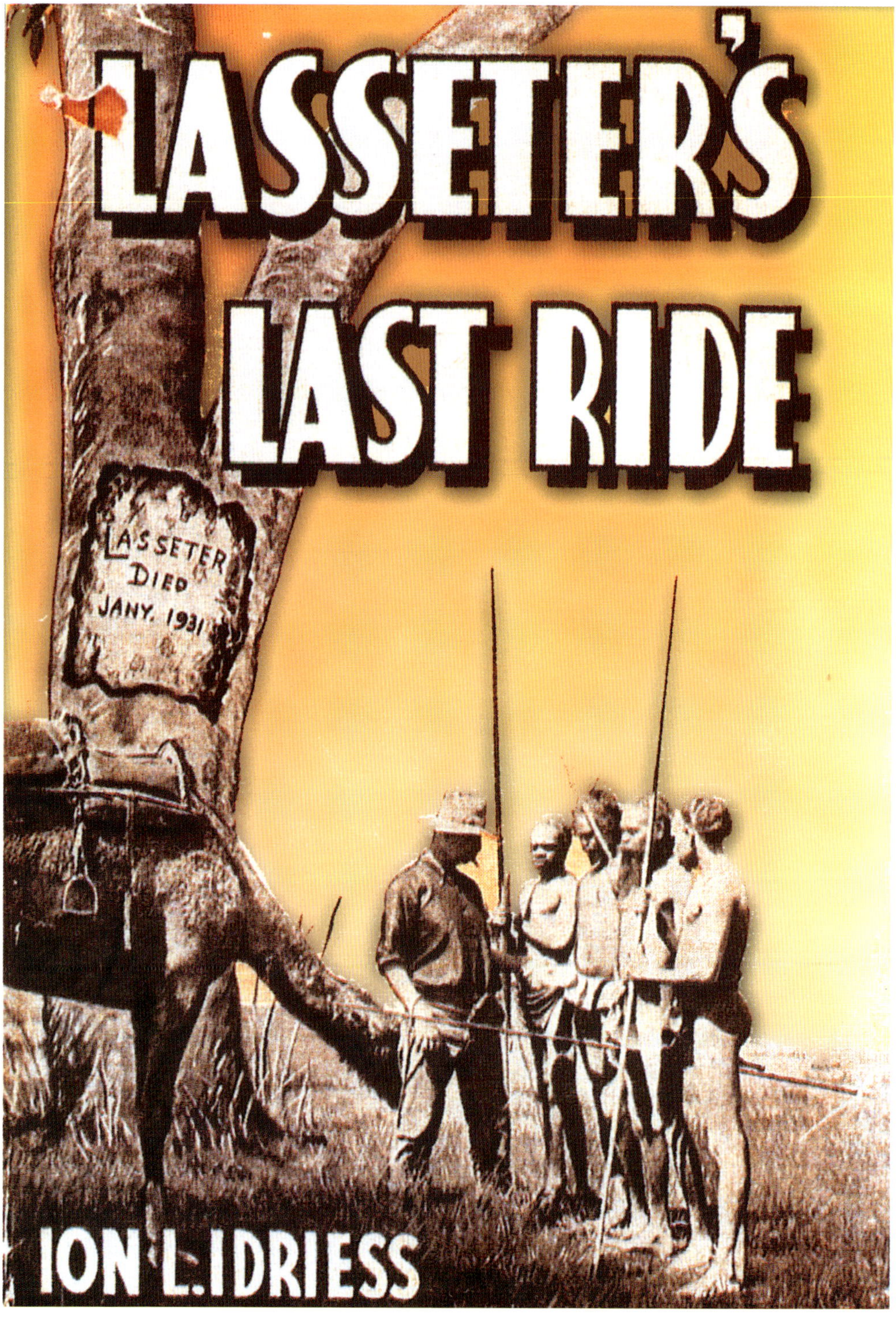


PLATE 11.1 Artist unknown, dust jacket for *Lasseter's Last Ride*, 1939 edition

taking part in the consumer economy, for living on 'the smell of an oily rag'. Whether or not this was true for the urban immigrant workforce, it certainly could be said of the other half of the goldfields immigrant workforce: the Italians and Yugoslavs who worked as timber cutters in the spur camps of the woodline.

Woodline work was poorly paid and costly for the new employee. He had to pay high prices for his own tools. Axes, files and shovels came from the company store but were set against his own wage. An axe, for example, could cost two and a half times more than the ordinary store price in Kalgoorlie.[8] Given his isolation, often sixty or seventy miles down the line from town, it was hard for a woodcutter to shop elsewhere. The woodline company store bound the worker to his situation. Cutters could draw items of food and clothing from the store, but these items were also set against their pay. Many workers ended their week owing their pay-packet to the store. In effect, they received weekly rations instead of wages, inevitably postponing the pay period when they could get real money back in their pockets. In some cases, married men with young families did not receive pay for up to a year.

If the breadwinner injured himself cutting timber, there was no safety net, no workers' compensation, and the only way to get medical help was by way of a returning woodline train, a long slow journey up the line to the main camp and, with luck, transport to the Kalgoorlie hospital. Women in labour made equally urgent journeys in open rail trucks, exposed to the heat of the sun or winter's drenching showers, and more than once losing a baby in what could be a twelve-hour trip.



Even the necessities of life could prove hazardous. At the end of the line, frequently a long way from the company store, men living in camps had a long wait for their rations. When they arrived, the food was often poorly or scarcely wrapped and, in the case of meat, sometimes badly fly-blown. In summer it frequently had to be thrown away. While meat might be cut one day and sent down the line the next, there was no refrigeration at the spur camp.

Single men lived in tents, often saving much of their pay to bring their family out from Europe, a process that could take a decade. At family reunions fathers sometimes saw their offspring for the first time, children they had conceived but not seen born, who were nearly adolescents. Those families which did eventually make the journey from the Mediterranean usually had little idea that their long voyage would end in a spur camp in seemingly featureless bush. Annie Grubisa, who had lived near Trieste, came to the woodline with her mother as a young child. Her mother was the only woman at her husband's camp:

> Well, when you come from a place like we did and come to the bush with nobody, well it was a shock. But after a while she said she wouldn't go home. At home women had to work out in the fields digging potatoes, etc. and planting.[9]

Woodline life for families was essentially transient. As the timber cut out, the main camps moved on, usually once every three months, laying down new

**Figure 8.2** John Wearmouth, *A Family at No. 5 Camp*, date unknown.

branch or spur lines deep into the bush to cut a new timber block. Everything went: the school, the butcher's shop, the store and the workers' homes were all loaded onto flatbed rail trucks and set down again at the new destination. Years later, Olga Bibich, a woman who had spent her childhood on the woodlines, recounted 'moving days':

> All the huts had to go onto the railway wagons to be moved, and, with the families, whoever moved first, the family staying behind would cook their breakfast on the morning they left. So when you came to the new camp they cooked your breakfast when you arrived the next time. That's the way we handled it because we had to break down the fireplaces before we could shift.

The ephemeral nature of hearth and home did not deter women from creating as much permanence as possible. Olga Bondi recalls her mother cooking on a wood stove 'in the old Slav way, mostly boiled soups, boiled meats, because we didn't have any refrigeration, only the Coolgardie safe, so food had to be cooked and eaten straightaway'. Olga Bibich retains the strongest memory of the woodline camps as 'the smell of basil. Where I lived they'd have their half drums with a few bits of parsley, basil and all kinds of herbs which they took with them when they moved.'

It was also a life where Italians mixed with Slavs and, by most accounts, well. Tension came later with World War II. In the main, the two groups formed a shared culture, playing cards or drinking together at night and working side by side each day.

The woodline's working divisions reflected the social and economic status of immigrants. The men who drove the timber trains or kept them running were mostly of English, Scots or Irish descent. They self-consciously described themselves as 'Silvertails'. The men who filled the wagons were Italians, Montenegrins, Croats or Serbs:

> As a matter of fact if an Australian came out to the head of the line, he didn't last too long. He might come out with his tent, stretcher, and pannikin but he'd be selling it within a month. Not too many stuck it. Most Aussies out there were either bosses or engine drivers.[10]



Repeated strikes during the sixty-year life of the woodline illustrate the working conditions of the non-silvertails. There were strikes in the early 1920s

**Figure 8.3** Jack Sonsee, *Camp Accommodation Being 'Manhandled'*, date unknown.

**Figure 8.4** Emil Grizich, *Sharpening Tools with a Little Music*, date unknown.

and again in 1937 when workers took issue with their employer, the West Australian Goldfields Firewood Supply Company, over the conditions of their award. The Australian Workers Union, acting for the cutters, sought a forty-hour week. Their long working hours made it almost impossible to get away from the isolated camps at weekends. The only way they could travel was, much as the timber they cut, in open wagons. Their employer did not meet their request for covered transport.

The need to escape the isolation was as important as any other industrial gain. In court the Australian Workers Union representative, Charlie Oliver, described the woodline working life as so monotonous and lonely that some men were known to 'put their hat on a stump and talk to it'.

The nature of woodline life also meant that the immigrants had little interaction with other Australians:

> We knew those people were out there, in a vague sort of way, but they seemed like a separate community. There was not much inter-mixing you might say between the Gwalia citizens, or even the workers on the mine, and the families out at the Halfway Camp, and, of course, the single Italians, or the foreigners,

> actually at the woodcutting terminus, which they shifted frequently. So really those people lived for many years in a very isolated existence out there.[11]

The union alleged that the isolated conditions alone affected some of the cutters badly. Some became mentally depressed, a condition aggravated, especially in the case of single men, by their social isolation from family and friends.

News from their homes in Europe in the late 1930s did little to reassure them. Hitler and Mussolini had already begun to threaten international peace. Yugoslavians, in particular, became very concerned for the security of their country and the safety of their families as the European powers continued to accede to Hitler's territorial demands.

World War I had produced disastrous consequences for emigrants from the Balkans. World War II, even without the same xenophobia, made life difficult for Italian workers on the woodline. Unless they could prove naturalisation they risked internment after 10 June 1940 when the Axis powers in Rome, Berlin and Tokyo proclaimed a mutual defence pact. Non-naturalised Italian civilians throughout Australia became direct casualties.

Few Italian woodline workers had any idea that the war in Europe would affect their lives so drastically. Mussolini's imperial adventures seemed unreal beside keeping their jobs and supporting their families. Bill McNally, an engine driver on the woodline, had the distasteful task of taking a batch of Italians to Kalgoorlie for internment. He knew one young man who had only just come out from Italy to join his father on the woodline. The father had died suddenly and unexpectedly and the son, without any other family in the camp and no naturalisation papers, was now liable for internment. As the train halted at his pick-up point, he raced up to Bill, threw his arms around him and begged him to tell him his fate: 'What are they going to do with me? Will they shoot me?'[12]



The strain told. Of sixty-two internees from the woodline, three were sent to Perth for medical treatment and three others to Claremont and Heathcote mental hospitals. There was a limited period in which Italian civilians could appeal against their internment, but, to judge from the military records, few were either aware of their right or felt able to exercise it. In the army records for November 1942, the names of many internees are recorded with the description: 'No appeal lodged and period allowed now expired. Eligible for transfer.'[13] Transfer meant a long journey across the Nullarbor to Australia's major internment centre: Loveday in South Australia. Once there, the only reprieve from imprisonment was local farm work.

In the same year, the Commonwealth government declared woodline timber production a vital national war industry because of its direct effect on gold production (also vital to the war effort). The loss of Italian workers to internment meant that production fell sharply. West Australian Premier J. C. Willcock (also Minister for Forests) asked the Commonwealth government to exclude Italian woodcutters from internment, in the national interest, and to return their workers. He was partially successful. The company eventually regained thirty or so Italian cutters, yet, because of potential tension with

Yugoslav workers after Hitler's invasion of Yugoslavia, they went to work from an isolated camp, well away from their former neighbours and friends.

George Progomet, himself from a Yugoslav family, recalls that a special police station was set up at the main woodline camp to ensure peace between the two peoples. He remembers little animosity, despite the authorities' concern. A few extremists whipped up bad feeling at times, but most were content to put the war aside.

The Yugoslav workers may well have congratulated themselves on not being on the 'wrong side' this time round, after their own internment in World War I. But World War II imposed restrictions on their movements too. Since the woodline was now a 'protected industry', nobody could leave his job. Many had hoped to return to Europe, join Tito's partisans and resist the Nazis. Absenteeism grew, possibly from resentment at being tied into the Australian war effort when they wanted to be elsewhere. Reports to the authorities had little effect on the practice. Again, woodline production went down.

The coming of peace in 1945 brought considerable change to the woodline. The disruption of Italian lives through internment caused many, on their release, to seek work elsewhere. Many now joined friends and relatives to go cane farming in Queensland. Many Slavs had also saved enough to buy a block of land somewhere else. Woodliners shifted to the rural edges of Perth to buy small vineyards or market gardens, or to the south-west to grow potatoes on small farms. The woodline had given them a hard-earned ticket to success.



Several Yugoslav families left Australia altogether after World War II. Marshal Tito had urged 'sons of Yugoslavia', now in Canada, New Zealand and Australia, to return and rebuild the country. Two ships were sent to Australia and carried away more than 1100 former immigrants. For many, the return was disillusioning. Post-war conditions in Europe were worse than they had been led to believe. Some tried, often unsuccessfully, to return to Australia.

Such an exodus might well have crippled the goldfields fuel supply had not Australia's own post-war immigration scheme, 'Bring Out a Balt', provided some relief. The timber industry itself was in slow but terminal decline by 1945 as other power sources became available to Kalgoorlie. Although it would cease operations altogether in 1965, for a time the woodline became once again the first work experience for a new generation of migrants. But, where their predecessors had been rural people used to manual labour, many of the post-war workers came from the urban centres of Poland, Latvia and Lithuania. Most had spent years in northern Europe's 'displaced persons' camps, awaiting acceptance to any country that would take them. A worldwide shortage of merchant vessels due to the war slowed down the process of resettlement. When they arrived in the late 1940s and early 1950s, they were bound by the terms of post-war reconstruction and development to complete a two-year period of labour, usually in civil construction works. Those who left the ship at Fremantle often ended up on the woodline.

This workforce was easily the most highly qualified to ever work the woodline – except in terms of the job itself. Many had a good education but had lost position, income, status and even members of their family in the political and racial purges of the 1930s and, later, from the war itself. Jack Sonsee was responsible for their induction into woodline life and work:

> At HQ we erected a demonstration camp, with a tent to show the new men the tools, the files, the hammer, etc. and the lamps which they'd use in their tent on the woodline. They had to buy their own tools before they went out there. The look on those men's faces. They were astounded. They'd never seen a cross-cut saw or an axe before, never even touched them. When they showed you their hands it was obvious they'd never done any manual work, they'd been a doctor or an accountant, but they were given a job and they had to take it.[14]

Recruiting 'displaced persons' could also prove wasteful. Jack noticed that

> their physique wasn't always suitable either. They had to cut three tons a day to make a living and pay back the money they owed the company for their tools and stores. They were on a two year contract but less than fifty per cent survived. Some only lasted a week. The turnover was ridiculous.[15]



There were plenty of exceptions. Some, like Sigurds Titmanis, fulfilled their two-year contract and went on to take advantage of other opportunities in post-war Australia.[16]

The 'displaced persons' were the woodline's final source of labour. From then on, numbers fell as the demand for timber fell and the company approached the end of the line. Changes in the nature of the work also affected the close, if isolated, social life of the woodline. Increasingly, the company sent out contractors with road trucks to bring in wood from the bush. It was easier to hire one man with a bulldozer to clear a bush track for access to the timber than to employ a rail gang to put down spur lines. This, in time, reduced the need for outlying camps and ended the isolated life of the labourers.

The woodline, in its sixty or so years of operation, was transient in more than one sense. Not only were its rail lines and camps temporary and portable, it was also a first step, albeit a hard one, for generations of workers who entered the Australian community. Many prospered subsequently. Others, less fortunate, were caught up in its isolation and hardship and never escaped.

The woodline story remains a significant chapter in Australia's environmental history and its engineering history, if we explore the improvisation and ingenuity required to sustain it. Above all, it tells a social history of the enduring impact of an isolated and largely immigrant community, strongly affected by forces beyond our own shores, on the economic and cultural development of this country.

## Notes

1 Letter from an unnamed captain (possibly Captain R. Collins) to Director of Military Intelligence, Victoria Barracks, Melbourne, Australian Archives, pp. 14/1 (WA) 16/1/29.
2 B. Bunbury, *Timber for Gold: Life on the Goldfields Woodlines*, 1997, Fremantle Arts Centre Press, Fremantle, p. 114.
3 Letter to Military Department in Kalgoorlie from solicitor R. D. Lane, Australian Archives, pp. 14/1 (WA) 16/1/29.
4 B. Bunbury, 'A Bad Blue: The Kalgoorlie Riots of 1934', in B. Bunbury, *Reading Labels on Jam Tins: Living through Difficult Times*, 1993, Fremantle Arts Centre Press, Fremantle, pp. 102–26.
5 Bunbury, 'A Bad Blue', pp. 102–6.
6 Bunbury, 'A Bad Blue', p. 122.
7 R. Gerritsen, 'The 1934 Kalgoorlie Riots: A Western Australian Crowd', *University Studies in History*, vol. 5, no. 3, 1989, pp. 4–5.

8 Bunbury, *Timber for Gold*, p. 141.
9 Bunbury, *Timber for Gold*, p. 98.
10 Bunbury, *Timber for Gold*, p. 95. (Original interview, Oral History Collection, Battye Library, Perth.)
11 Interview with Colin Turnbull, Battye Library Oral History.
12 Bunbury, *Timber for Gold*, p. 141.
13 Australian Archives, MP 508, 255/717/65 (VIC).
14 Bunbury, *Timber for Gold*, p. 175.
15 Mario Battaglia, interview with Bunbury on 'Timber for Gold', ABC Radio National *Hindsight* feature, 1996.
16 Sigurds Titmanis from Latvia, who studied for a degree while still working in the goldfields and later became head of CSIRO's analytical laboratory in Perth.

# Part III White Gold/Black Gold

SETTLER AND INDIGENOUS HISTORIES OF GOLDMINING

Chapter 9

# Eyewitness?

## DRAWINGS BY OSCAR OF COOKTOWN

*Kim McKenzie and Carol Cooper*



Posters of north Queensland depict waterfalls and butterflies and bright parrots amid luxuriant vegetation leading down to jewelled seas. These are pictures from the narrow eastern flank of the dividing range. Eager miners trekking inland from Cooktown in the 1870s soon found themselves in quite different terrain. The approach to the Palmer River necessitated travelling through country that is steep and rocky and, for much of the year, harsh and dry. Rivers are its lifeblood; away from them, it is hard country indeed. The annual monsoon season would have given Aborigines greater freedom of movement, but for the remainder of the year their lives must have been based in the river valleys.

Before 1873 the only foreign incursions into this area had been the handful of small exploration and survey parties passing through. Unlike most Australian goldfields, the rush into the Palmer preceded the take-up of land by Europeans for pastoral purposes. The year 1873 saw a sudden invasion of thousands of Europeans and Chinese into these same river valleys, scouring the beds for alluvial gold. A baby born during the tumult of these years would, growing up, be witness to a particularly dramatic clash of cultures and fierce competition for limited resources, with all the dislocation and conflict these would bring (Plate 9.1).

In March 1899 Augustus Henry Glissan, manager of Rocklands cattle station in far north-western Queensland, posted a small book to friends in Melbourne. He said in an accompanying letter that he was sure his friends' children would 'enjoy it immensely'. By odd circumstance, the book and letter survived unnoticed through most of the following century until they became, by further chance, part of the collection of the National Museum of Australia.

Each of the book's forty pages is taken up by pencil drawings, and the letter describes their making:

> [They were] executed by one of my boys named 'Oscar', this boy I got at Cooktown in 1887 & he has been with me ever since & when I got him he was a thorough Myall – could not speak many words of English & had never crossed a horse, he was somewhere about 9 or 10 years of age then – but could not be certain to a year or two, the police got him for me . . . Ever since the boy came here he has had a liking for drawing & some of his etchings on stone with charcoal and chalk are so really well done, that I thought some day I would get him a book and some coloured pencils & let him have a good try at it.

With Glissan's letter is a set of captions, written in his hand and correlated by number to the drawings; 'this the boy gave me in his own way, & I have put it as well as I can'. As well as he could, perhaps, but not without his giving way to some temptation for embellishment. Captioning a depiction of police killing Aborigines as 'Dispersal usual way. Some good shooting' rings more true as the words of a nineteenth-century manager of a remote cattle station than those of a teenage Aboriginal boy of the time. A number of such captions sound a warning that some of Glissan's interpretations may be free to the point of being quite misleading of Oscar's own intentions in the drawings. Nevertheless, the captions and the text of the letter seem to be in broad accord with the subject matter of the drawings and provide a basis for understanding them and their context. In so doing, they create an argument for their importance as historical documents.



First, they establish that the drawings concern two areas: Cooktown and the Palmer River goldfields of north-eastern Queensland; and the cattle country around the town of Camooweal in the north-west of that state. The Palmer River was the most famous, indeed notorious, north Queensland gold rush of the late nineteenth century. In line with our cautious approach to these captions, it is worth noting that by the time Glissan wrote his letter in 1899 'Palmer River' had come to be used as a term embracing a number of the goldfields in the ranges inland from Cooktown, and even those, like the Etheridge fields, to the west of those ranges. As with many other goldfields, the rush to the north-east saw the creation of populations and infrastructures in towns such as Cooktown, Palmerville and Maytown (Plate 9.2). They were built with an expectation of longevity, even permanence, but in most cases fell into decline or ruin once the mining ceased to be profitable and the miners moved on. By then the damage had been done to the former owners of the land.

Glissan's text reveals that Oscar was Aboriginal. In his description of Oscar as 'one of my boys' he is, without doubt, using the word 'boy' in the argot of the cattle station as meaning a male Aboriginal worker, regardless of age. He uses the term in the same way in some of his captions, even clarifying it in the case of the caption 'Rocklands Boys (stockmen) and gins'. ('Gin' was a widely used contemporary term for an Aboriginal woman.) His further

**Figure 9.1** Oscar, *Some Maytown Swells Doing the Block*, 1890s. (National Museum of Australia)

description in the letter of Oscar as 'a thorough Myall' confirms beyond dispute that he was Aboriginal. Used in this way and at that time, 'Myall' was equivalent to other expressions of the day like 'wild blackfellow', though originally it was a term for hostile strangers used by the first inhabitants of the Sydney region.

A further accord between the text and drawings is the identification of the particular police force which figures in both. In fact, Oscar's drawings require little corroboration on this point, as they accurately detail the uniform and accoutrements of the Queensland Native Mounted Police.

In 1859 Letters Patent were issued which proclaimed Queensland as a separate colony of Britain. The new colony inherited a special mounted police force consisting of white officers and black troopers, with an established strategy for dealing with disturbances on the mining and pastoral frontiers. With a policy of recruiting troopers from districts well away from areas in which they would be deployed, this force used Aborigines against Aborigines and became a major agent of suppression of Aboriginal resistance. Rarely did it take action against other wrongdoers, a well-known exception being the use of a north Queensland detachment in pursuit of Ned Kelly's gang in Victoria in 1879. As the frontier expanded northward and inland through Queensland, the Native Police followed a pattern of punitive reprisals for Aboriginal 'depredations' of those who came to occupy their lands and assume their resources. Glissan's word 'dispersal' is a chilling euphemism, used again and again in the records of the Native Police (Plate 9.3).

The conduct of the Native Police was far from without criticism. Their history is marked by a succession of official inquiries and by numerous depositions of concern to newspapers of the day. These include, interestingly, vehement letters from a number of pastoralists. However, the fact that they were most often active in locations beyond official scrutiny left the Native Police relatively unrestrained.

Few records remain of Aboriginal accounts of the history of the Native Police. The drawings of them by Oscar might well be the most accessible and graphic record made by an Aboriginal person. If it can be determined that he was drawing his scenes from direct personal experience, then his work is that much more valuable as a historical document.



Analysis of Oscar's forty drawings has revealed that there is a greater level of possible interpretation than is at first evident, and certainly more than that which Glissan conveys through his captions. Perhaps the single most remarkable result of this analysis has been the discovery that, both within certain images and between groups of images, Oscar was using his drawings as a means of telling stories. While some drawings seem to be self-contained 'scenes', others appear to be parts of narratives, some even extending over several pages. The most dramatic of these is also the only one to date which we have been able to correlate with other documented accounts. It appears to be Oscar's account of the killing by Aborigines of Glissan's predecessor as manager of Rocklands and of the subsequent reprisals by Native Police. A newspaper report of the manager's death suggested coyly that it may have come about as a result of his 'interfering in the domesticities of the camp'. These events occurred in 1884, three years before Glissan's letter describes Oscar's coming to Rocklands from north-eastern Queensland. Thus, it is most likely that in this instance Oscar's drawing is of a story told to him, not events to which he was witness. Other drawings, however, seem to point to the possibility that Oscar is not only witness to his stories but includes himself within the drawings.

In 'Palmer Blacks preparing for battle' (Plate 9.4), a caption Glissan gives to two of Oscar's drawings, it seems as though there are two groups of men armed with spears and spearthrowers. On the extreme left and right, Oscar has drawn trees: presumably, he is placing the men within an open space surrounded by vegetation. The way the groups are drawn facing each other suggests that they are about to fight. However, in the extreme top left, there is a head which, at first glance, seems to be a drawing of another warrior begun but discontinued. Closer inspection suggests that Oscar has drawn an observer looking out from behind the cover of foliage, similar to that of the trees, at the scene within the clearing. This figure appears to lack the pronounced beard of most of the other figures in this drawing. It is of course only conjecture, but could this be Oscar depicting himself in a situation he remembers from his childhood?

Further conjecture can be made by looking at the way in which Oscar draws children. First of all, hardly any of the drawings throughout the book reflect a use of perspective to create a sense of depth. Rather, Oscar treats the page as a vertical, flat plane in which he places his figures. Overall, it can be said that he uses the comparative size of individuals within a group as indicative of his perception of their importance relative to each other. In a drawing such as the one captioned 'Camooweal white Ladies' (Plate 9.5), it is most likely that the figure on the left is that of a girl, drawn smaller in stature and elevated in comparison to the level on which the other figures stand. There are a number of other drawings in the book that suggest this as the most consistent way in which Oscar drew children.

What then is to be made of 'Native Police Boys, Palmer River' (Figure 9.2)? The figure on the right is, just like the girl with the Camooweal ladies, smaller and elevated above the level at which the others in the group stand. This figure is the only one with no evident facial hair, and also he seems to be holding onto the trooper next to him. Remembering Glissan's account of his acquisition of Oscar ('the police got him for me'), might this be a drawing of Oscar himself in company with those police?



In his *Reminiscences of Queensland*, W. H. Corfield describes the situation of one young boy taken up by Native Police during the early years of goldmining on the Palmer River:

> Sub-Inspector O'Connor's camp was at the Laura River. On one occasion, when dispersing some blacks, the troopers, who were all Fraser Island natives, saw the shiny, black skin of an aboriginal hiding in the bush some distance away. They fired, and a little fellow about six years of age got up and ran towards them. The troopers picked him up, and he became a favourite with them. They delighted in instructing him in drill and discipline, and he proved an apt pupil. O'Connor and myself became great friends, and many a happy hour I've spent at his barracks when passing to and from the Palmer. Knowing I had no blackboy, he gave me the little fellow he had so well drilled. I bought a pony for him to ride, and it was laughable to see him, if we happened to meet the troopers on the road, straighten himself up and salute the officer.[1]

There is no suggestion that Oscar was the boy Corfield describes in this passage. The age of that boy at the time Corfield was at the Palmer (1873–74) would make him far too old to be the 9 or 10 years Glissan gives as Oscar's age in 1887. However, the passage does give some clue as to a possible way in which

**Figure 9.2** Oscar, *Native Police Boys, Palmer River,* 1890s. (National Museum of Australia)

a young Aboriginal boy might come to be in the hands of Native Police. It is not known from Glissan's letter or any other source how long Oscar might have been with the police between the time of his presumed capture until he was sent to Glissan at Rocklands. It may be that he was with them long enough in some capacity to have come to see himself as linked to them in the way the above drawing suggests.

Conjecture again. But such an association between Oscar and the Native Police, especially if it led to an affinity on his part which was maintained even after he went to north-western Queensland, would help to explain one of the disquieting aspects of his drawings.

In the most dramatic of the drawings of the Native Police, they are shooting down fleeing Aborigines, or hanging them from trees, or executing them while they are bound. These are highly disturbing images. One of their most striking features is the consistent disproportion in size between the police and their victims. The troopers are drawn tall, with exceptionally long legs that convey superiority and control, even elegance (Plate 9.6). Oscar takes pains to draw their uniforms in detail – stripes, leggings, epaulettes and lanyards – and he gives their faces an individuality lacking in those of the victims. These differences clearly represent Oscar's perception of the differences in the relative power of the figures in the scenes. One possible interpretation is that Oscar regards the Native Police as figures to be feared. It is, however, difficult to escape the sense that he is less moved by the plight of the victims than he is impressed by the stature, bearing and power of the troopers. In one drawing of police from, presumably, north-western Queensland, Glissan has even written names below each figure – Dick, Harry, Albert and Dickey. Whether these identifications were suggested by Oscar, or were Glissan's own idea, is unknown. Regardless, the drawings have an accuracy and immediacy that place Oscar close to the scenes they depict.

In many ways, Oscar was himself in the kind of situation from which the Native Police drew its recruits. He was from a society which had just experienced invasion and massive disruption. In a few short years, most of the social structures and ties to land had been broken. Young men such as Oscar were, perhaps cruelly, offered a meagre share of the victors' spoils – a wage, a carbine, a horse, a uniform, some power and excitement.



Oscar's drawings may not document the past with the weight of detail of the European written record, but they do serve as a reminder of the complexity of the past and that it was a complexity made from the individual lives of people. If only more had been given a book and some coloured pencils, '& let have a good try at it'.

## ote

1 W. H. Corfield, *Reminiscences of Queensland 1862–1899*, 1921, A. H. Frater, Brisbane, p. 59.

# Golden Reflections

## ARTISTS, ENTREPRENEURS AND ABORIGINES ON THE NORTH-WEST AUSTRALIAN GOLDFIELDS



*Ian Coates*

164 At the end of the nineteenth century Western Australia was in the grip of gold fever. In the words of one visitor:

> The air in Perth is full of the yellow fever. Its germs, in the shape of talk of reefs, leases, claims, yields, trial crushings, camels, syndicates, stocks and Company flotations, are as thick as a London fog. In the smoking-room, the bar, the club, the exchange, the bank, and even in the drawing-room, the chorus goes up in praise of Mammon, of the bright yellow metal.[1]

The author of this evocative description, A. F. Calvert, was a London-based mining engineer, entrepreneur and journalist who vigorously promoted the goldfields of Western Australia.[2] In the few years between 1894 and 1900, West Australian gold was the centre of attention of the British stock market. Investors sank their funds into a few successful companies and many more ill-fated ones.[3] Such an environment created opportunities for entrepreneurs such as Calvert to flourish. Not all of the colony's residents were impressed with his activities. During his fourth visit to Western Australia, one newspaper noted:

> Since the West Australian boom set in, Mr Albert F. Calvert has used every possible means to puff his own person and position. He has expended very large sums of money upon sensational advertisements, which were so set up in the papers that they might very well have been mistaken for reading matter.[4]

Gold, and the promise of its wealth, created an environment in which illusion thrived. It was a time and place where potential and promise featured more than solid results, a world where opportunities and individuals often turned out differently from the way they first appeared.

Images, whether in the form of photographs or drawings, also belong to the world of illusion. They can mask and/or simplify the intricate relations between an artist and a subject. They can serve to hide as much as they reveal. This chapter is an examination of two sets of visual representations of Aboriginal people from the goldfields of north-west Western Australia during the 1890s, which at one level illustrates this point. Yet I will argue that, despite the deliberate avoidances in some of these depictions, examining them carefully can give an insight into the local social dynamics between Aboriginal people and Europeans on the goldfields at this time. They sit within a web of representations produced in the context of intensifying European settlement, competition with the original inhabitants over resources, the introduction of



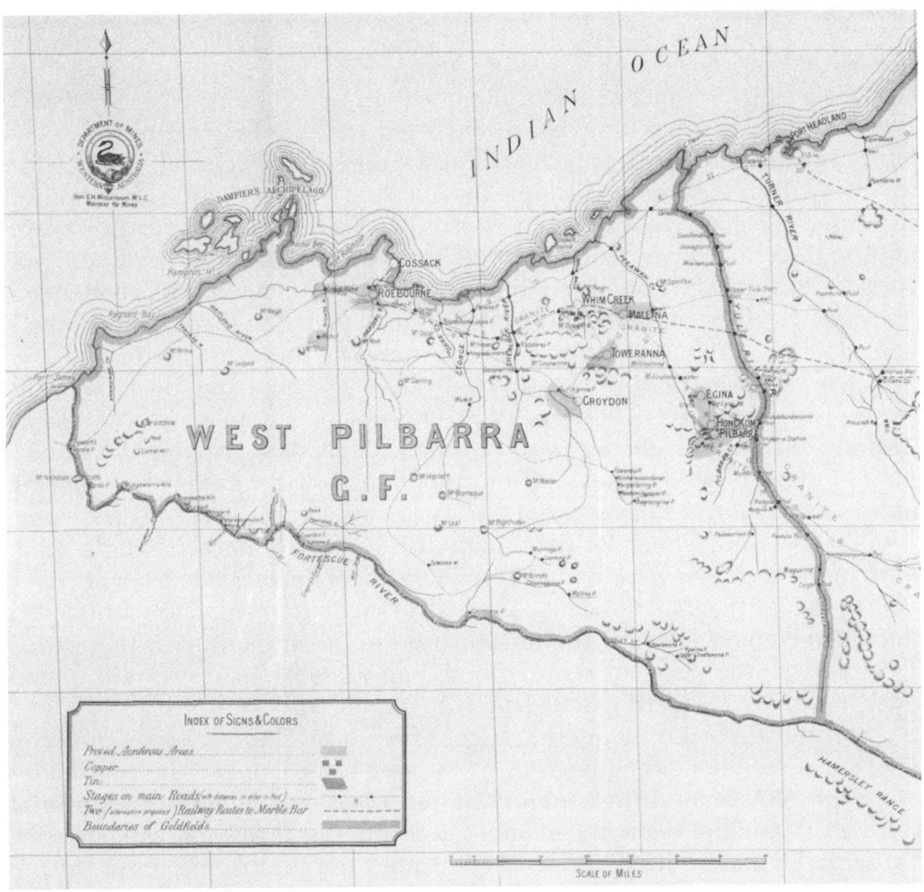


**Figure 10.1** Map of the West Pilbarra [Pilbara] Goldfield, Western Australia, in *Report of the Department of Mines for the Year 1896, 2nd Session 1897.*

Aboriginal people into the labour force, and the settlers' desire to attract international investment to the goldfields. In very different ways these works are testament to a complex history of ethnic cohabitation in the region. The first set was created by the ethnographic collector and photographer Emile Clement, and the second by the artist Walker Hodgson. Both of these European men were present on the West Pilbara goldfields, which flourished briefly in the 1890s.[5]

By the end of the nineteenth century the pursuit of gold by Europeans was having a significant impact on the lives of Aboriginal people in the north-west district. European settlement had begun in the region during the 1860s. It was initially centred around pastoralism and later pearling. Unlike the gold industry, both of these activities had drawn substantially on Aboriginal labour. Labour shortages were a frequent problem in the West Australian colony and led to the introduction of convict labour in 1850 – after other Australian colonies had stopped taking convicts – which lasted until 1868.[6] The shortage of European labour was further exacerbated in the northern parts of the colony by regulations banning convicts and ex-convicts from working in areas north of the Murchison River, because of the harshness of the climate.[7] The labour shortage in the north created an opportunity for Aboriginal people to work on pastoral stations. Aboriginal labourers were not usually paid in cash, instead receiving food, rations and other goods. Despite being alienated from their traditional lands by the introduction of pastoral livestock and limitations on access, many Aboriginal workers and their families continued to live on those stations which had taken over the country for which they were the traditional owners.[8]



The discovery of gold near Roebourne in 1882 dramatically altered the relationship between Aborigines and Europeans in north-west Western Australia. At this time there was an economic depression gripping the eastern states, and the promise of wealth led to a sudden and large influx of Europeans, mostly male, to the north-west. This flow of new gold seekers also created another source of labour for pastoralists, as unsuccessful miners sought other work in the area. Conversely, for Aboriginal people the influx led to a decrease in their employment on the stations. As well as swelling an alternative workforce, the change coincided with the introduction of fencing on many properties. This reduced the need for shepherding, one of the primary types of station work in which Aboriginal people had been employed.

On the goldfields themselves, there were no specific legal restrictions on Aboriginal people holding licences or working claims, unlike the restrictions on people of Asian or African origin.[9] Despite the fact that there are few official records of Aborigines working on mines, it is clear that many Aboriginal people, particularly women, lived and worked around the flourishing towns on the Pilbara goldfields, such as Mallina, Roebourne and Whim Creek, working as cooks, looking after horses and fossicking for gold.[10] This was the background against which Emile Clement and Walker Hodgson entered the goldfields of the north-west in the 1890s and created two separate depictions of Aboriginal people.

**Figure 10.2** Emile Clement, untitled photograph, Towranna, 1890s. (Pitt Rivers Museum, University of Oxford)

Like Calvert, Dr Emile Clement was an entrepreneur attracted by the allure of gold in the north-west. In addition to his involvement in gold exploration and mining, Clement created and marketed representations of Aborigines for overseas museums. These took the form of ethnographic collections and photographs obtained from the West Pilbara fields.

Clement first arrived in Roebourne in 1895. His journey had been long and digressive. He was born in Prussia in the early 1840s. By the age of 26 he had acquired a doctorate and was teaching natural philosophy and chemistry at Bramham College in Yorkshire. Over the next ten years he taught in a number of English schools, as well as spending extended periods in Germany, publishing in the process two geology and chemistry textbooks.[11] In addition to teaching, Clement raised money by selling large collections of Bronze Age pottery to a number of British museums for substantial sums.[12] By the 1890s his entrepreneurial interests had turned to Australia. Utilising his knowledge of geology, he travelled to the rapidly expanding West Pilbara goldfield to assess the potential for establishing a goldmine.

Between 1895 and 1901 Clement made three trips to Australia, during which time he spent three years in the north-west. Most of his time was spent establishing a mine at Towranna on behalf of a British-based company.[13] Clement was keen to promote the prospects of this mine. In 1898 he predicted to a London shareholders meeting that there was 'enough quartz on the [Towranna] field to supply not only a 10-head battery, but a 20-head battery for many years to come'.[14] Despite Clement's optimism, within two years many of the Pilbara goldfields were in decline, including Towranna. In early 1901, on the occasion of the last crushing of ore from the mine, the local paper wrote:

> This crushing, we are sorry to say, is the last which the Towranna Company is likely to have, the mine having been closed down last Saturday. After his first inspection of the mine, some four or five months ago, Mr. Field [the mine manager] came to the conclusion that the property had been over rated.[15]

During the time that Clement was in the north-west, he frequently used his scientific credentials to promote both himself and various mining ventures, none of which had a long life.



At a time of some investment uncertainty, Clement's views on the British perception of the north-west mining industry were highly sought after in the Roebourne district. It appears that his scientific credentials gave him the status of a savant in the region. Late in 1896 the lead story in the *Northern Public Opinion* began:

> It is not often we have an opportunity of hearing, direct from a reliable source, the standing our Northern goldfields occupy in London, and owing to this fact the few curt observations made by Dr Clement will be all the more appreciated.[16]

From the style in which Clement's activities were reported, he apparently encouraged such interest by keeping the press informed of them: 'Dr Clement informs us that he is undecided as to where he will erect his two head stamper battery, but at present he favours either the new find near the Lower Nickol, or Weerianna'.[17]

In addition to his mining activities, Clement acquired large collections of Aboriginal objects and natural history specimens during his time in north-west Western Australia. These he later sold to museums throughout Britain and Europe.[18] Although he wrote at one point that he had 'visited every native camp within a hundred miles of Roebourne',[19] it is apparent from the provenance of his ethnographic objects that he acquired most of them within the immediate vicinity of the Towranna mine. His collecting activities were well known in the district, as the local press eagerly reported: 'The native weapons and collections of Nor'-West moths which the doctor took to England with him were landed in splendid condition, and excited many econiums [*sic*] from savants'.[20] His practice of building up confidence and trust in his mining ventures resembled the techniques he employed when negotiating the sale of objects to museum curators in Britain.

Clement's involvement in collecting and selling such material raises the issue of his relationship with Aboriginal people. Despite his voluminous correspondence with museum curators about the collections he was selling, Clement rarely reflected upon his relations with the original inhabitants of the region. Yet something of his attitude to them is revealed in his contribution to a lively debate which took place in the British press in 1899, during one of his sojourns in England. At this time, the London newspapers had raised allegations that two West Australian Justices of the Peace, convicted of cruelty to Aboriginal people in the north-west region, had been treated unduly leniently by the authorities in Western Australia. This occurred against the background of Western Australia's campaign for self-government, and concerns in Britain about the treatment of Aborigines in the colony.[21] The *St James Gazette* published a number of editorials questioning whether the incident was evidence of more systemic violence towards Aboriginal people on the part of white settlers. The issue provoked a series of letters to the newspaper. As one reader put it: 'This matter is worthy of the attention of every Englishman who has the prosperity of the Empire at heart. If we are to continue as a great colonising Power, it can only be by a strict adherence to the principles of equity and justice.'[22]

Provoked by a letter which criticised the practice of indenturing Aboriginal labourers as 'worse than slavery',[23] Clement entered the debate, suggesting that such views were naive and accusing the writer of having visited only those Aboriginal people living near the major towns where 'the lazy good-for-nothing black population congregated'.[24] Instead, he favoured an indenture system, citing his own experiences in the region: 

> I have had black servants, and speak from experience. I visited every native camp within a hundred miles of Roebourne, and asked hundreds of blacks how they liked the life on the stations, and in only one single instance a man complained. This black had been a chief, and he asked me to tell the Queen, on my return to England, that the white-fellow had taken away his land, that he could no longer hunt the kangaroo, and bustard at his own sweet will; that he had plenty of waterholes, plenty of meat, plenty of 'gins' (wives) before the white man came; but that now he was a shepherd at a station.[25]

Clement continued sarcastically: 'And a very fat and sleek shepherd he looked, and his three wives and numerous children well nourished and merry'.[26] Such views and language show clearly that Clement was at odds with the burgeoning humanitarianism developing among the British press (and among some West Australian clergy).[27]

In Clement's extensive correspondence with museums, he never mentioned that the reason for his presence in the north-west was to pursue goldmining and exploration. It is probable that this information would have made the material on offer less attractive, reducing its economic value. Clement took pains to represent himself as a 'scientific' naturalist and anthropologist.[28] Such a guise helped him to represent the local Aborigines, and the collections of their objects which he sold, as being removed from the influence of Europeans. The growing metropolitan obsession with collecting only authentic artefacts made this ruse essential to his livelihood.

In addition to selling innumerable Aboriginal objects to museums, Clement also sold copies of photographs which he had taken of Aboriginal people near Towranna. Here, again, we find him cultivating an 'ethnographic' representation of his subjects, situating them in a manner designed to maximise their spatial and temporal remoteness from Europeans. By his own account, Clement took at least eighty photographs, but a cyclone which hit Towranna in April 1898 supposedly destroyed all but two.[29] Both these images are similar in content, probably taken on the same occasion. They depict male and female figures of a variety of ages, wearing body paint and ornaments and standing outside a bough shelter (Figure 10.2). The omissions are as striking as the inclusions. All visual reference to goldmining has been carefully avoided, despite the fact that it was a dominant feature of the Towranna landscape and the very reason for Clement's presence there. Apart from some items of clothing, nothing in the photographs refers to the existence of a relationship between Europeans and Aborigines. The selectivity of Clement's work becomes more apparent when it is compared with contemporaneous images of social life from the region.



One such set of images was created by Walker Hodgson, an artist who accompanied the aforementioned Calvert on his tour of the West Australian goldfields in 1895. Hodgson's sketches were included in Calvert's published account of the trip and they include many depictions of Aboriginal people. This work, published in England, was one of a number of pamphlets, books and newspapers produced by Calvert. Although he used some of these to promote British investment in the goldfields, he also appears to have had a genuine interest in Australia, particularly Western Australia.

Few details are known of Hodgson's life and career.[30] However, it is apparent that, unlike Clement, he did not seek to make a career out of creating and marketing a particular type of image of Aboriginal people. At one point Calvert described the activities of Hodgson in the field:

> Our artist, who has a natural aptitude for lighting upon curiosities that are overlooked by the casual observer, was usually out of his blanket an hour or more before the rest of us had stirred, and when he once got loose on his rambling excursions it required considerable ingenuity to find him again. Whenever a halt was called on our journeys he would tuck his sketch book under his arm and leave us.[31]

Using his 'natural aptitude for lighting upon curiosities', Hodgson produced a wide range of images showing an extraordinary diversity of relationships between goldminers and Aboriginal people. The variety is highlighted by the contrast between the paranoia of 'Surrounded' (Figure 10.3), an image of

Aborigines encroaching upon a lone miner loading his dry-blower, and his image of 'cricket on the goldfields' (Plate 10.1).[32] The latter documents a complex cross-cultural interaction involving both objects, such as cricket bat and ball, and concepts, such as sport and game. The inclusion of the Afghan cameleer indicates a further dimension to the cultural heterogeneity of the goldfields.

**Figure 10.3** W. Hodgson, *Surrounded,* in A. F. Calvert, *My Fourth Tour in Western Australia,* Heinemann, London, 1897. (National Library of Australia)

Considered as a whole, three styles of imagery are evident in Hodgson's work containing Aboriginal themes. First, his drawings include some depictions of Aboriginal people in an 'ethnographic' style, similar to that of Clement's photographs. These contain scant evidence of European influence. There is an absence of European figures or clothing, and they often depict people undertaking 'traditional' activities. In contrast to these 'ethnographic' images, some of Hodgson's other drawings show Aboriginal people engaging with European objects in diverse ways. Examples include an image of an Aboriginal person riding a bicycle, and another depicting an Aboriginal boy sporting a helmet, pipe and gun (Figure 10.4). Obviously, many readings of such images are possible; however, at one level, they clearly document a complex black–white relationship on the goldfields, in contrast to the restricted style of Clement's photographs, discourse and collections of objects.

**Figure 10.4** W. Hodgson, *Borrowed Plumes*, in Calvert, *My Fourth Tour*, 1897. (National Library of Australia)

**Figure 10.5** W. Hodgson, *'Home, Sweet Home', 12,000 Miles Away*, and *I Say, You'll Catch Cold, My Lord!*, in Calvert, *My Fourth Tour*, 1897. (National Library of Australia)

A third feature of Hodgson's images is the way in which he showed Aboriginal people, particularly women and children, interacting with European goldminers in culturally complex ways, suggesting frequent contact between goldminers and Aborigines which is not documented in 'ethnographic'-style works. His depictions of women are particularly interesting and touch on an issue which greatly concerned colonial authorities. Henry Princep, the Chief Protector of Aborigines, was particularly anxious on this issue: 'I cannot say that the condition of the natives altogether on the Pilbarra Goldfields appears to be satisfactory. There is a great deal of immorality. We may expect, however, more of this sort of thing in a shifting community than in a purely pastoral district.'[33]

Hodgson has captured something of this 'immorality' in a series of images representing interaction between miners and Aboriginal women (Figure 10.6). 'A tin of salmon', in which a miner beckons to an Aboriginal woman, standing some distance away, with the offer of food, suggests the commodification of women, an exchange of salmon for sex. Interestingly, the framing of this work limits the exchange to two individuals.[34] 'A letter home' depicts a remote miner sitting in his tent, writing to 'home'. In the doorframe, but outside, stands an Aboriginal woman. Hodgson suggests that a relationship exists here but one which cannot be acknowledged in the letter. 'A little flirtation' shows a miner leaning on his shovel, talking to a smiling Aboriginal woman. The equal positioning of both figures here portrays a situation in which significant agency is ascribed to Aboriginal women.

Hodgson created these three works as separate images, and they appear in different parts of Calvert's book, but they can be easily juxtaposed to show three aspects of Hodgson's interpretation of relations between European miners and Aboriginal women. When brought together, they suggest a largely predatory relationship, though one in which there is certainly some degree of genuine intimacy. They hint at the existence of sexual relationships, something rarely recorded in textual sources.[35] Annette Hamilton, writing about the Northern Territory, has reflected on the discrepancy between official records and the oral history of Aboriginal people. She noted

> the disparity between the great significance Aborigines placed on the sexual and domestic role of Aboriginal women in the interaction with white men, and the lack of prominence accorded them in official records . . . there is no frank acknowledgment that practically every white man, whether boss or worker, lived with an Aboriginal woman, or a series of them, throughout the early period of settlement.[36]



Through his deft imagery and dry irony, Hodgson has documented an interpretation of this pervasive reality as he encountered it on the north-west goldfields. It is a reality which had no place in the ethnographic style of Clement's representations.

Aboriginal people are often absent from European accounts of the discovery and mining of gold in Western Australia. At best their presence is limited to being either a threat to goldminers, as on the Kimberley goldfields, or a curious, colourful feature of the landscape.[37] Rarely are the complexities of the day-to-day interactions and relationships between miners and Aborigines described and acknowledged.

The representations created by Clement and Hodgson are two very different non-textual takes on this enterprise. Clement's ethnographic collections and photographs sought to show Aboriginal people isolated from the complex gold society of the West Pilbara in the late nineteenth century. His photographs show nothing of the heterogeneous cross-cultural society of the region, a society which included British, American, Asian and Aboriginal men, women and children. Clement certainly did not acknowledge the existence of relationships between Europeans and Aborigines. By contrast, more diverse and narrative-based, Hodgson's range of images, and the interactions contained in them, suggest that complex and often contradictory relationships developed

**Figure 10.6** W. Hodgson, *A Tin of Salmon*, *A Little Flirtation*, and *A Letter Home*, in Calvert, *My Fourth Tour*, 1897. (National Library of Australia)

between black and white on the north-west goldfields. Ironically, such relationships are masked by the 'ethnographic' style of Clement's work.

Deborah Bird Rose and Daryl Lewis have written of the difficulty in trying to maintain separate black and white histories in Australia when so often the two are intertwined: 'Rather than white history or black history, what we keep finding . . . is the history of human beings in association (positively, negatively, ambivalently, absurdly, tragically, but almost invariably side by side) with each other on this continent'.[38]

Clement, an entrepreneur of both gold and anthropology, marketed a vision of Aboriginal people as ethnographic types, completely outside the society of the European gold seekers on the West Pilbara goldfield. By contrast, Hodgson's work, created outside of the scientific and ethnographic paradigm, brought together black and white in action, catching something of the ambivalence, the absurdity, the tragedy and the occasional poetry of the goldfields. In doing so he has provided us with a rare and revealing record of our colonial past.

## Notes



1 A. F. Calvert, *My Fourth Tour in Western Australia*, 1897, William Heinemann, London, p. 15.
2 Like many of the characters involved with gold, Calvert was a complex figure whose life was full of twists and turns, success and failure. During the 1890s he made four visits to Western Australia. He published numerous pamphlets and articles on Australian themes, ranging across Aborigines, history, travel and mineralogy, and from 1893 to 1894 he published the *Western Australian Review* in London, which promoted the West Australian mining industry. Calvert also promoted the establishment of a portrait gallery, museum and library in Perth, and he pledged financial support to what became the Calvert Exploring Expedition. It was this last venture which led to his being declared bankrupt in 1898, after which his interests turned to Spain. See G. Blackburn, *Calvert's Golden West – Albert Frederick Calvert: A Biography and Bibliography*, 1997, Hesperian Press, Carlisle, WA; B. Nairn and G. Serle (eds), *Australian Dictionary of Biography*, vol. 7, 1979, Melbourne University Press, Melbourne, pp. 528–9.
3 See J. W. McCarty, 'British Investment in Western Australian Gold Mining, 1894–1914', *University Studies in History*, vol. 4, no. 1, 1961–62, pp. 7–23.
4 *Northern Public Opinion*, 30 November 1895.
5 The increase in goldmining in the west Pilbara region during the 1890s led to the proclamation of a separate goldfield in 1895 which included the diggings at Mallina, Whim Creek and Towranna. See A. Maitland, *The Mineral Wealth of Western Australia* (Geological Survey Bulletin no. 4), 1900, Richard Pether – Government Printer, Perth, p. 36.
6 See R. T. Appleyard, 'Western Australia: Economic and Demographic Growth, 1850–1914', in C. T. Stannage (ed.), *A New History of Western Australia*, 1981, University of Western Australia Press, Perth.
7 P. Biskup, *Not Slaves Not Citizens: The Aboriginal Problem in Western Australia 1898–1954*, 1973, University of Queensland Press, Brisbane, p. 18; and J. S. Battye

(ed.), *The History of the North West of Australia Embracing Kimberley, Gascoyne and Murchison Districts,* 1915, V. K. Jones & Co., Perth, p. 18.

8 See S. Hunt, *Spinifex and Hessian: Women's Lives in North-Western Australia 1860–1900,* 1986, University of Western Australia Press, Perth, pp. 96–102.

9 Under the West Australian Goldfields Act 1886, 'Asiatic or African aliens' were prohibited from obtaining a miner's permit. However, the employment of Aborigines within mines was discouraged by the regulations attached to the Act, under which a mine operator could not include Aboriginal labour as part of the minimum number of employees that each mine was required to use.

10 See Hunt, *Spinifex and Hessian,* p. 103.

11 E. Clement, *Symbolic Chemistry: Showing the Decompositions of Inorganic Substances with Text Tables Annexed,* 1870, Edward Baines & Sons, Leeds; and E. Clement, *Tabular View of the Geological Systems, with Their Lithological Composition and Paleontological Remains (Especially Adapted for Students Preparing for the Royal Military Colleges of Woolwich and Sandhurst),* 1882, W. Swan Sonnenschein & Co., London.

12 See I. Coates, 'Lists and Letters: An Analysis of Some Exchanges between British Museums, Collectors and Australian Aborigines (1895–1910)', PhD thesis, Australian National University, 1999, Appendix 5.

13 Towranna, or Toweranna as it is now known, is located inland, between the Sherlock and Yule Rivers, 100 kilometres north-east of Roebourne.



14 *Financial Times,* 19 August 1898, p. 5.

15 *Northern Public Opinion,* 22 February 1901, p. 2.

16 *Northern Public Opinion,* 28 November 1896, p. 2.

17 *Northern Public Opinion,* 23 September 1899, p. 2.

18 Between 1896 and 1901 Clement sold collections containing approximately 900 Aboriginal objects to the British Museum, the Museums of Science and Art in Edinburgh and Dublin, the Pitt Rivers Museum, Bristol Museum, and Kew Museum of Economic Botany, as well as major museums in Leiden, Hamburg, Moscow and Basel. During this period he also sold important collections of natural history objects to the British Museum (Natural History), the Oxford University Museum, Kew Gardens and the Leiden Herbarium. See Coates, 'Lists and Letters', 1999, pp. 114–41.

19 *St James Gazette,* 6 April 1899, p. 4.

20 *Northern Public Opinion,* 3 June 1899, p. 2.

21 See Biskup, *Not Slaves,* p. 25, and B. K. de Garis, 'Self-Government and the Emergence of the Political Party System, 1891–1911', in Stannage, *A New History of Western Australia,* 1981, p. 337.

22 'Outis', *St James Gazette,* 29 March 1899, p. 4.

23 'Worse than Slavery: The Testimony of a West Australian Official', *St James Gazette,* 30 March 1899.

24 *St James Gazette,* 6 April 1899, p. 4.

25 *St James Gazette,* 6 April 1899, p. 4.

26 *St James Gazette,* 6 April 1899, p. 4.

27 See S. Hunt, 'The Gribble Affair: A Study in Colonial Politics', *Studies in Western Australian History,* 8, 1984, pp. 42–51.

28 For example, Clement wrote to the British Museum: 'I am leaving again for North West Australia . . . and write to ask what I can do there to further the interests of Anthropology' (letter to C. H. Read, 29 September 1896, Department of Medieval and Later Antiquities Archive, British Museum).

29 See letter from Clement to E. B. Poulton, 29 November 1898 (Oxford University Museum Library – Poulton MSS Series). Copies of these photographs are held by the Pitt Rivers Museum, Oxford, and the National Museum of Ireland, Dublin. Again, the environment of entrepreneurship which existed on the goldfields causes one to be suspicious of Clement's claims about the destroyed photographs, possibly serving as they do to enhance the value of the 'surviving' images.

30 Hodgson was apparently something of a bohemian figure. During the period 1891–92 he produced a series of nearly seventy portrait drawings of people prominent in the British art world, twelve of which are held in the National Portrait Gallery in London. It appears this portfolio was to be used as illustrations for a series of articles. Apart from these works very little is known of Hodgson's life (Christopher Wood, pers. comm.).

31 Calvert, *My Fourth Tour*, p. 132.

32 Many of Hodgson's works included in Calvert's account have been given ironic titles. It is not clear whether these are the creation of Hodgson or added later by Calvert.



33 Aborigines Department, 'Report for the Financial Year Ending 30 June 1899', *Western Australian Parliamentary Papers and Votes and Proceedings*, Paper no. 40, p. 6.

34 See M. A. Jebb and A. Haebich, 'Across the Great Divide: Gender Relations on Australian Frontiers', in K. Saunders and R. Evans (eds), *Gender Relations in Australia: Domination and Negotiation*, 1992, Harcourt Brace Jovanovich, Sydney. Much of the literature describes the sexual relations between European men and Aboriginal women as occurring within broader contact involving group relations, rather than individuals, and European men.

35 See Jebb and Haebich, 'Across the Great Divide'.

36 A. Hamilton, 'Bond-Slaves of Satan: Aboriginal Women and the Missionary Dilemma', in M. Jolly and M. Macintyre (eds), *Family and Gender in the Pacific: Domestic Contradictions and the Colonial Impact*, 1989, Cambridge University Press, Cambridge, p. 237.

37 J. Gooding, 'Presence and Absence in the Western Australian Landscape', *Westerly*, 3, 1999, pp. 99–108.

38 D. B. Rose and D. Lewis, 'A Bridge and a Pinch', *Public History Review*, 1, 1992, p. 28.

PLATE 13.1 Old Stamp Battery – Showing the Granites in the Background. (Normandy Mining Limited)

PLATE 13.2 Tanami Mine Joint Venture – Aerial View. (Roger Potts, Otter Gold NL)

PLATE 13.3 Granites Treatment Plant Operating by Night. (Normandy Mining Limited)

# Lasseter's Stories

## TENDING THE GHOSTS OF DESERT GOLD



*David Raftery*

As Geoffrey Blainey remarks in his history of Australian mining, *The Rush That Never Ended*, 'Australia's map is criss-crossed with the spidery tracks of the metal seekers'. The lure of gold, and other precious metals, impelled the fashioning of routes of transport across country as diverse as it was inhospitable. The network of railways that Australia developed owed much to finds on the new goldfields. Every port north of the Tropic of Capricorn was instigated, or at least expanded, as a result of gold discovery.[1]



Lasseter is one of many names that are synonymous with gold mania and Australian prospecting. Unlike Ballarat, Kalgoorlie or Charters Towers, which can all boast iconic and tangible strikes of gold – nuggets, reefs or other spectacular successes – Lasseter's gold remains a ghost, or more properly a series of ghost stories. Today, Lasseter stands as shorthand for the history of gold prospecting in the nation's most remote mineral field: Central Australia. Despite the absence of the gold, the man who claimed its existence, Harold Bell Lasseter, is memorialised nonetheless. Those 'spidery tracks of the metal seekers' across the inhospitable landscape of Central Australia are well trodden, not least by other prospectors seeking his reef. The history of prospecting in the Petermann Ranges, where Lasseter finally met his fate, can also be read as a history of transport. The technological developments that spawned railways, cars, trucks and the aeroplane underscored different generations of desert gold-seeking expeditions. It was the desire for gold, though, that brought both human and mechanical resources together in a quest for the elusive strike.

While Lasseter's individual role in the expedition of 1930–31 is consistently emphasised, he was by no means a lone prospector. In fact, he was officially employed by the Central Australian Gold Exploration Company Limited

(CAGE), a hastily convened company with bold dreams of exploiting the rich reserves of minerals believed to be waiting in Central Australia. Aerial travel had introduced a new dimension to prospecting. Sir Charles Kingsford Smith, for one, saw the aeroplane as the key to unlocking the interior and ensuring that Australia continued to march 'onwards to her destiny as the Great Nation of the South':

> No other country can, like Australia, claim that its real foundation as a nation was based on the discovery of gold. The gold rushes of the Yukon and the Klondyke [*sic*] in Alaska were outrivalled by the stampedes of Bendigo, Ballarat, Ophir and Coolgardie, where men from the four corners of the world concentrated on the frenzied search for the precious yellow metal, which has so upset the equilibrium of the world today.
>
> Where the icy vaults of Alaska imprisoned that country's vast gold deposits, the dreary sun-baked solitudes of our little-known interior beckon adventurous souls on and on, with the gold 'that must be there', the irresistible lure. Lack of water and the uncertainty of supplies have hampered operations up to within the last few years, but the humble beginning with a tiny Gipsy [*sic*] Moth, piloted by the author on the Lasseter expedition in 1930, immediately drew attention to the possibilities of the aeroplane in the geological survey field.[2]



However, it was not aeroplanes and motor cars, the technological trophies of the time, which propelled CAGE forward in its quest for gold. It was the testimony of one Harold Bell Lasseter that prompted the formation of the company, and the expedition's eventual departure from Alice Springs on 21 July 1930. Errol Coote, who accompanied them, was a Sydney journalist before he found fame as author of *Hell's Airport and Lasseter's Lost Legacy*, chronicling his experience as pilot on the 1930 expedition.[3] As part of his regular political rounds he would often talk with John Bailey, then president of the omnipotent Australian Workers Union (AWU). It was in Bailey's Sydney office that Coote first met Harry Lasseter. Lasseter was aware that the union had recently cut a deal with Prime Minister Scullin to procure a Rolls Royce engine from the Imperial government so that 'Wizard' Smith could attempt a world land-speed record. While the attempt had not been successful, it did bring the AWU to the attention of Lasseter and, as a result, brought forth his incredible story.

In the AWU office, Lasseter regaled Bailey and Coote with a startling account of his previous mineral discoveries in the Central Australian region. He claimed to have located a reef of gold some thirty years earlier, somewhere near the border of Western Australia and South Australia. According to the story, he had in the late 1890s been sailing a vessel off the Queensland coast. Upon hearing of ruby deposits in the MacDonnell Ranges, just east of Alice Springs, he took leave of his boat at Cairns, procured some horses and made the long journey to Central Australia. The gems that he found, however, proved not to be rubies but ruby-coloured crystalline quartz. Disappointed, he decided to head west in the hope of finding work in Kalgoorlie or further afield. Believing that the mountain range extended to the west coast of the continent, he planned to follow it to the port towns of Fremantle or Carnarvon.

The ranges soon gave way to dry desert plains, punctuated by spinifex and sandhills. Stopping at one of the few waterholes in this region, Lasseter happened upon flakes of gold. Further investigations suggested that the flakes emanated from rock within a small range. He excitedly followed this rock outcrop over sandhill country, gathering several samples for later testing. This enthusiasm, however, did not sustain his horses, which soon became weary of the travel and died. Alone in the desert, with no horse and little water, bearing samples of what he believed to be gold, Lasseter must have been a desperate man. Fortunately, he was rescued by an Afghan cameleer who took him to a nearby camp presided over by Harding, a government surveyor. Having seen the gold samples, Harding begged the prospector to take him back to the country of their origin. Still traumatised by his desert ordeal, Lasseter refused both to return or to divulge its whereabouts.

Lasseter was soon back working in Coolgardie and remained in contact with Harding. They later staged an expedition to the reef together, travelling  inland from Carnarvon. Samples of gold taken were later found to yield three ounces to the ton. Inspired by such results, both men attempted to raise capital for an exploration project. But with Kalgoorlie booming, it was difficult to arouse interest in a remote place and in the claims of two obscure men. Still convinced that 'his' reef held untold riches, Lasseter claimed to have attempted another journey to search for it in 1911. This time he travelled up through Oodnadatta, taking men and camels with him. His crew, however, lacked the motivation required for such a trip. They soon abandoned their quest for gold in favour of the comforts of Oodnadatta's Transcontinental Hotel. Frustrated by the faithless once again, Lasseter enlisted in the army and fought in World War I. Upon return to Australia after sundry travels, he worked first in Canberra, then on the Sydney Harbour Bridge. Ion Idriess claimed that the bridge was viewed by Lasseter, in typical fashion, less as a job than a project of passion:

> An unusual man in several ways. One of his dreams was the Harbour Bridge. Long ago he had drawn plans for such a bridge, and these plans have been pronounced by competent engineers perfect down to the last rivet. Highly educated men with whom he came into contact have expressed their surprise at his knowledge of unusual subjects, never suggested by the man's everyday talk. His mates on this expedition found him a good camp-mate, but liable to be impatient at even unavoidable delay, and headstrong over difficulties. They had an idea, too, that though the finding of the reef occupied his mind, his ultimate object embraced something greater. As a matter of fact, at this time a subdued excitement boiled within Lasseter, as day by day every forward lurch of the trucks brought him a little nearer his lifetime's desire.[4]

By 1930, when he approached the AWU, almost twenty years had passed since Lasseter last claimed to have sought his reef, but his desire to find it again had obviously not waned. From the AWU he sought only sponsorship for camels and provisions. Though it was the middle of the Great Depression, John Bailey managed somehow to ensure a much more grandiose provision. Entranced by the man's story, Bailey immediately engaged Lasseter at £5 per week: that way his knowledge could not be harnessed by rival prospecting interests. He also set about assessing Lasseter's credentials by hiring two experienced mineral men, George Sutherland and Fred Blakeley. Sutherland was a veteran of New South Wales mines, and Blakeley was a prospector with considerable experience in the Coober Pedy region of South Australia. While these men interviewed Lasseter and checked his connections, Bailey instructed Coote to investigate the aerial logistics of a Central Australian expedition and, once Lasseter had been sufficiently scrutinised, to use his journalistic skills to publicise the scoop.

Coote's headline, 'Fokkers Search for Desert El Dorado', heralded the birth of the Lasseter expedition, and a meeting was soon convened to introduce Lasseter to prospective shareholders for the new venture. A mere twenty-four hours after the registration of the company name, more than £5000 had been

subscribed, notwithstanding Depression conditions. Bailey, the unanimously elected chairman of directors, told fellow shareholders the further good news: that free rail transport for the expedition party had been offered by the New South Wales government, and that the Atlantic Union Oil Company would supply the fuel to power the trucks and planes over remote expanses of desert at no charge. Food and provisions were also to be provided at low cost. Mr A. G. Hebblewhite, representing a large British firm, provided CAGE with a state-of-the-art six-wheel truck for six months. Coote recalls Bailey's grateful reflections on the importance of the expedition:

> Australia, as a part of this great Empire, must bear her burdens of responsibility. How else can she emerge successfully from the mire of depression? What can she do to fulfil her obligations, her financial obligations, to the mother country? Can anybody at this meeting tell us of a better way out of our difficulties than by backing this man, Lasseter?[5]

Only 'Peter the Pest', a compulsive heckler at public meetings, replied in the negative, suggesting that Australia would be better served by trying to balance its budget. Bailey's deft rejoinder was well calculated to tap populist sentiments:

> I have reason to believe that the man who passed that remark is a spy for the mining interests of Pitt Street who, because they have been excluded from this show, are determined to break up the meeting. That man has no right to be here![6]

The 'Pest' was rapidly escorted from the building by two hefty and excitable shareholders, but they would have done better to heed his warning. A few simple cross-checks on Lasseter's life would have made the story of the lost reef look very shaky indeed.

In the story that Lasseter had told John Bailey and Errol Coote, the prospector had stressed that he had found his reef around thirty years previously, but he was publicly producing a rather different version. In a letter to the West Australian Minister for Mines in October 1929, Lasseter had stated that the reef had been discovered *eighteen* years earlier, in 1911.[7] A subsequent letter to the same source claimed that he had actually found the reef thirty-three years earlier. However, most damaging to the credibility of Lasseter's story was its striking similarity to accounts of previous prospecting expeditions in the same region. Max Cartwright, in *From Bush Tracks to Highway*, and Michael Terry, in *Untold Miles*,[8] describe, for example, a series of prospecting expeditions by Messrs Earle, Stock and Lamb in 1895 and 1896, which bear an uncanny resemblance to the story spun by Lasseter in the AWU offices. Earle, a Coolgardie-based bank official who had previously enjoyed some success in a gold syndicate, had shown Lamb, an experienced Oodnadatta prospector, some quartz gold samples. Lamb and Earle had then travelled west together towards the Rawlinson Range, but, after a disagreement, Earle had returned to Oodnadatta, showing several passengers on the southbound train the quartz specimens he had brought from Coolgardie. This ore he falsely claimed to be from a recent reef strike in the Rawlinson Range.



In the aftermath of the rumour, a syndicate led by Ted Warman and Mr W. Cockrum and son left Marree in 1900 to locate this supposed reef. Travelling through the Mann Ranges to the south of the Petermann and Rawlinson Ranges, the party was attacked by Aborigines, who killed a camel driver. Despite this setback, a second Warman and Cockrum expedition set out from Marree the following year. Leases were pegged in the Mount Davies area before ill health forced the party to return to safety. Another group argument developed, whereupon the party's camel driver murdered Warman and fled. Earle never returned to the Mount Davies region, but the story of 'Earle's Reef' entered Australian gold mythology. Cartwright suggests the possibility that Lasseter borrowed its details to construct his story. Whatever the case, at least seventeen prospecting expeditions to the 'Earle's Reef' area were mounted between 1895 and 1930. None of these missions yielded gold.

Seemingly, too, Lasseter was as suspicious in the present as he was creative with the past. Despite the substantial resources at CAGE's disposal, he was not satisfied with the transport arrangements being coordinated by Fred Blakeley, the team leader. While awaiting the arrival in Alice Springs of the *Golden Quest*, with its pilot Errol Coote and mechanic Phil Taylor, Lasseter made a secret and private arrangement with a dogger (dingo scalper), Paul Johns, who was to bring a supply of camels to a point near the West Australian border.[9] Lasseter planned to find a reason to separate himself from his expedition colleagues and join up with Johns and the camels in order to rediscover the reef. His obsession about locating the lost gold reef had swelled into a paranoid fear that someone else might beat him to it.

By the end of July 1930, the CAGE expedition party had left Alice Springs for Ilbilla, a nine-day journey north-west of Alice Springs. Seven white men made up the party. Sutherland and Blakeley, the prospectors, led the expedition. Taylor, from the New South Wales town of Parkes, had been recruited as the expedition's mechanic. His skills were often tested as the trucks struggled to cope with the rugged conditions. CAGE proposed to stock Ilbilla with fuel and provisions as a base for launching aerial surveys. Coote was employed to fly Lasseter on reconnaissance missions. Fred Colson, a well-travelled local bushman, was employed for his regional knowledge and his mechanical and logistical expertise. Completing the party was Captain Blakiston-Houston, an Irish army officer with surveying skills and a thirst for adventure. Yet even this assemblage of expertise could not compress the huge distances or overcome the rugged terrain that dogs all exploration efforts in the Central Australian region. Coote's, Idriess' and Cartwright's accounts of the expedition are replete with details of incapacitated trucks, fuel shortages and the continual need to create new motor tracks through inhospitable country.[10] Despite all the excitement about the technological sophistication of the expedition, the prospecting party soon came to realise what hunter-gatherers had long known: reliance on water in the desert enforces a nomadic lifestyle.[11]



Under instructions from its board of directors and fuelled by its own exploratory zeal, the CAGE party struggled determinedly on, using whatever water they could find. Having stripped the differential drive on Colson's Chevrolet truck while crossing the Aiai Creek, Coote and Colson had to travel back to Alice Springs to procure spare parts. Taylor, Sutherland, Blakeley and Lasseter remained at Ilbilla. A temporary airstrip was cleared near Aiai Creek, where Coote landed the *Golden Quest*, but he was soon back at Alice Springs for treatment to injuries incurred after crashing on take-off. When they heard of this accident, the Sydney directors of CAGE organised the supply of another plane, the *Golden Quest II*, which Pat Hall piloted to Alice Springs. Coote and Hall then flew back to Ilbilla to rejoin their fellow expedition members.

It was precisely at this time that Paul Johns arrived at Ilbilla from the Finke River Mission at Hermannsburg with several camels. Meanwhile, Lasseter had become convinced after a reconnaissance flight that the expedition was 150 miles too far north and would have to negotiate numerous sandhills before the reef could be reached. The CAGE party attempted to fashion crude motor tracks from Ilbilla but soon returned after limited success. Conscious of the strains being placed on the trucks, Blakeley led the party back to Alice Springs for more supplies and equipment. Lasseter, however, refused to leave the area. He was particularly keen that Johns and his five camels should remain involved in the search.

The pair travelled through Mount Udor, Putardi, and across Lake Amadeus to Mount Olga before reaching the Petermann Ranges. It is believed that Lasseter left Johns for a couple of days, during which he claimed to have refound his reef. Cartwright estimates that they returned to Ilbilla sometime in October, when their fuel supplies were running short.[12] Although Ilbilla was well

**Figure 11.1** H. B. Lasseter and His 1930 Expedition Party at Pantas Well, MacDonnell Ranges, Lasseter in front and others (l to r): Coote, Sutherland, Colson, Blakeley and Taylor. (National Archives of Australia)

stocked with food, the five camels were in poor condition and Johns wished to return to Alice Springs or Hermannsburg to acquire fresh ones. Again, Lasseter was obstinate in his refusal to leave the area. As a compromise, he retained two of the camels, while Johns returned to Alice Springs with the other three. The lone prospector again headed south from Ilbilla in search of the elusive reef.

Although the original CAGE expedition had already disbanded, several of its members made attempts to locate Lasseter and Johns. With the *Golden Quest II* once again in working condition, Coote flew to Ayers Rock, which was to serve as the new base for the company's rescue and support efforts. Taylor travelled to Ayers Rock with a camel team, hoping to synchronise his arrival with Coote's. Delays prevented this happening; Coote damaged the aircraft's propeller while landing and was forced to wait nine days until help arrived. When Taylor did finally arrive, the plane was repaired and Coote flew back to Alice Springs. There he was ordered by CAGE's directors to fly the *Golden Quest II* back to Sydney, where two men, the pilot Les Pittendreigh and an original CAGE shareholder, Mr Hamre, would replace him.

What happened to Lasseter from this point on is a matter of much conjecture. He could well have travelled as far west as the Rawlinson Range and Lake Christopher in Western Australia. Michael Terry's expedition of 1932, which is documented in his book *Sun and Sand*,[13] found and photographed Lasseter's Lake Christopher camp. It was here that Lasseter had etched the words '*dig under fire*' in the earth. From this point it is probable that he retraced his tracks in an easterly direction back to the Hull River, stopping for some time in a cave that now bears his name. From there, he probably headed east for ten or twelve miles before losing his camels. He then walked back to the cave, where he was assisted by several Aboriginal people in his attempts to return to health.

In the meantime, efforts to rescue Lasseter were themselves proving dangerous. After returning from Ayers Rock, where Coote had been rescued and the *Golden Quest II* repaired, Taylor was instructed by his CAGE directors to lead a camel team to Ilbilla, where a base would be set up from which Pittendreigh and Hamre would attempt to locate Lasseter from the air. They were not even able to make it to Ilbilla, their plane crashing at Dashwood Creek after running out of fuel. Eventually, the air force was called in and the two men were found after a seven-day air search. Johns and Taylor then made a final attempt to locate Lasseter with camels. Taylor soon became seriously ill and was forced to return to Hermannsburg for treatment. Becoming increasingly desperate, he, on behalf of CAGE, contracted a local pastoralist, Bob Buck, to find Lasseter – dead or alive.



Buck left the Finke River Mission on 24 February 1931, following camel tracks to the west. He eventually found Lasseter's body, on 29 March. After discussions with local Aborigines, Buck estimated that Lasseter had been looked after by Aborigines for something like sixteen weeks and had died around 30 January 1931 (Plate 11.1).

At first sight, the sequence of events that made up the CAGE expedition of 1930 and 1931, and which culminated in the death of Harold Lasseter, appears a tragi-comedy of errors and chimeras. This is especially so in the light of the initial excitement over the supposed potential of aerial prospecting to produce new gold discoveries. Faith in the benefits of new technology pervaded the shareholders and the many who followed the expedition's progress through Coote's reports in metropolitan newspapers. Behind this technological optimism lay a still greater hope, invested in the discovery of gold itself – the precious metal that could deliver Australia from the mire of the Great Depression. Lasseter had come to embody all of these desperate hopes.

Despite the incredibility of Lasseter's claims, as well as his obstinacy, stubbornness and fatalism, it is too easy simply to dismiss him as a lone fool, crazed by the promise of gold. For if he was a dreamer, he certainly had his partners. The Central Australian Gold Exploration Company launched their second and final expedition in September 1931. Different personnel were employed: this time Bob Buck acted as leader and principal guide. Notably, neither trucks nor aeroplanes were used – camels were to provide the sole means of transport. Motion picture of the expedition was also taken. After prospecting around the Western Australia/Northern Territory border regions in October, and finding Lasseter's diary, this party returned to Alice Springs in late November, carrying no specimens of gold.

Expeditions did not end with CAGE either: Morley Cutlack led several from 1935 onwards; and Hummerston went out in 1934, and even published an article in which he claimed to have found Lasseter's reef. From Cartwright's account it seems that fraudulent claims were often made in order to lure capital from distant investors.[14] In 1936, Hummerston and Cutlack staged a joint motor expedition; this too ended in controversy, with both reporting that they had been the victims of attack by local Aborigines. Ted Strehlow, then a patrol officer for the Native Affairs Branch, was sent to investigate these claims. He found no evidence of any such attack. The promise of a rich reef of gold seems to have been perpetually fuelled by the lies of desperate men.

Lasseter's diary documents something of his life during his last three to four months. Alienated from his expedition mates, Lasseter died among the Pitjantjatjara of the Petermann Ranges area. The final horror of his ordeal is revealed in the diary letter he wrote to his wife, Rene, which was found by Bob  Buck in 1931. Blinded by sandy blight, and delirious beyond repair, he feared a gruesome death:

> I don't suppose I've an earthly chance of surviving. I can carry 2 gallons and 3 pints of water but that is hardly likely to take me the 80 miles to Mt Olga and on no food, whatever. I've brought this all on myself by going alone . . .
>
> Good Bye and God bless you Rene, darling wife of mine and may God bless the children.
>
> Harry.[15]

Rather than another round of conjecture as to the circumstances of Lasseter's demise and the whereabouts of his reef, the last word should go to someone who witnessed his presence in his final days. I do not mean Johns, the dingo scalper, who is readily acknowledged as the last European to see Lasseter alive, but, rather, Eddie Etamintja, a Pitjantjatjara man well into his seventies. As an infant, he met Lasseter sometime late in 1930, at the cave now bearing the prospector's name. I mentioned in passing to Eddie Etamintja that I understood that Lasseter had once travelled through the areas in which we were having a conversation. Etamintja confirmed this enthusiastically, joking that it was Lasseter that had brought the camel (now in the Petermann Ranges area in great numbers) to Central Australia. However, he turned grave when he insisted that Lasseter be 'left alone, we can't dig him up!' He went on to worry that Lasseter would not be looked after by anyone else: 'Who's going to look after him when we're gone?' Lasseter's desperate dreams of gold, which brought

foreigners to the interior, and the search for minerals and their production now dominate foreign investment in the country. In 1878 Australians joined the first rush to a New Guinea goldfield.[3] About 100 miners and two notorious camp followers, Annie Smith and Jessie Ormiston, landed in Port Moresby. It was just five years after Captain John Moresby had told the outside world of the 'many-bayed' harbour, 'surrounded by round-topped grassy hills, having the Australian gum-tree scattered over them, between'.[4] The miners arrived before any internationally recognised government had claimed the area, but they carried their own law with them, declaring that they would try before their peers any men who 'used violence wantonly' against the local villagers.[5]

The miners had responded to a report that Jimmy, a New Caledonian who had worked on Australian goldfields, had picked up quartz spotted with gold. Having prospected the Goldie, Brown and Laloki Rivers inland of Port Moresby, the miners believed they were in good gold country but found little sign of it. By the time the wet season set in at the end of the year, hardly any of them were left in the area. Even Annie and Jessie had returned destitute to Thursday Island. The miners had learnt that even the most experienced north Queensland bushmen had trouble travelling in the foothills of the Owen Stanley Range. Horses were useless beyond the grasslands around Port Moresby, and white men were incapable of carrying the stores needed to sustain them. Malaria was a 'fearful reality': nearly all had suffered, many had been debilitated and a few had died. They were puzzled by the people whose lands they had crossed and whose lands they hoped to mine. Because Papuans had settled villages, kept pigs and planted food crops, they appeared to be landowners, and many miners thought they would have to negotiate with them before entering their lands or turning it over in their search for 'colour'. Some villagers were prepared to work for the miners or to sell food to them, so a range of relationships was possible. But, while some groups of people were found 'agreeable', others were said to be hostile and well armed, and prospectors faced ambush or open aggression.



In the event, the first payable fields were not found under the rainforest canopy or in the headwaters of the wild rivers, but close to the beaches and palm trees of the islands that trail off Milne Bay in the south-east. The islands were more accessible to foreigners, and the islands and their seas contained much that the foreigners wanted: whales, trepang, pearls and men.

Those encounters between ships' crews and islanders were marked by violence. Between 1878 and 1887 there were some seventeen incidents in which foreigners were killed in the islands of the south-east. More islanders than foreigners died in these affrays and some of those deaths took place off shore. In 1885 seventy-one men who had been illegally recruited for work in Queensland were returned to their homes on Sudest (or Tagula) Island. They carried with them thirteen bundles of trade goods to compensate the relatives of the men who had died. That same sad return of young men and bundles took place on other islands too.

Most of the Australians who went north to prospect and mine in New Guinea had already tried their luck on the Queensland fields. From the early

1870s the Palmer was the dominant field and was known for its intense race consciousness and violence.

Given such experience and the recent history of conflict in the islands of south-east New Guinea, sporadic fighting, and even battles, between miners and islanders seemed inevitable. There was not much chance of government officers enforcing peace: Sir William Macgregor and his small staff were arriving to set up the new administration of British New Guinea just as the first finds were being made on Sudest. But when the miners responded to reports that David Whyte, a pearling captain, had discovered gold on the island in 1888, the meeting was generally peaceful. The Sudest sold some food to the miners, although they had little to spare, and worked for them, at first carrying from the beach to the alluvial fields and later working on the claims.

Before miners arrived on Sudest, the violence between islanders and foreigners was only one of the conflicts in the area: a pattern of raid and counter raid between islanders had long prevailed. Indeed, with the greater general turbulence and the introduction of guns, there may have been an increase in warfare between islanders. In this extension of traditional warfare, the Sudest were losers. By 1888 many had retreated inland, but their food gardens and ocean-going canoes had been destroyed, and they were being preyed upon by neighbouring islanders. The Sudest were in no position to resist the miners, and they seem to have realised that if the foreigners stayed ashore the raids would cease. By working for the foreigners and trading with them, the Sudest also had a chance to repair their wealth and strength relative to other islanders. It was in the self-interest of the Sudest to tolerate, even welcome, the miners.



Over the next two years, significant changes took place in the culture of both miners and islanders. In Australia, the miners thought of themselves as individuals, workers and pioneers. In New Guinea, they could continue to claim to be both on and beyond the Australian frontier. They still wanted to think of themselves as workers, but now they were also employers. They were dependent on the villagers to carry for them, and, on the claims, each miner had a 'team' – about ten New Guineans – who dug the water races, shovelled the alluvium into the sluice boxes and worked around the camp. As employers, the miners had to be concerned about wage rates, recruiting policies, and relations between masters and servants. All of this took them a long way from the diggers of Ballarat; but most still thought of themselves as battlers.

Having learnt the techniques of mining, some Sudest began mining for themselves. At first they sold gold to white miners at less than its value, but by 1891 they were entering the stores, watching the storekeeper place it on his scales, and then getting their full value in goods. Some white miners resented the competition and were said to have holed the panning dishes of those Sudest who wanted to work for themselves, but a combination of government officers working to protect Sudest rights, and other factors, ensured that the Sudest continued to mine. Sudest Island was a poor alluvial field: at its height there were only two or three hundred white miners on the island, and within a few years there were only thirty or forty. Miners, who had to meet the costs of transport as well as imported food and wages, needed to recover a lot of gold to cover their expenses; the Sudest, who could live off their subsistence gardens, had few costs. As the amount of gold that white miners sold in the stores declined, the small amounts brought in by the Sudest became more important, and the storekeepers began stocking more of the products that the Sudest wanted – fish hooks, cooking pots, kerosene for lamps, stick tobacco, and the cloth they turned into *laplaps*. Here there could be no alliance of white miners and storekeepers to stop the Sudest from mining the gold on their land.

Even when all of the white miners had gone, the Sudest kept mining. An Australian government officer in 1911 watched a group opening up a new face on a hillside, and talked to other groups who were reworking old ground. They told him that they were finding a few pennyweights – not worth much in cash but more than they could make as labourers on plantations. That transfer in alluvial mining skills was an important accidental consequence on several of the early minor New Guinea goldfields and continued on most of the later mainland fields. It gave an independent source of cash income to villagers, and in the period between the wars and through to the 1960s some of the highest income earners in Papua New Guinea were indigenous goldminers.



Just as Australian miners were deciding that there was not much gold for them on Sudest, prospectors opened a slightly richer field on nearby Misima Island. Sudest and Misima became part of the Louisiade goldfield proclaimed in 1889. Six years later, gold was found on Woodlark Island, and Marua, the second of the island goldfields, was proclaimed. At their heights, both Misima and Woodlark attracted around 400 white miners – few by the standards of Australian goldfields, but a high concentration of foreigners in the islands. One result was a sharp and variable impact on the island populations. For example, at the time that the miners arrived, both Woodlark and Misima were home to around 2500 people. By the 1930s the population on Misima was increasing markedly and reached 3000 before World War II. It was 6000 by 1970. By contrast, on Woodlark the population had fallen to just 800 in the 1920s and showed little increase until the 1940s. Recovery was then slow, but the population had reached 1200 by 1970. That wide variation in population decline and growth was probably repeated on other islands, and possibly on the mainland, although it is not so easily measured. Just why some peoples were more vulnerable to the introduced diseases – dysentery, venereal disease, influenza, whooping cough, and German measles – is uncertain, but the demographic changes had a profound impact on the political and economic balances between communities.

While most miners followed the reports of gold to the islands, prospectors still believed that the big finds would be made on the mainland. After traces of gold were found on the Mambare River on the north coast, George Clark led an expedition inland, but he was ambushed and killed. By 1896 payable gold had been located on the upper Mambare on what was to be called the Yodda goldfield, and Kokoda, the first inland station, was established to administer the area. Few miners reached the Northern Division fields by what later became well known as the Kokoda Track; most used the northern rivers as their points of entry. To reach payable ground they had to walk for several days through difficult, rain-drenched country, and they had to have teams constantly carrying to keep them supplied with basic stores. The labourers who carried or worked on the claims experienced a high death rate. Appalling in the early days, it was still 10 per cent by 1903 when there were around 1000 labourers on the field. Yet what really marked the Gira and Yodda fields was the violence.

Apart from the killing of Clark, at least four miners, the government officer at Tamata on the Mambare, John Green, and his police were killed. The miners' labourers, so much more vulnerable, died in greater and unknown numbers. But the most reported deaths occurred when government officers and their police attempted to extend British and then Australian administrative control across the area. For over ten years from 1895 there were savage encounters as different groups fought and then made their settlements with the official patrols. The officers often made frank reports. In 1901 Alexander Elliott, Assistant Resident Magistrate, wrote to Port Moresby that he and his police and a miner had killed thirty-six men and left another seventeen with their legs broken. And, he admitted, 'there must have been a few more wounded of those who got away'.[6] In such clashes, when twenty, thirty or forty warriors were killed, no officers or police met the same fate. Beyond government control, the miners imposed their own law. In 1908 two experienced miners on the Waria, Frank Pryke and Arthur Darling, retaliated after labourers had been killed. Pryke wrote that they attacked at daylight: 'I had a Lee Enfield, automatic Winchester, and two ordinary Winchesters and a shot gun and Darling was even better fitted out'. Pryke gave no indication of the number of casualties, but he was astonished at the speed with which the people disappeared – he thought that either they were 'used to being surprised or else they train for it'.[7]



In the Northern Division, the miners had confronted the Orokaiva and the related Binandere peoples. These were rich, numerous, aggressive and expanding at the expense of their neighbours. They were constantly forming alliances of clans who joined together in attack and defence. At maturity a young man was presented with his fighting weapons: the long spear, the stone club and the wooden shield wrapped in cane. Most of all, his reputation as an adult depended on his prowess as a warrior, and the welfare, even the very existence, of the group depended on its fighting men. Imbued with belief in the importance of warfare, confident of their own strength, and seeing outsiders as a threat, the Orokaiva were ready to fight intruders. They ambushed and skirmished, but their beliefs, customary battle practices and weapons inclined them to form lines and, armed with axes, spears and shields, charge across open ground. Invariably, they were mown down by rifle fire, but to the surprise and admiration of government officers the survivors would sometimes re-form and charge again.

After the early clashes, the miners and government officers were prepared for violence and they placed less value on the life of a Melanesian than they did on that of a European. The miners and their labourers came suddenly and spread out rapidly, giving officials no chance to extend control gradually; and the beliefs and customs of the Orokaiva compelled them to make disastrous attacks on the well-armed foreigners. Moreover, all through the years when the Orokaiva were fighting the foreigners, they continued with their own warfare. At any given time, they probably thought that their most important long-term relationships were not with the government or any foreigners but with their traditional allies and enemies. Clans attempted to divert the foreigners' rifles away from themselves and at their enemies. They could do this openly with the police, who readily understood the ambitions of particular clans and the significance of the delicate alliances. They could also sometimes manipulate unwitting officials. As soon as clans reached an accommodation with the government patrols, they usually tried to get some of their own group recruited into the police – that was a ready transfer of the warrior tradition into the new ways. Once a group had a representative in the police, it could more easily protect itself from government vengeance and make sure that the destructive power of the foreigners was aimed at its enemies.

When the miners opened up a new field in 1909 on the Lakekamu River,  inland from the Gulf of Papua, they entered the southern edge of the country of Anga-speaking peoples. To their Papuan neighbours and the miners, they were known as Kukukukus. Small rainforest-dwellers, they wore a characteristic bark cape and seemed always to be armed with bows, arrows, bone knives and stone clubs. They roamed widely, often making sudden surprise attacks on distant peoples – a quick dash into a sleepy village, killing and plundering and disappearing before any counterattack was possible. The Kukukukus were rightly feared.

The miners knew of their reputation and were inclined to fire a few shots over their heads if they saw any of them. The Kukukukus responded by sometimes launching arrows from concealed positions into the miners' camps – but the bowmen were rarely seen. Isolated prospectors were vulnerable to the Kukukukus, who ambushed them or made dawn raids on their camps, clubbing, knifing and looting. Over the next twenty years they killed at least three miners and inflicted severe wounds on others. But because of their style of warfare, and because they quickly disappeared into distant mountainous country, the Kukukukus did not suffer the battle deaths of the Orokaiva.

In 1902 a few miners began working claims in the Keveri Valley, about three days walk inland from Cloudy Bay on the south Papuan coast. High enough to be cool in the evening, Keveri with its timbered grasslands, ridges, fast-flowing clear streams and surrounding mountains became the miners' happy valley. Although Keveri was another poor alluvial field, experienced miners with a good 'team' of labourers could work in the area for several months and perhaps collect a hundred ounces or more of gold – enough to cover expenses and leave something for a trip 'south'. The Bauwaki language people of the area, usually known simply as the 'Keveri', lived in small stockaded clusters of houses on the ridge-tops. They were constantly feuding, but those involved in raid and counter raid were often small parties, and

**Figure 12.1** A Kukukuku: Dreaded Warriors of the Mountains, September 1942. (National Library of Australia)

warriors might choose an isolated victim, rather than attempt an open attack on another war-party or a defended village. A miner's labourer who wandered off was at risk, and so was a labourer who chose to desert, but generally there was little conflict between the miners and the Keveri. As the miners worked the stream beds and the Keveri were on the ridges, there was no immediate conflict over resources; and, more importantly, the miners' self-interest required peace with the villagers.

Because there were rarely more than ten white miners in the valley, storekeepers did not set up in the area. The miners therefore had to provision themselves and their labourers. They could bring in basic stocks, but to stay for long periods they simply had to be able to buy fresh garden foods from the Keveri. The miners had no choice but to establish and maintain a reputation as trustworthy trade partners. If any miner or labourer disrupted the commerce between camp and village, all miners were likely to suffer. When government officers made their rare visits into the valley and attempted to arrest those Keveri men accused by their neighbours of murder, the miners condemned the officers for harrying the villagers and disrupting trade. George Arnold, a lonely miner in the valley in 1912, took a noble if misguided stand on behalf of



the village where he had been trading and shot one man in a raiding party. This was the only recorded shooting of a Keveri by a miner.

Many of the same men mined in the Northern Division, and on the Lakekamu and the Keveri. On all of those fields they encountered peoples with strong warrior traditions. In the Northern Division, this produced open clashes and the Australian and British officers, especially those who had been in the Boer War, used the language of war to describe what was happening; on the Lakekamu, surprise attacks and sniping were sustained over twenty years; and in the Keveri Valley, the people traded with the miners, sometimes agreed to carry for them and fought each other. This is a simplified statement of complex relationships, but it reveals a pattern in which two interacting factors were critical: the modes of customary warfare among the villagers, and the economic self-interest of the miners. Where miners needed villagers, they established mutually beneficial and peaceful trading relations. And the Sudest, finding it in their own interests to secure an alliance with the miners, were as effective as the most mercenary of Europeans in securing a peaceful meeting. In spite of spectacular clashes with foreigners, many villagers thought that relationships with traditional allies and enemies were more important. In most cases, they were right: the foreigners went away and the well-being of local peoples was largely determined by the strength of their group, by its capacity to obtain and hold resources and to gain access to the new opportunities opening in the cash economy.

The miners, like many people encountering the unfamiliar, often reduced the people they met to stereotypes, and this intensified the differences in their various local relationships. They came to think of the Orokaiva as warriors, a people whose aptitude for war might one day make them valuable assets of empire, but they believed the Kukukuku were deceptive and unpredictable in their violence. A Kukukuku who approached a miner's camp hoping to trade, or even to be given work, was likely to have the dogs set on him or to receive a blast from a shotgun. The labourers, who shared the miners' presumptions about the Kukukuku, were inclined to join the general alarm and hostile reaction.

In the same way, realities exaggerated by prejudice influenced the miners' relations with labourers and with Papua New Guinean women. The labourers were indentured – hired on one-year contracts and paid only 10 shillings per month plus all costs of clothing, food and shelter. At the end of his contract a labourer could hope to take back to the village £4 or £5 worth of trade goods, and returning labourers were important in transforming the economies of the villages from stone to steel. The miners recruited labourers from the western border with Dutch New Guinea to the islands of the south-east, but they concentrated recruiting on the D'Entrecasteaux Islands, particularly Goodenough Island. By the 1920s, in some of the Goodenough Island villages, all of the men had spent time away as indentured labourers. The Goodenough, known as Gosiagos, believed that a good man was one who went early to his garden, worked hard, consumed little and demonstrated his skill and generosity when he and his family brought in a rich harvest.[8] They needed no instruction in the work ethic. It became almost a rite of passage for young Gosiago men to go away before marriage to work on the mines or the plantations. By contrast, men from the nearby Trobriand Islands rarely put their mark on a contract to go and work for a miner. Yet when miners were looking for women to live with them in the camp, they could sometimes persuade a Trobriand woman. Women from Milne Bay, Misima and the Trobriands conformed more to the miners' sexual aesthetics; these women also came from communities that allowed them some freedom and which were tolerant of their forming sexual relationships with men from other places.



When diseases such as dysentery swept through mining camps and killed many labourers, or when the labourers were the victims of violence on the goldfields, government officers took boxes of trade goods and notices of death back to the villages. Few of those officers remained untouched by the villagers' questions about the young men who were dead or had disappeared, and by the distress of the relatives. Especially in the early years, the Gosiagos paid a high price for their reputation as hard workers, while the Trobriand Islands, the 'islands of love' in the romantic literature of the south seas, suffered a high incidence of venereal disease.

Little gold had been found in German New Guinea before World War I, but, once the country came under Australian control, prospectors were keen to try the new ground. In 1926 at Edie Creek they made the rich strike that they had always believed was concealed by rainforest, mist and mountain. Upper Edie was 7000 feet above sea-level, and trees and rocks were moss-covered. New Guinean men from the coast, who had never known cold, wrapped themselves

**Figure 12.2** Miners Putting in a Sluice Box on One of the Workings at Edie Creek. (H. L. Downing Collection, Mitchell Library, Library Council of New South Wales)

in blankets and crouched over fires. Edie was a six-day walk from Salamaua, the port that grew rapidly to service the new goldfield, and by the end of the year more than 200 white miners and 1000 New Guinean labourers were on the field. In the thatched hotel at Salamaua, the champagne corks were nailed to the ceiling, souvenirs of the record shouts made by successful diggers. Recruiting fees for labourers rose from around £5 or £7 10s a man to £20 or £25, but the labourers themselves rarely benefited from their scarcity – their wages remained at around 10 shillings a month. Errol Flynn was one of the adventurous, opportunist Australians who tried his luck as a recruiter on the Morobe field.[9]

The rich alluvium on Edie Creek, the first of the New Guinea goldfields to give real wealth to the successful diggers, was just the beginning of the development of the Morobe field. Shafts and drives were soon sunk into valuable lodes, and crushers installed. Also, the gold-bearing creeks – Edie, Koranga and others – had long been washing silts and gold into the Bulolo Valley, leaving millions of cubic yards of alluvium that was too poor for individual miners but rich ground for dredges. Placer Development, a Canadian-based company with much Australian capital, set up Bulolo Gold Dredging to exploit the valley leases.[10]

**Figure 12.3** Indentured Mining Labourers Wearing Their Sou'westers.
(H. L. Downing Collection, Mitchell Library, Library Council of New South Wales)

From the opening of the Morobe goldfields, transport was the basic problem. As returns were high and the demand immediate, aeroplanes were seen as a short-term solution. In 1927 the first aircraft began flying regularly from Lae to Wau, the town that was developing to serve the growing community on the field. When Bulolo Gold Dredging finally committed itself to mining the river flats, it took the bold decision to rely on aeroplanes to fly in all parts for its dredges and hydroelectricity plant. The giant dredges were dismantled so that no part weighed more than 7000 pounds, and then the tri-motored, all-metal Junkers aircraft lifted them across the mountains where they were reassembled. Nearly all other supplies – cars, motor bicycles, furniture, kegs of beer, bags of rice, even the horses for a race meeting – travelled by aeroplane into Wau and Bulolo. Guinea Airways, the main airline operating out of Lae, carried 2000 passengers in 1930 and nearly 10,000 in 1935. By then, Lae could claim to be the busiest airport in the world.[11]

Miners spreading out from Edie found gold in other creeks and were soon testing ground well beyond the Morobe District. Prospectors, such as the Leahy brothers, made pioneering patrols through the valleys of the central highlands,[12] but no strike challenged the richness of Edie and the Bulolo. By 1940 the Territory of New Guinea was exporting over £3 million worth of gold, nearly all of it from Morobe. Gold shielded New Guinea from the impact of the world depression that devastated the plantation industry. By 1940 more than 7000 New Guineans were under indenture on the goldfields. Working in Bulolo, Wau, high on Mount Kaindi or on the claims spreading out along the Watut to Surprise Creek became a common experience for young men, often recruited from the distant Sepik District.[13] Most were flown from the coast to the goldfields: the early airlines charged one shilling per pound weight for the human cargo.

In the post-war period, some of the dredges began churning again, but Bulolo Gold Dredging never reached pre-war levels of profitability: production began to decline after 1954 as the richest alluvial was worked out. The World Bank's comprehensive report on the economic development of Papua and New Guinea in 1964 concluded that 'the outlook for the development of mineral resources [was] not encouraging'.[14] It then looked as though most of the gold would soon be produced by Papua New Guinean miners. By the mid-1960s some 3000 of them were panning and sluicing on the goldfields, many supporting themselves as subsistence farmers just as the Sudest had done seventy



years earlier.

Yet, even as a draft of this report reached the Australian government, geologists employed on Bougainville by Conzinc Rio Tinto of Australia were testing silts and sinking drills into ground close to where miners had worked small gold claims in the 1930s. Before the mine at Panguna eventually began production in 1972, a workforce of more than 2000 foreigners and 5000 Papua New Guineans was employed, a port and a town built, roads laid down, and thirty million tons of overburden lifted to expose the ore in the open-cut crater.[15] The gold and copper mine at Panguna was then bigger than any mine in Australia. The Ok Tedi mine, high in the Star Mountains near the western border of the mainland, began production in 1984.[16] Misima, redeveloped on the old sites as a modern open-pit mine, opened in 1989, followed by Porgera in Enga Province in the highlands in 1990, and Lihir off the coast of New Ireland in 1997. In addition, there was the spectacular chaos of Mount Kare.

On the border of the Enga and Southern Highlands provinces, on the wet, cloud-covered slopes of Mount Kare, too high and swampy to garden, CRA geologists established a camp. But from early 1988 Papua New Guinean miners began moving into the area, and, as news spread that the field was rich, concentrated and shallow, the last great Australasian alluvial gold rush began. Papua New Guineans, equipped with shovels, crowbars and panning dishes and standing knee-deep in slush, worked tiny claims, sometimes with mates guarding the margins against interlopers. At the height of the rush, somewhere between 7000 and 10,000 people crowded the field; there they found around $3 million worth of gold. Where Edie had been serviced by aeroplanes, Mount Kare had helicopters, the hopeful and the successful coming and going by chopper taxis. Some visitors thought they were looking at a great democratic harvest, a triumph of small and local miners; others saw anarchy and

**Figure 12.4** Old Mess Hut at Edie Creek Showing the Warden's Quarters, the Courthouse and Police Headquarters in Foreground, 1929. (H. L. Downing Collection, Mitchell Library, Library Council of New South Wales)

violence, where the strong triumphed over the weak. Nearly all of the easy gold was worked out within two years: the blue plastic sheets that had provided temporary shelters were furled; the gold buyers, prostitutes and food sellers closed their businesses; and the gold-bearing lands along Pinuni Creek and on the upper Gewa River consisted of churned mud and exposed rocks.[17]

The mines, large by international standards, have generated much investment in Papua New Guinea (Lihir had a start-up cost of a billion dollars). Minerals have been providing some 70 per cent of exports, and much government revenue has been derived from direct investment in the mines and from various taxes. Many individual Papua New Guineans have benefited from the mines. They have been trained to high technical levels, and some have moved into senior management positions. Subsistence farmers have found markets for their produce, and villagers have been given access to services that governments have been unable to deliver to remote areas. Measured by basic indicators such as literacy, life expectancy and child mortality rates, life for people close to mines is often better than it is for those in isolated areas without mines.

**Figure 12.5** Guinea Airways advertisement, in *Pacific Islands Monthly*, 15 August 1939.

But the mines are also enclaves of massive capital investment and the world's most advanced technology. Relative to their significance in exports and government revenue, they employ few staff. And those goldmines come at great environmental, social and political cost. The scale of the transformation of the landscape has been beyond imagination. The Ok Tedi mine annually sheds some thirty million tons of tailings and forty million tons of rock waste into the Ok Tedi River. Such quantities inevitably overwhelm the country immediately below the mine, having an impact over hundreds of kilometres beyond the junction with the Fly River and down to the sea. Add to this the pollution from metals and refining chemicals released into waterways, the threat and reality of lawsuits, and the clamouring demands for compensation payments, and the balance between costs and benefits becomes less clear.

From the early attempts to regulate mining in British New Guinea and Australian Papua, governments recognised that Papuans were landowners with rights. The transferred Queensland Mining Act was amended to include conditions compensating 'native' owners for damage to their lands. If mining was likely to cause damage to a village, the owners were not only to be compensated but were also required to give their consent. Although there is little evidence that any Papuans or New Guineans were ever consulted or compensated on any fields before World War II, those clauses recognising indigenous ownership north of Torres Strait were significant. By the time that the transnational companies began developing the mines that led to the resurgence of the industry in the post-war period, greater consciousness of indigenous land rights had emerged and Papua New Guineans were moving into positions of authority and could influence negotiations.



The contested and revised agreement for the Panguna mine on Bougainville was, for the time, a generous settlement. In fifteen years of operation, the mine paid in taxes and dividends more than 1000 million kina (the kina was then worth close to one Australian dollar) to the central government, 75 million to the provincial government, and 24 million in royalties and compensation to the local landowners. But the two million kina that the company paid annually to the landowners had to be divided among over 5000 title holders and other claimants, allowing just a few hundred kina to an average beneficiary.[18] Perhaps more importantly, local people claimed that mistakes were made in the recording of landholders, families had difficulty ensuring that the money received was distributed equitably and used efficiently, and people living just outside the designated area received nothing. In practice, the compensation payments fuelled resentment. The impact of the mine on the environment, the bringing of black and white outsiders into the area and rewarding them with high wages, the apparently small and arbitrary payments against reports of millions in total earnings, and the incapacity of the clans to disperse funds fairly, all influenced the sense of isolation and difference felt by Bougainvilleans. In 1988 a small group of local people stole explosives from the mine and used them to destroy mine installations. The next year the mine was closed and a ten-year sporadic war was under way: Bougainvilleans fought

**Figure 12.6** A Papuan Miner, photograph in government report, 1950s. (Source unknown) (Collection of Hank Nelson)

for secession from the central Papua New Guinea government, and they fought each other for the direction and leadership of Bougainville.

For at least 500 years, gold has been important in attracting foreigners to the island of New Guinea. It still is. Where the world's most advanced techniques and giant machines are used in unstable, rain-drenched country, rarely touched by more than axe and digging stick, where billion-dollar investments are made among people so lightly connected to cash economies, where transnational companies deal with governments with less cash and expertise, then the contrasts between the costs and benefits must be extreme. Without a mine, women may continue to die in childbirth, infants may remain below expected rates of growth, illiteracy may stay high, and few people may gain opportunities to earn cash and buy the goods that the rest of the world consumes as a matter of course. With a mine, they may face ecological, social and political disaster.



## otes

1 The island of New Guinea has been divided into various political entities. In 1884 the south-east became British New Guinea and the north-east, German New Guinea. In 1906 British New Guinea became the Australian Territory of Papua, and in 1914 the Australians captured German New Guinea which was administered by an Australian military force until the League of Nations gave to Australia the mandate to administer the territory. After World War II, Australia combined the territories of Papua and New Guinea under the one administration, and just before independence in 1975 all of eastern New Guinea became Papua New Guinea, and Papua New Guinea became the name of the new nation. Western New Guinea was Dutch or Netherlands New Guinea until 1962, and from 1963 it was effectively West Irian and a province of Indonesia. In 2000 the name of the province was changed to West Papua.

2 J. L. Whittaker et al., *Documents and Readings in New Guinea History: Prehistory to 1889*, 1975, Jacaranda Press, Brisbane, p. 183. See also G. Souter, *New Guinea: The Last Unknown*, 1963, Angus & Robertson, Sydney. De Saavedra had probably landed on Biak.

3 H. Nelson, *Black, White & Gold: Goldmining in Papua New Guinea 1878–1930*, 1976, Australian National University, Canberra, gives a general account of the early strikes and provides notes on the sources used.

4 Captain John Moresby quoted in B. Jinks, P. Biskup and H. Nelson (eds), *Readings in New Guinea History*, 1973, Angus & Robertson, Sydney, pp. 8–9.

5 Nelson, *Black, White & Gold*, p. 78.

6 Nelson, *Black, White & Gold*, p. 123.
7 Nelson, *Black, White & Gold*, p. 140.
8 Michael Young has written extensively on the Goodenough Islanders, for example *Fighting with Food: Leadership, Values and Social Control in a Massim Society*, 1971, Cambridge University Press, Cambridge.
9 J. H. Moore, *The Young Errol: Flynn before Hollywood*, 1975, Angus & Robertson, Sydney, pp. 42–4. He kept, or probably made up, a diary about recruiting (pp. 108–22).
10 A. M. Healy, *Bulolo: A History of the Development of the Bulolo Region, New Guinea*, 1967, New Guinea Research Bulletin no. 15, New Guinea Research Unit, Australian National University, Canberra. There are popular accounts of the gold strikes, for example I. Idriess, *Gold-Dust and Ashes: The Romantic Story of the New Guinea Goldfields*, 1933, Angus & Robertson, Sydney.
11 I. Willis, *Lae: Village and City*, 1974, Melbourne University Press, Melbourne; and J. Sinclair, *Wings of Gold: How the Aeroplane Developed New Guinea*, 1978, Pacific Publications, Sydney.
12 B. Connolly and R. Anderson, *First Contact*, 1987, Viking, New York, and film of the same name.



13 The annual reports (*Report to the Council of the League of Nations of the Administration of the Territory of New Guinea*) give annual production of gold and the numbers of people employed in the industry.
14 *The Economic Development of the Territory of Papua and New Guinea*, 1965, International Bank for Reconstruction and Development, Johns Hopkins Press, Baltimore, p. 222.
15 D. Denoon, *Getting under the Skin: The Bougainville Copper Agreement and the Creation of the Panguna Copper Mine*, 1999, Melbourne University Press, Melbourne, p. 146.
16 R. Jackson, *Ok Tedi: The Pot of Gold*, 1982, University of Papua New Guinea, Port Moresby; and G. Banks and C. Ballard (eds), *The Ok Tedi Settlement: Issues, Outcomes and Implications*, 1997, National Centre for Development Studies and Resource Management in the Asia-Pacific, Australian National University, Canberra.
17 P. Ryan, *Black Bonanza: A Landslide of Gold*, 1991, Hyland House, Melbourne; and C. Filer, 'The Melanesian Way of Menacing the Mining Industry', in B. Burt and C. Clerk, *Environment and Development in the Pacific Islands*, 1977, Australian National University and University of Papua New Guinea, Canberra, pp. 91–122.
18 Denoon, *Getting under the Skin*, p. 193; and S. Dorney, *Papua New Guinea: People, Politics and History since 1975*, 1990, Random House, Sydney, ch. 5.

# *Jukurrpa* – Golden Dreams

Chapter 13

*Derek Elias*



In the Tanami Desert of the Northern Territory, more than $3 million per year is currently distributed to Aboriginal communities and individuals. This money, primarily paid out by mining companies, flows from legal agreements made with traditional landowners through their representative body, the Central Land Council. Under the Aboriginal Land Rights (Northern Territory) Act 1976, there are provisions (sections 35 and 64) for financial payments to be made to the traditional owners of areas associated with, or affected by, mining and exploration. In the Tanami Desert the majority of traditional owners identify themselves as Warlpiri, or a sub-group thereof, and exploration and mining are focused exclusively on gold. 'Golden Dreams' refers to a common equation made by Aboriginal people in the larger communities (Lajamanu and Yuendumu) of the Tanami Desert between gold (transformed into royalty money) and certain sacred places or 'Dreaming tracks and sites' (*jukurrpa*). It is the confluence of goldmining, royalty monies and sacred places that constitutes the background of this chapter. It traces the historical development of the search for gold in the Tanami Desert, fuelled by the dreams of wealth of the prospectors and miners who went there.

The encounter between the Warlpiri and the predominantly European explorers and prospectors in the central Tanami Desert was an unusual one in the broader context of the exploration and colonisation of Australia. The reasons for this are related to the fact that, because of their relative geographical remoteness, the Warlpiri were not to fall under the direct authority of the Australian state until the middle of the twentieth century. The Tanami Desert was the scene of some of the last incursions by colonisers into the heart of the continent. The arrival of these explorers and prospectors marked the culmination of a long process of 'opening up' Australia at the turn of the last century. From that time, the pursuit of gold played a key role in the development of relations between the Warlpiri and the rest of Australia.[1]

In the Northern Territory, a large-scale and comparatively rapid expansion of the pastoral industry was under way on the better grasslands to the north of the Tanami Desert. One of the South Australian Government Residents' reports on the Northern Territory emphasised the strained and violent contact between Aboriginal peoples and pastoralists and indicated that occupation of the land by both groups was 'hopelessly irreconcilable'. Its author wondered how the 'real property' interests of the Aborigines could be preserved.[2] The Warlpiri were to avoid this collision over land use because of the unattractiveness of the Tanami Desert, which had been reported to be an area of limited potential for stock grazing. The Tanami had few permanent water supplies and scant, unpredictable rainfall, even though the land on the fringes of the desert was taken up by pastoral properties.

The first contact in the Tanami Desert between the Warlpiri and the outside world commenced with Colonel Peter Warburton's exploration expedition in the 1870s.[3] The party was driven by the colonel's interest in finding and using Warlpiri water supplies, the lack of which nearly cost him and his men their lives on several occasions. The primary objective of such exploration parties was to assess and report on economic resources, in particular water, while surveying routes through the uncharted interior. The next significant exploration of the Tanami Desert was led by Allan Davidson in 1900, who was accompanied by two prospectors, a camel tender, an Aboriginal guide, nine bull camels and six months of rations.[4] The lasting consequence of this expedition, almost thirty years after Warburton's journey, was the discovery of gold at both Tanami and the Granites.[5]



Although Davidson reported finding gold at Tanami and the Granites, he was of the opinion that it was not in sufficient quantities to make it a viable proposition. This qualification did not discourage numerous small parties of prospectors, many of whom had given up on the Halls Creek field, over the border in Western Australia, from travelling to the area. Between 1902 and 1908, prospectors who visited both Tanami and the Granites obtained some good results, and a makeshift well was sunk at Tanami in 1908. In 1909, the area attracted interest from the South Australian government, which decided to send a geologist, Mr H. Brown, to the area to comment on potential gold reserves. Brown's favourable report was followed by a small discovery at Tanami, and the combination of these factors led to a mild rush in that year.[6] W. Braitling, who was present at the rush, told Mervyn Meggitt that at one time there were 500 men on the field.[7] The presence of such large numbers of non-Aboriginal people in the Tanami Desert proved to be short-lived.

As a result of increasing interest in both Tanami and the Granites, Lionel Gee was appointed as Mining Warden and stationed at Tanami in 1909. By 1911 he was overseeing the prospecting of about sixty men, and the search for gold had extended to all of the country throughout the region. The best results were obtained at the Granites, and the government encouraged further development by having wells sunk along the tracks to Tanami from both Halls Creek and Wave Hill. The government well was sunk at Tanami in 1911. Gee's report on the Tanami goldfield commented that the prospectors were encouraged by the results and that they were keen to remain as long as there was water available.[8]

Little interaction with Aboriginal people was reported during this initial period of consolidation by the prospectors, but in 1910 a prospector, Stuart, was killed at a soakage (Aboriginal well) near the Granites by an Aboriginal man, and his equipment was stolen. This incident did little to encourage more than the already limited interaction between the Warlpiri and prospectors, but oral histories collected by Dick Kimber indicate that violent reprisals in the wake of the death of Stuart were far more extensive than officially reported.[9] In discussions about the incident at the Granites, Meggitt's informants in Lajamanu indicated to him that 'most of the tribe avoided the place for years to come'.[10]

The initial interest in the Tanami–Granites region soon waned. In 1914 the Reverend Wilkinson reported that there were only a few prospectors scattered about the field (mainly at Tanami).[11] There can be little doubt that the violence that had occurred at the Granites was also a major reason discouraging a more permanent presence of Warlpiri people in the area. Other factors that contributed to their departure were related to broader demographic shifts brought on by the development of the pastoral industry. In the 1880s, pastoral settlement was taken up in the Sturt Creek and the Stirling, Ord and Victoria River areas to the north and west of the Tanami Desert. In this region the killing of Aboriginal people, and the heavy death toll inflicted by introduced diseases, produced acute shortages of labour on pastoral stations. The combination of dwindling interest in the gold workings at Tanami and the Granites, and the severe drought of 1924–29, exacerbated this situation. During this period many Warlpiri were forced to leave the centre of the desert to search for food and water. Parties travelled to the cattle stations to both the north and east, increasing their contact with the pastoralists with whom they had had little interaction, but only to provide their labour for very brief periods in order to acquire goods.

It was at this time that the relatively well-watered country around the Lander River immediately to the east of the survey area was the scene of the last two punitive expeditions led by police in Central Australia. At least 100 people were killed in the Cockatoo Creek and Lander River area, an incident that was to become known as the Coniston Massacre.[12] This catastrophe, and its subsequent demographic impact, triggered a wholesale uprooting process that began in 1928 for the Warlpiri of this area and which ended only with the establishment of the communities.[13] The profound changes caused by the drought drew the majority of Warlpiri out of the Tanami Desert and into prolonged, frequently disastrous contact with pastoral and settler society on its fringes. For those who remained, the limited resources of Tanami and the Granites that could be obtained from the remaining prospectors became of even greater social and economic significance.

**Figure 13.1** An Aboriginal Man from the Granites, in Chains for Alleged Theft of a Bag of Flour, in *Adelaide Advertiser*, 9 December 1932. (National Library of Australia)



The demographic shift signalled the beginning of a twenty-year process that was to concentrate the majority of the Warlpiri at Tanami and the Granites and fundamentally change their day-to-day interaction with all of their other important sites. The prospectors' pursuit of economic wealth in the form of gold resulted in the sudden infusion of new meaning of a distinctly economic character into certain Warlpiri places. These subtle shifts were brought about by the attention given by both Warlpiri people and the wider Australian society to the Granites (*Yartulu Yartulu*) and Tanami (*Jarnami*). For Australians, these regions that possessed gold deposits became significant in what was, for them, the vast background emptiness of the Tanami Desert. Yet these were just two places, among thousands, of religious and social significance for the Warlpiri. In the aftermath of white occupation in the region, the Granites and Tanami became important centres for the Warlpiri because of the access they afforded to highly desirable new goods, such as tobacco, flour, blankets, tools, tinned meat, tea, sugar and, most significantly of all, a permanent water supply.

In 1926 the first large-scale attempt at mining gold in the Tanami Desert was made. The Tanami Gold Mining Company formed in Perth and brought the first mechanical stamp battery and gas engine to Tanami by donkey teams (Plate 13.1). The battery was to operate for almost two years before the combination of lack of water and poor results led to the abandonment of the venture. The venture was in part stimulated by the reported discovery of a rich lode by a lone prospector, Jimmy Wickham, who was scouring the land to the west and south of the Tanami–Granites region. 'Wickham's Find',[14] as it became known, rapidly achieved legendary status among prospectors as well as catching the attention of the wider public. According to Michael Terry, Wickham 'came to Wyndham in 1925 with a sugar bag full of specimens carrying such rich gold that the story book tales fade out of the picture'.[15] The myth of 'Wickham's Find' was to inject substantial energy into the flagging activities of prospectors in the Tanami Desert.

Thomson's 1926 expedition was the most organised and well documented of the post-Wickham prospecting expeditions and was to name many significant sites in the Tanami Desert. Yet the closest the party ever got to 'Wickham's Find' was to locate some trees marked and initialled by the man himself. Thomson reported that all of the 'natives' encountered were friendly and that his party made frequent use of Warlpiri assistance: one man was given a tomahawk for helping to find water, and on another occasion men were given tobacco for demonstrating the use of a bullroarer.[16] Apparently, the Warlpiri people they met were quite familiar by this time with the intruders' obsession with gold and the party was encouraged to head south. They were informed that 'white fella catchem gold one time'. As one of Thomson's party observed, this comment was most likely a reference to wolfram mining on Mount Doreen: 'Blacks always associated anything in the prospecting line with gold'.[17]



Thomson's expedition was the first of many in search of 'Wickham's Find'. They were all to end in often-fatal failure. At the Granites, Thomson met Terry, and both men concluded that the Tanami–Granites area needed extensive prospecting and bulk sampling. A short time later, in 1930, Terry was to make the original Tanami Road between Halls Creek and Alice Springs through the Granites and Tanami. Older Warlpiri men at Lajamanu refer to the Tanami Road as the 'white man's Dreaming track' because it links up the white man's major places – the towns and the mining areas.[18]

The drought finally broke in 1929–30 and, with the assistance of the new track and the reports of 'Wickham's Find', there was an increase in the number of prospectors working both Tanami and the Granites. Despite a few good finds, by 1932 there were only a small number of men left once again. This situation altered dramatically in mid-1932, when one shaft at the Granites suddenly produced very high returns, sparking the 'Granites Rush' and swelling the number of people on the field to over 200 men. Within the broader economic climate of the time, it has been noted that the period was ripe for the atmosphere of a gold rush, given the worldwide depression, high unemployment and an increase in the price of gold.[19] The Granites rush generated the second serious large-scale goldmining venture in the Tanami Desert: this one was led by Charles Henry Chapman from Queensland and chronicled by journalist F. Eric Baume. Baume's accounts of the Tanami Desert were popularised in a book and a series of articles that recounted stories of prospecting, tragedy, adventure, life in the desert, encounters with Aboriginal people, and, most importantly, the search for gold.[20]

The Granites rush, like all previous gold rushes in the central deserts, was soon over. Yet it was important because it brought the first official policy that addressed the presence of the Warlpiri in the Tanami Desert. The circumstances surrounding the Coniston Massacre and the actions of the Warlpiri on the fringes of pastoral settlement during the drought in the 1920s meant that they were now feared as violent and unpredictable cattle spearers. It was also reported that many of the Warlpiri were now suffering from introduced venereal disease, which suggests that sex may have been exchanged for access to goods and rations. For such reasons, the Warlpiri were officially required to remain at least ten miles from the miners' camps.[21] Up until this time there had been very limited involvement of the Warlpiri in the goldfields of Tanami and the Granites, and their interaction with the miners had been confined to occasional exchanges of goods for labour.

The impact of the Granites rush on the Warlpiri and their environment was profound. The policemen stationed there at the time pursued a policy of open harassment of particular groups.[22] The prospectors, who were having little luck, wandered throughout the region with their camels and exhausted a great deal of the Warlpiri's valuable water supplies in order to maintain their prospecting activities.[23] The Warlpiri were increasingly resentful of this tremendous disruption of the fragile economy of the Tanami Desert, and it is clearly documented that they were under the impression that they had been invaded and it was high time for the interlopers to depart.[24] This attitude of the Warlpiri was to alter markedly in the years following the Granites rush.



By 1933 only the Harris and Chapman families and a handful of other prospectors remained on the goldfield. They worked among several thousand Warlpiri people in the Tanami Desert. The Warlpiri had become increasingly desirous of the goods that could be obtained and the safety net provided by the permanent water supplies to be found at Tanami and the Granites. The void left by the departure of the majority of prospectors and administrative personnel opened up employment opportunities for the Warlpiri that were taken up willingly. Doubtless, the diseases and reports of violence that accompanied such association were discouraging, yet there was a subtle shift in attitude by younger Warlpiri men in particular.

It was at this time that the current eldest generation of Warlpiri people remembers moving to the Granites to work as miners. Victor Simon Jupurrurla's father, Pilala Jakamarra, was encouraged by Chapman to bring Warlpiri people in to stay at the Granites and work. The Warlpiri exchanged their labour for such goods as flour, tea, clothes and tobacco and established permanent camps next to both the Granites and Tanami mining operations. Men such as Henry Anderson Jakamarra and Abie Jangala recall a life of hard work in the mines, interspersed with long visits to their relatives who preferred to remain in the desert.[25] In summertime the men would take goods out to their families and then participate in ceremonial cycles before returning once more to the

Granites or Tanami in order to work digging for gold, gathering wood and carting water in order to accumulate new goods.

Although the Warlpiri worked hard for the new commodities, they had little idea of the purpose of extracting gold, as Ronnie Lawson Jakamarra explained:

> They cooked gold in a fire and made it square like they cooked bread. We couldn't understand, what is that one? Are they going to eat it? We didn't understand gold in the early days, we know now. I worked right up until dawn, I don't know what for, just for tucker, trousers and shirt, no money, we worked flat out, people on both sides, Granites and Tanami. We never understood the kardiya (white) law, he never told us – see this is gold, we really like it – Chapman and Harris, they never told us like that.

It was during this decade after the Granites rush that the Warlpiri of the central Tanami Desert were finally to congregate in such numbers that the government was forced to intervene. A major factor that fostered the growth of the relationship between the Warlpiri and prospectors was the amendment in 1933 of the Aboriginals Ordinance 1918. This ordinance aimed to protect the  welfare and employment of Aboriginal people in the Northern Territory with respect to minimum wages for maintaining a person and dependants.[26] It also addressed health and moral issues. In the area under consideration, permits to employ Warlpiri at the Granites were issued to prospectors from 1933. Chapman, who finally left the Granites in 1953, reported in his newspaper that during his period of twenty-one years at the Granites he had used mainly 'native' labour.[27]

The use of Warlpiri labour by the miners was overseen by patrol officers of the Aborigines Branch, later to become the Native Affairs Branch. Part of the officers' brief was to inspect the welfare and working conditions of Aboriginal employees. Strehlow's and Sweeney's patrols regularly noted that there were often serious abuses by the miners regarding the proper issue of rations in return for labour in the Tanami Desert. The activities involved in goldmining were heavily labour-intensive. The paradox was that, despite the exploitative nature of a relationship that weighed heavily in favour of the miners, the Warlpiri nevertheless wanted them to remain because of the rations and water supply that could be had.

Sweeney's 1944 patrol observed about 100 Warlpiri living in the Tanami–Granites region and noted that, with respect to the Tanami Desert, the younger Warlpiri wanted to stay and 'have no desire to return – they speak of it as "hungry country"'. He acknowledged, however, that in the mining area the older people are still drawn to their old tribal hunting grounds when 'seasons are good'.[28] Sweeney also recommended that rations be issued by the Harris family at Tanami. The decision by the government to issue rations was to prove the catalyst that would culminate in the forced removal of the Warlpiri from the Tanami Desert to permanent settlements on its fringes.

The Harris family distributed rations at Tanami for approximately a year, and when they left the Tanami Desert in 1945 these duties were taken over by Patrol Officer McGarry of the Native Affairs Department. McGarry found the Warlpiri people living at Tanami in 'appalling conditions' and he predicted that the issuing of rations would attract up to 250 people, whose demands would outstrip the resources of the local area.[29] Almost immediately, the number of Warlpiri people gathered at Tanami swelled to 163, putting enormous pressure on the water supply. Frightened of losing the water, McGarry soon moved to the Granites and its more reliable bore. This relocation resulted in a massive influx of Warlpiri to the area. The simultaneous and competing demands on the water supply by the miners and the Warlpiri meant that both interests could not be accommodated. This would prove to be the crucial moment of history in the Tanami Desert and for the Warlpiri.

In 1946, a decision was made to transport all of the Warlpiri people living at the Granites to a newly established ration depot that was to become the community of Yuendumu. The settlement began with an initial population of 377, with another large group of Warlpiri living at Mount Doreen. The economics of the situation were clear enough to the government. It was decided that the miners, who required only the water supply and needed little management, could remain in the Tanami Desert where they were producing gold. The Warlpiri on the other hand, who supplied only labour and whose occupation of the desert produced nothing else of economic value, required food and medical care. Food and care could be provided on the fringe of the desert to the south, where the people would be closer to services and a more reliable water supply. The physical dislocation of Warlpiri people from their places in the Tanami Desert was complete. Within the space of fifty years, people who had occupied, travelled and lived in an area of around 70,000 square kilometres were now all located together in one place, on the extreme southern edge of their land.

The government succeeded in ensuring that people remained on settlements by the rationing policy and isolation from the cash economy. Nevertheless, the establishment of settlements had some unintended effects that actually fostered the perpetuation of cultural difference – difference that the authorities had presumed would disappear with the drastic alteration to the traditional lifestyle of Aboriginal people. Jeremy Long noted that, in Papunya, an assured food and water supply actually worked to favour continuing ritual traditions and culture by allowing ceremonies to be held more frequently and for longer periods of time.[30] Such benefits were also enjoyed at Yuendumu and Lajamanu, and the community situation also reproduced a continuing social and religious attachment of people to place – in spite of their dislocation.

There is no doubt that, among the Warlpiri, sedentary settlement engendered a richer ceremonial life for women, who were afforded more opportunities to display and learn designs, songs and dances for *jukurrpa* than would have been possible before contact.[31] Beyond this, the physical separation from place experienced by the Warlpiri served to heighten the emphasis of other kinds of links with respect to ceremony and identity, although this period of absence from the land meant that their ability to locate accurately the entire complement of their places was declining. To some extent this situation was reversed by the change of official policy reflected in the Land Rights Act and by a new period of intense exploration and goldmining in the Tanami Desert.

What were the consequences of the movement of Warlpiri people to cattle stations, mines, ration depots and, finally, welfare settlements? These processes rendered the Tanami Desert empty of Aboriginal people and, by implication, empty of place. Yet memory, experience and the continued vitality of religious and social activity meant that place remained central to the Warlpiri's lived experience of the world. Moreover, developments that have occurred in the political and economic positioning of Aboriginal people within Australian society have culminated in a partial reappropriation of the Tanami Desert by the Warlpiri, if not specifically on their own terms. The introduction of the cash economy, improved access to vehicles and the passing of the 1976 Land Rights Act have all combined to afford the Warlpiri enhanced opportunity to once more visit and live on their land after a separation of thirty years.



The current politics of place for both the Warlpiri and development interests is one in which the project of mapping is of paramount importance. At the present time, Warlpiri places are at the centre of debates that seek to determine their location and physical boundaries. This determination is of great significance for mining companies and for the conservative federal and Northern Territory governments, which want to maximise companies' ability to access land in order to discover and exploit mineral reserves. The Land Rights legislation provides the Warlpiri with a considerable amount of power over access to their land.

The advancement of the Northern Territory's mineral base has long been a key element of the ruling Country–Liberal Party's agenda. The Northern Territory government's opposition to land claims by Aboriginal people was deemed necessary in order to remove an obstacle to the development of economic infrastructure. For the government, the defeat of Aboriginal land claims would have allowed mining and exploration to go ahead unimpeded. Not only did the government fail in this respect, its vehement contesting of land claims actually helped the Warlpiri to successfully relocate places of significance, thus strengthening the latter's position. The Warlpiri, through the Central Land Council, quickly established a number of agreements with mining companies. Through the procedures of exploration and contingent 'site area clearances' provided for in these agreements, people were rediscovering and locating many of their places. As more exploration tracks were made, they provided the means for people to access their places more easily and facilitated the preparation of subsequent land claims and clearances as people began to travel more frequently through their old country.

There have been significant social and economic benefits that have flowed to both traditional owners and the mining industry through the Central

Land Council's operation of the statutory processes of the Land Rights Act in the Northern Territory. In particular, this process has been assisted by the more progressive mining companies that recognise the special nature of the link between Aboriginal people and the land.[32] The success of the agreements made between mining companies and Warlpiri people in the Tanami Desert hinges on a company's acceptance of, and respect for, this special link. It is the foundation of a workable and mutually beneficial relationship.[33] Such a position, adopted by certain elements in the mining industry, is based on a very different premise from that which governed mining in the Tanami Desert in earlier years.

The advent of the Land Rights Act signalled an escalation of interest in Warlpiri places and land tenure, not only for Warlpiri people and their representatives on the Central Land Council but also for the Commonwealth and Northern Territory governments and the mining industry. The steady transfer of land back to Warlpiri hands was followed closely by a rising interest in gold and mineral exploration in the area during the 1980s. By the early part of the next decade, the central area of the Tanami Desert was, literally, being held under the microscope.



During this period, both NormandyNFM (previously NFM) and TJV (Tanami Joint Venture) were allocated exploration licences over the Granites and Tanami, respectively, that had lain dormant during the 1970s. An exploration licence was given to NFM in 1975, and subsequent mineral leases were offered by the Northern Territory Department of Mines and Energy in 1980. An agreement was made between the Central Land Council and NFM in 1983. This agreement was necessary because of the need for NFM to secure more land in order to set up its processing plant and other support infrastructure for the mine, including a reliable water supply. In 1987 mining recommenced at Tanami and in 1991 control of the mine passed to Zapopan. Zapopan held the exploration licences negotiated with the traditional owners around the TJV mine site. TJV had soon exhausted the deposits within the existing lease. By 1995, further deposits were found in close proximity to the mine and a joint venture was formed by Otter Gold Mines Ltd and Acacia Resources. The mine now operates as Tanami Mine Joint Venture (Plate 13.2).

There have been numerous companies and licences involved in the development of goldmining in recent Tanami Desert history. Agreements between Warlpiri people and mining companies have successfully reintroduced the industry to the region. The Tanami operation relies on production from numerous small pits and has remained a smaller, yet increasingly viable, operation than that of NormandyNFM at the Granites. The attraction that the Tanami Desert region holds for gold exploration has been bolstered by the success of NormandyNFM in the discovery of a remarkable deposit west of the Granites (Plate 13.3).

From the early 1990s the Warlpiri of the Tanami Desert have been involved in day-to-day consultations regarding access to land, in large part prompted by the Callie find which is widely considered to be 'one of the great modern-day Australian gold discoveries'.[34] NormandyNFM's Callie deposit is by far the greatest reserve of gold so far discovered in the Tanami Desert, and it is the first goldmine in Australia that has been developed with the consultation and permission of the Aboriginal owners. The Land Rights Act has certainly not restricted the access of mining interests to Aboriginal land. Indeed, since the Act's inception in 1976, annual exploration expenditure in the Northern Territory has gone from $1 million to $40 million and is showing little sign of slowing.[35] The expenditure on gold exploration in the Tanami Desert alone was in excess of $12 million in 1997.[36] Currently, there are more than 100 exploration licences in the Tanami Desert awaiting consideration. More than 160 have already been granted within fifty-three agreements between the Central Land Council and various mining companies, and there are currently seven mining leases.

By and large, government programs for the economic development of Aboriginal people in remote areas have met with little success. In contrast, mining companies have been remarkably effective in exploiting the few opportunities in remote Australia. The leasing initiatives in the Tanami Desert are of great significance because of the mutually beneficial relationships they create. Mining companies benefit through the extraction of gold on Aboriginal land; the Warlpiri receive considerable financial benefits through royalty payments as well as regional and community development and employment.[37]



The mining industry has returned with a vengeance to the Tanami Desert after an absence of almost thirty years, yet there are important points of difference between the mining situation as it currently exists and as it occurred earlier in the century. The fact that the Warlpiri now live in communities such as Yuendumu and Lajamanu means that there is a minimum of day-to-day interaction with the miners, unlike their closer physical association in the past. At the same time, however, a closer working relationship has developed in which respect for the Warlpiri's relationship with their places plays a role in determining access to land as much as the exigencies of mineral exploration. As Doug Johnson Japanangka has said, it allows Warlpiri people to realise a critical role in a process from which they have long been excluded:

> We take our people into the bush because we care about our *jukurrpa*, our dreams, and our sites, they are very important to us. We can't give away our sacred sites, we are part of the *jukurrpa*, it is our ceremony, country and ritual. We check our families land before mining can start, not like those early days where they mined on sites, poor buggers . . . we make proper agreements between both parties . . . before they put pits where they liked and worked them blackfellas like hell.[38]

The 'Golden Dreams' in this chapter's title simultaneously refers to several phenomena that contain different meanings for both the Warlpiri and Euro-Australians in their unique encounter in the Tanami Desert. On the one

**Figure 13.2** Top to bottom, l to r: Abie Jangala, Henry Anderson (Cook) Jakamarra, Ronnie Lawson Jakamarra, Victor Simon Jupurrurla, Doug Johnson Japanangka. (Photos: Ian Bryson, reproduced with permission of the subjects)

hand, it refers to the continuing hope of mining companies in their search for gold, a dream that for many early prospectors proved a tragic illusion. On the other hand, gold is imagined by the Warlpiri in terms of substance travelling and running under the ground in a manner not unlike their own Dreaming tracks *jukurrpa*; indeed, it is spoken of as 'white man's Dreaming'.

The current context of goldmining in the Tanami Desert involves three fully operational mines and pits with processing facilities located at both Tanami and the Granites, with more discoveries highly likely. For the Warlpiri, gold represents an infusion of economic rewards in the form of royalties paid from its extraction at or near places that have been the sources of profound social and cultural significance from time immemorial.

The search for gold in the Tanami Desert continues today after exactly 100 years. In that time it has spawned displacement, exploitation and disillusionment, as well as wealth, knowledge and cultural exchange. The quest for gold in the central Tanami Desert has lain at the heart of an instructive chapter in the history of black–white relations in Australia.



## Notes

Thanks to Chris Anderson of NormandyNFM for his insightful comments on a draft of this essay and to Paul Hynch who directed me to many valuable references.

1 Chris Anderson drew my attention to this point of looking at 'culture contact' with respect to how frontier industries may have played a role in the cultural maintenance available to indigenous people.

2 Government Resident, *Report on Northern Territory*, 1890, South Australian Parliamentary Papers, no. 28, South Australian Government Printer, Adelaide, p. 9, quoted in H. Reynolds, *Dispossession: Black Australians and White Invaders*, 1989, Allen & Unwin, Sydney.

3 For a detailed account of this expedition, by its leader, see Colonel P. E. Warburton, *Journey across the Western Interior of Australia*, 1875, Sampson Low, London.

4 A. Davidson, *Journal of the Explorations in Central Australia by the Central Australian Exploration Syndicate 1898–1900*, 1905, South Australian Parliamentary Papers, no. 27, South Australian Government Printer, Adelaide.

5 Tanami refers to the goldfield at the site of *Jarnami*, which is distinguished from the Tanami Desert which takes its name from the site.

6 M. Cartwright, *Lasseter's Gold, Tanami Gold: A Story of Two Highways*, 1996, Flinders University Press, Adelaide, p. 65.

7 M. Meggitt, *Desert People*, 1974, Angus & Robertson, Sydney, p. 21.

8 L. Gee, *General Report on Tanami Goldfield and District*, 1911, South Australian Parliamentary Papers, no. 31, South Australian Government Printer, Adelaide.

9 Dick Kimber, 'Notes on Land South of Tanami', n.d. (unpublished ms), quoted in J. Stead, *A Claim to Chilla Well Pastoral Lease by the Warlpiri*, 1985, Central Land Council, Alice Springs, p. 30.
10 Meggitt, *Desert People*, p. 21.
11 Cartwright, *Lasseter's Gold, Tanami Gold*, pp. 70–1.
12 M. C. Hartwig, 'Coniston Killings', BA Hons thesis, University of Adelaide, 1960.
13 T. Strehlow, 'Geography and the Totemic Landscape in Central Australia: A Functional Study', in R. Berndt (ed.), *Australian Aboriginal Anthropology*, 1970, University of Western Australia Press, Perth, pp. 92–140.
14 Also spelt as 'Wykham's Find'; see C. Goodall, 'The Granites Gold Rush', *The Australian Lapidary Magazine*, September 1972, p. 5.
15 M. Terry, *Hidden Wealth and Hiding People*, 1934, Putnam, London, p. 167. (A complete history of this incident is in press: P. Hynch, 'Exploration History of the Tanami Desert', Hesperian Press, Perth.)
16 *Daily News* (Western Australia), 23 March, 6 April 1935 (newspaper did not receive any expedition reports until 1935).
17 *Daily News*, 23 March 1935.
18 The track assisted the spread of cattle stations from Alice Springs to the south of the survey area. Most notably, Braitling took up and occupied the Mount Doreen pastoral lease in 1932.



19 Goodall, 'The Granites Gold Rush', p. 7.
20 F. E. Baume, *Tragedy Track: The Story of the Granites*, 1933, F. C. Johnson, Sydney, and F. E. Baume, 'The Heart of Australia', *The Steering Wheel and Society and Home Magazine*, article series in ten parts, 1934–35.
21 Baume in Meggitt, *Desert People*, p. 25.
22 Baume, *Tragedy Track*, p. 56.
23 Baume, *Tragedy Track*, p. 134; C. Madigan, *Central Australia*, 1944, Oxford University Press, Melbourne, p. 240.
24 Baume, *Tragedy Track*, p. 134; Terry, *Hidden Wealth*, p. 233.
25 Interviews with Henry Anderson Jakamarra, Abie Jangala, Ronnie Lawson Jakamarra, Victor Simon Jupurrurla and Doug Johnson Japanangka conducted by Derek Elias at Lajamanu, 1 July 2000. See videorecording by Ian Bryson, Australian National University, Canberra. All quotes from these men are taken from these interviews.
26 T. Rowse, *White Flour, White Power: From Rations to Citizenship in Central Australia*, 1988, Cambridge University Press, Cambridge, p. 57.
27 *Centralian Advocate* (Northern Territory), 10 April 1953.
28 Patrol Officer Sweeney quoted in T. Rowse, 'Enlisting the Warlpiri', in T. O'Regan (ed.), *Communication and Tradition: Essays after Eric Michaels*, Continuum, vol. 3, no. 2, 1990.
29 F. O'Grady, *Francis of Central Australia*, 1977, Wentworth Press, Sydney.
30 J. Long, 'Papunya: Westernization in an Aboriginal Community', in M. Reay (ed.), *Aborigines Now*, 1964, Angus & Robertson, Sydney, pp. 72–82.
31 B. Glowczewski, *Du rêve à la loi chez les Aborigènes: Mythes, rites et organisation sociale en Australie*, 1991, Presses Universitaires de France, Paris; M. Laughren, 'Review of Glowczewski', *Australian Aboriginal Studies*, 2, 1993, 74–80.

32 Central Land Council, *Annual Report 1997–1998*, 1998, Commonwealth of Australia, Canberra, p. 49.

33 T. Ireland, 'Exploring for Gold on Aboriginal Land in the Northern Territory', in *Building International Partnerships in Environmental Management for the Minerals Industry*, third international and twenty-first annual Minerals Council of Australia environmental workshop, Newcastle, New South Wales, 14–18 October 1996, Papers vol. 1, Minerals Council of Australia, Canberra, pp. 302–13.

34 S. Bell, 'SOG Pulls Out of Tanami Search', *Australia's Mining Monthly*, December 1998 – January 1999, p. 67.

35 Ireland, 'Exploring for Gold'.

36 Central Land Council, *Mines and Myths: The Truth about Mining on Aboriginal Land*, 1998, Central Land Council, Alice Springs, p. 6.

37 I. Manning, *Native Title, Mining and Mineral Exploration: The Impact of Native Title and the Right To Negotiate on Mining and Mineral Exploration in Australia*, 1997, Aboriginal and Torres Strait Islander Commission, Canberra, p. 33.

38 See Note 25.

# Part IV Daily Life and Domestic Culture

Chapter 14

# Mrs Charles Clacy, Lola Montez and Poll the Grogseller

## GLIMPSES OF WOMEN ON THE EARLY VICTORIAN GOLDFIELDS

*Margaret Anderson*



Between 1851 and 1860 nearly 600,000 immigrants arrived in Victoria. Among them were almost 164,000 women and girls.[1] Not all of them were bound for the goldfields, but there can be little doubt that the lure of gold was paramount in the minds of those who embarked on the hazardous voyage to Australia in those years. Men dreamed of making quick fortunes, of escaping the world of wage labour, of perhaps becoming landowners and employers in their own right. But of what did the women dream? Could they share in these dreams of independence? If so, what did independence mean in the context of mid-nineteenth-century strictures about women's nature and women's place?

David Goodman, in this book and elsewhere, has written powerfully of the masculinity of gold-rush society and its historiography. Although he by no means excludes women from his account, they are, of necessity, secondary to his main theme, which is gold getting and its attendant mythology. And yet these women are far from passive observers. Through the 'fragments' of their writings, he suggests, 'we get glimpses of an outsider's view of the masculine landscape . . . a view so alienated from the dominant ethos that it amounts to a critique of it'.[2] Alas, he is probably right, and glimpses of these women's lives are all that we can hope for, but they are illuminating glimpses for all that and not only for their alternative perspectives on the ways of men. A female world, on the diggings and in gold-rush Melbourne, also emerges from these sources, a world with its own ethos and its own dreams. It deserves consideration in its own right (Plates 14.1 and 14.2).

But let us stay with the world of men for one moment and visit that holy of masculine holies – the Shamrock Hotel in Bendigo. Charles Thatcher, popularly dubbed the Colonial Minstrel, is in a sentimental mood this night, as with true Victorian pathos he sings to his patrons of 'Mary's Dream'. Mary is indeed every digger's dream – the serene wife and mother, cheerful in adversity, ever hopeful of her man's ultimate success. Although ill-fortune has dogged her path, she is reminded in a nightmare of the grim realities of poverty in England, and comforts herself with the knowledge that in Australia she can at least dream of prosperity to come:

> 'Come, tell me Mary, how it is that you can look so gay,
> When your husband at Mount Blackwood for weeks has been away;
> Come when I will, your countenance is radiant with a smile,
> and yet I know you've had bad luck for a tremendous while.'
>
> Says Mary then, 'it is because
> Hope animates my breast,
> Although ill luck and troubles
> On us have sorely pressed;
> But Fortune does not always frown,
> But smiles sometimes, you see,
> And with lots of nuggets my good man
> May soon return to me'.[3]



Thatcher's genius lay in capturing just enough of the reality around him to strike a chord with his audience, who greeted his songs of the diggings with particular delight. They laughed with him when he satirised the commissioners or the police and cried with him too, on occasion, when he sang especially sentimental songs of home.[4]

But does Thatcher's imaginary Mary have anything to do with the thoughts and feelings of real women on the goldfields at the time? I think it does. In fact, there are uncanny echoes of Mary's dream in the contemporary letters and reminiscences of two working women who came to Victoria in the early 1850s. These writings lack the eloquence of those composed so carefully by their more educated sisters, the spelling is often erratic and the punctuation minimal, but their matter-of-fact narratives of trials and triumphs are sometimes more moving than those other more elegant, but also more contrived, accounts. The narratives are bald, but they speak clearly down the years of dogged determination and of hope – not of winning great riches necessarily, but of achieving some security and a respite from constant labour.

Sarah Davenport's life was probably closest to that of the imaginary Mary. Sarah was the wife of a skilled tradesman, a cabinet-maker, whose skill might have been expected to cushion her against poverty, but his chronic ill health meant that she was constantly forced to take in additional work, just to support their four children. With few prospects of a better life in England, Sarah and her husband at last decided to follow his sister to New South Wales. Her laboriously penned memoirs were probably written in about 1869.

'As maney pepole leving home for new south wales and Port phillip sent gloing accounts of the country both for good wages and no scarcity of work and the healtheness of the climate . . .', she began, 'the news to us was from them that we shold come out . . . we was all in good hopes that we was coming to beeter our selves tho sorry to leve our friends'.[5] Their venture did not begin well. They had barely left port when the ship struck a sandbank and had to be abandoned. Many of their belongings were lost and Sarah was forced to return to Manchester to seek help from her parents. After such an inauspicious beginning they obviously considered abandoning the project, but Sarah held firm:

> i did not wish to begin Life again in old England i wanted to make a fresh start in a new country my husband was a cabinet maker by trad and he used to suffer with the sick headach almost every week i had to work very hard my self to keep our familey and i found my strenth getting very low i concluded the best to try a new country.

Fresh disasters were to follow. Sarah lost her youngest child after an accident on board the second vessel, then suffered a miscarriage.[6] Characteristically, she refused to give in to her 'tryals', but prosperity continued to be elusive. The Davenports eventually decided to try Port Phillip, still in the hope of 'bettering' themselves, and there, finally, they had some luck, although Sarah was still at work, labouring alongside her husband and sons, washing clothes and washing gold.



Lucy Hart's motives were similar, although she was more fortunate in her husband. John Hart was, by his wife's account, 'very saveing and very fond of money'.[7] He worked hard for the family in Adelaide, but still sickness meant that they seemed to take 'on step forine and to back'.[8] Lucy took in washing, and later managed a cow and a fowl yard, selling the extra milk and eggs. Slowly they began to make some headway. As men began leaving for the goldfields, Lucy persuaded John to join some friends and try the diggings. He agreed, confident that she could support herself if need be without touching their savings. Within a few months he had sent for her to join him and she wrote in triumph to her mother:

> Should I have been so well off in England NO work hard and be half starved. Australia is the place to live I would not come back to England unless I had enough to keep me without work on no account neither would my husband I am speaking now the sentiments of our hearts but people must be *saving industrious* and *persevering* we have deprived ourselves of many things we might have had but what was it for all to try to do something for ourselves so that my husband should not allways work under a *Master* and happy am I to inform you that we gained that point he is now his own Master.

Better still, Lucy no longer needed to work so hard herself, as she proudly told her mother:

> He was then determined that I should not work so hard any more. I have no need to now, for with the money that I brought with me, we could raise 300 pounds, and now I consider myself well off in the world, so I do nothing but my own work now.

Here was prosperity indeed!

These accounts differ substantially from those left by more prosperous women. Most obviously, both Lucy Hart and Sarah Davenport were full participants in the decision to emigrate: Sarah may even have been the initiator of the idea, while Lucy claimed to have encouraged John to try the diggings. England held no prospects of a better life for either of them. This was not the situation for many middle-class women. Some, like Georgiana McCrae or Elizabeth Ramsay-Laye, might not even have been consulted by their husbands about emigration, and, while their subsequent lives were undoubtedly far more comfortable than even the successful Lucy Hart could imagine, they were still a far cry from the comforts of home.[9] Elizabeth Ramsay-Laye was more reconciled to her lot than Georgiana McCrae (she may also have known that her stay in the colony was not to be permanent), but she was quite frank about her initial



feelings. Only a rigid sense of lady-like propriety prevented her expressing them openly:

> It was just at the moment when the gold fever was at its height that I was unexpectedly called on to accompany my husband to Melbourne. Disgust, indeed, is not a word too strong to express my feelings at the moment, particularly as I had to wear a calm face and not distress loving friends by any exhibition of feeling.[10]

Elizabeth soon overcame her disgust and thoroughly enjoyed her eight years in the colony. Even two years living in a tent on the Castlemaine diggings did not dampen her enthusiasm. She was to write almost apologetically in her reminiscences, published on her return to England in 1861: 'Perhaps many may think my descriptions too *couleur de rose*; but if I felt a roseate tint every-where shining on my path, could I cast shadows where there were none?'[11]

Indeed, there were few shadows in the Ramsay-Layes' privileged existence. Elizabeth was young, recently married, and wealthy. She had no children to worry over, or to keep her from indulging in her passion for horse riding. If the continued absence of children was a disappointment, she did not say so. The life she led from her tent at the Castlemaine diggings was much the same as her subsequent life in Melbourne, or her previous life in England for that matter – she visited friends, attended balls and picnics, went for long gallops in the bush and accompanied her husband and his friends as they shot their way from one end of the colony to the other. It is difficult to imagine a greater contrast with the experience of Sarah Davenport. Indeed, if Elizabeth's reminiscences contribute anything to our general understanding of life on the diggings, it is to highlight the vast gulf that separated rich from poor, even on the goldfields.

Within the democracy of 'diggerdom',[12] the lady and the gentleman still set themselves apart – so long as the money lasted. Perhaps this awareness was more significant for women, although the many reminiscences of men betray a similar consciousness.[13]

In material terms, such distinctions meant a great deal, even if everyone was living in tents. Shortly after she arrived at the diggings, Elizabeth and her husband were invited to tea. She described her neighbour's domestic arrangements in detail and they make extraordinary reading:

> The party was numerous; the tent one of the largest in the camp, was lined with green baize; one end of it was fitted up with sofas, arm-chairs, and a grand piano. Small, round tables were tastefully dispersed, on which some very pretty ornaments, books, and portfolios of drawings were placed. At the other end there was a large table with cups and saucers of every size and pattern, a large mud and stone fireplace, with a blazing fire, on which two immense kettles were singing. Loaded pistols decorated the mantelpiece.

Apart from the loaded pistols, the mud-and-stone fireplace and the mismatched crockery, she might have been describing a substantial drawing-room in Melbourne, or even London. The elegance of the company, however, was more than equal to the surroundings:



> The ladies – for ladies they were in every sense of the word – were well dressed, some even elegantly, in the newest fashions. After tea had been disposed of and the things taken away, we had some good vocal and instrumental music, and for a time forgot we were in the land of the diggers.[14]

We can only guess at the determination which saw these things, particularly a grand piano, transported by bullock dray over the colony's notoriously dreadful roads to Castlemaine! And it would be easy to laugh at the little oasis of civilisation these women undoubtedly saw themselves creating in the midst of the desert of diggerdom. With their visiting networks preserved as carefully as those in Melbourne, they maintained the social distinctions which defined 'society' amid all the chaos of the goldfields.[15] Caroline Chisholm would probably have approved, although her determined campaign to bring order even to the goldfields of Victoria through the civilising hand of woman concerned fellow reformers like William Howitt, who feared for the women in such a society.[16]

But for most women on the diggings, possessions such as grand pianos were the stuff of fantasy. Their worldly goods were far more meagre and makeshift. Martha Clendinning, who, with her little girl, accompanied her doctor husband to Ballarat in 1853, described living in a tent, divided in half, with a sleeping apartment at the back: 'We had iron bedsteads, mattresses and blankets, a couple of chairs, and a few cooking utensils. Such was the furniture we provided for our first Australian home!'[17] She mentioned no bed linen, although sheets on the beds were allegedly one of the first signs of a woman's presence in a digger's tent. Neither did she list a table among their possessions, but perhaps this was because most goldfields tables were home-made constructions, built from local saplings, with their legs driven into the dirt floor of the tent.[18] The Clendinning tent was a far cry from the Ramsay-Laye 'establishment',[19] but according to Martha it was far superior to the 'very rough shelters' most diggers called home. 'It was generally where they had respectable wives', she observed, 'that an effort was made to have their tents tolerably comfortable'.[20]

Women like Elizabeth Ramsay-Laye embraced what she described as the 'gypsy life' of the goldfields,[21] and looked back fondly on their experiences. But for others, who lacked the comforts (or the domestic staff) of the Ramsay-Laye establishment, life in a tent or in a bark hut held little romance. Emily Skinner arrived in Melbourne in 1854 at the age of 22 to join her digger fiancé (Plate 14.3). She was the daughter of a London jeweller. Not a wealthy jeweller evidently, because Emily travelled out as an intermediate rather than a first-class passenger, but wealthy enough to have provided his daughter with a comfortable home.[22] She recorded rather ruefully that she had never ironed anything before she came to Australia. Emily married William Skinner in Melbourne and then travelled sometime later to join him at the Ovens diggings (Beechworth), where

he had moved after being 'moderately successful' at Forest Creek (Castlemaine). Although pleased to be reunited with William and anxious to look on the bright side, Emily was still dismayed by the bark hut she found waiting for her:

> Our hut was built of bark – roof, sides and all . . . I must confess I was rather disappointed in the house. I thought some of the Melbourne houses strange after home, but this was worse and worse. I soon found though, that it was unusually good for the time and place, as most people lived in tents.[23]

Like so many of the digger families described in Suzanne Hunt's chapter, Emily and William established a garden. Emily also learned from an American baker how to make bread. 'The bill of fare was not much varied', she recorded, 'imported fish, meats, fruits, etc. were luxuries; however we had got beyond the mutton and damper stage of the Colony'.[24] Her first child was born safely and all seemed settled, when William and his partners decided to go to the Woolshed diggings. Although she was to regret the decision, Emily decided to leave the hut, with its garden just beginning to bear fruit, and join her husband in tent living. She considered staying behind. 'Had I been wisely regardful of my own and the babe's comfort, I should have done so', she reflected sadly, 'but no, I had had enough of being left to myself'.[25] In the humid heat of that first summer at the Woolshed, Emily and William both fell ill with fever. The Woolshed was prone to fevers, including typhus.[26] During this illness, their little son, then five months old, died. It may be that he had caught the same fever, but reading between the lines suggests the likelihood that he died because Emily was too ill to feed him. She was so ill, in fact, that she was barely aware of his death: 'It was only when partly restored to health that the anguish of the loss came upon me', she recalled.[27]

After persevering through a bitterly cold winter and continuing bouts of illness, with little gold to compensate, the Skinners returned to Beechworth. Here their health improved, but William's luck did not:

> Rather hard times were these, and sometimes I was so poor that I neglected to write my usual home letters because I could not afford the stamp . . . Little do women who have come here in later years know of the hardships their predecessors had to encounter. Sometimes in lonely places where no other woman was within reach, perhaps sick and weak, I had to sit down beside the tub to do the necessary washing or attend to the children. At another time, finding myself on the floor with the poor child crying beside me, I was dimly conscious of having fainted away from weakness. How one longed for mothers and sisters at such times, and envied the poorest woman at home who in sickness generally have some relative near.[28]

Emily Skinner paid dearly for her husband's love affair with gold. She tried to understand it,[29] but it is clear that his quest for independence doomed her to a decade or more of poverty, loneliness and hard domestic labour in primitive conditions. And in the end he was forced to retreat to the city as a wage-earner. The dream of independence proved illusory for William.



Even so, Emily was better off than thousands of others and she was well aware of it: as Graeme Davison's chapter shows, most observers of the goldfields and gold-rush Melbourne in the 1850s pointed to the squalor that lurked beneath the glittering surface. 'In Melbourne the extremes of wealth and poverty meet', Ellen Clacy wrote,[30] while William Howitt warned would-be emigrants of the 'lottery' that was gold seeking.[31] Canvas Town,[32] a sprawling warren of tents pitched on the southern bank of the Yarra, housed many an unhappy wife whose husband preferred to move around the goldfields unhindered, but even in Canvas Town there was rent to be paid, and the food prices in rapacious gold-rush Melbourne were horrifying to newcomers and old colonists alike.[33] Ellen Clacy illustrated her portrait of gold-rush Melbourne with the heart-rending tale of one educated young woman who died in poverty in Canvas Town, just as her husband finally returned for her: 'Sad as this history may appear, it is not so sad as many, many others, for often, instead of returning with gold, the digger is never heard of more'.[34] In desperation, abandoned wives and vagrant children sought the protection of the Benevolent Asylum in Flemington Road, while in neighbouring South Australia the colonial government instituted an urgent inquiry in 1855, alarmed by the numbers of destitute seeking relief.[35] These are the stories which need to be balanced against the better-known tales of wild drinking sprees and of diggers' wives decked out in extravagant silks and satins (Plate 14.4).[36]

In fact, marriage was as much a lottery for women as gold seeking was for men, but colonial women still had few realistic alternatives. There was reputedly plenty of work available, but all in traditional female occupations and at rates of pay which were never intended to sustain an individual independently, let alone a family.[37] Not surprisingly, those writing home to encourage friends or family to follow them to Australia cited high wages or business opportunities (or gold, of course) to men, and good prospects of marriage to women. Charlotte Perottet was one such. She joined her husband Louis at the Ovens diggings in 1856, and in one of her regular letters home to her mother reflected on the contrasting opportunities colonial life might offer her brother and two sisters: 'I sometimes wish George was here, as I think he would do very well . . . the gentlemen are very numerous, plenty of chances for the 2 girls, if they were to come out'.[38]

Plenty of chances indeed, for it was certainly assumed that the young women who emigrated to the colony in such numbers at this time came principally in search of husbands. Charles Thatcher sang happily of the 'scrumptious young gals' adorning the diggings in greater numbers:

> Scrumptious young gals, thanks to free emigration,
> Out to this country for husbands you fly;
> Mrs Chisholm we look on with great veneration,
> Because she's kept up such a constant supply.[39]



What Mrs Chisholm would have made of a character reference from Charles Thatcher we can only imagine, but the broader purpose of her work was clearly understood and apparently generally supported. The editor of the *Bendigo Advertiser* was fulsome in his praise of Mrs Chisholm's endeavours, and her fleeting visit to the Victorian goldfields in 1854 was reported in the local press in some detail. Although he was obviously in some awe of what he termed her 'masculine mind', the editor could only applaud her efforts to improve Victorian society. 'The absence of the benign influence of female society is severely felt', he wrote, concluding: 'That Mrs Chisholm may be enabled fully to carry out the mighty and excellent ends she contemplates must be the desire of every wise and worthy member of society throughout the colony'.[40]

Nevertheless, disquiet was expressed from time to time at the distressingly mercenary tendencies displayed by some of these young women. In the context of the goldfields and a society which, according to most observers, was almost entirely given over to 'Mammon', the hypocrisy of this seems ludicrous, but it was a recurring theme in Thatcher's songs and in the satirical newspapers. In one new song, 'The Disappointed Girl', which was apparently greeted with 'immense applause', Thatcher warned women who were too choosy that, even in the colony, they could still end up as old maids if they persisted in settling only for successful diggers:

> So, young girls, just ponder this song in your mind –
> Don't treat every suitor with rigour;
> A willing though poor man's as good, you will find,
> P'raps as any flash lucky gold digger.

I wonder if Emily Skinner would have agreed with him. Nevertheless, if the masculine anxiety reflected in these songs concealed even a grain of truth, we might conclude that women were indeed able to manipulate the favourable colonial marriage market to their advantage, at least some of the time.

Thatcher returned to the theme in the 'Faithless Nursemaid', but there is a darker note here, for although the heroine, Jane, married her rich digger all right (having by implication forsaken the poor, but devoted, singer), they quickly ran through all their cash, and her husband turned into a wife beater:

> Poor Jane, alas! brown she is down,
> But it serves her right, you must agree;
> She drinks, and he beats her like fun –
> So she'd better by half have had me.[41]

Almost every chronicler of the diggings commented on the drunkenness and violence they saw around them, with women frequent victims, despite their scarcity and apparent desirability. That earnest Quaker, William Howitt, had no doubt that the demon drink was the source of the problem. 'So long as men drink alcohol', he warned, 'brutality, beatings and murders of wives will follow',[42] and follow they did, if the court records from the diggings are any indication. Every month saw several men charged before the bench with assaulting their wives but, as always, these were probably the tip of the iceberg. It might be thought that the close proximity of goldfields living would have afforded wives some protection, but this does not seem to have been the case. Martha Clendinning was the unwilling observer of her neighbour's assault on his wife: 'Next evening I heard loud and angry voices and peeping out I saw the butcher fling her out of the tent and kick her savagely till the blood streamed from her face'.[43]



Such women could resort to the courts, but there was little the courts could do for them. Mostly the men were fined, bound over to keep the peace and sent home to their wives.[44] In extreme cases a husband might be gaoled, in which case the wife had to support herself until his release: it was a two-edged sword.[45] In Bendigo in 1854 the Police Magistrate, Mr McLachlan, expressed his determination to 'put a stop to the constant instances of women being abused on the diggings by their husbands and others',[46] but there is no indication that he was any more successful than anyone else in stamping out the practice. What he thought of two other cases – one of wife-stealing and the other of wife-selling – that came before him in the same period, is not recorded.[47] Early feminists were to shy away from the issue, but some of these women knew only too well that the root of their problems lay in economic dependence.[48] The divided McCrae family had, perhaps, more than its share of troubles over the years, but they were united in sympathy for their sister, Margaret Thomas, and her daughter, trapped in an unhappy marriage to a tyrannical husband and father. A family conference was called to decide whether Margaret should leave her husband, but the McCrae men reminded her that she could not hope to support herself alone. As Thomas Ann Ward Cole confided to her diary, 'these poor Ladies are in [Dr Thomas'] power for when people marry it is for better & for worse'.[49]

Much has been made of the opportunities for men in gold-rush Victoria, both for economic advancement and later for political participation. But were there openings also for women, apart from the qualified advantage of easier access to marriage partners? And did any of the more general freedoms of goldfields life extend to women? Here the evidence becomes even more patchy, snippets here and there, which are tantalising but elusive. What does seem clear is that actual gold digging was more or less an exclusively masculine occupation. It was probably far too strenuous for women and, in fact, stories abound of men who abandoned gold digging, defeated by the sheer, back-breaking labour it entailed.[50] There is no record of a goldmine worked by women. Women could fossick for gold, though, and wash gold dirt, sometimes with useful, even spectacular results. Sarah Davenport confounded her husband by discovering a



particularly profitable claim in Ballarat, while Margaret Kennedy, a local overseer's wife, was one of the first to find gold in Bendigo. But in each case these women were swiftly supplanted by their men.[51] Nevertheless, there were enough women washing gold to attract the attention of the licensing police and to interest artists like S. T. Gill, who drew them alongside their men, sometimes with babes in arms, rocking the gold cradle (see Plate 14.2).[52]

Of course, there were other employment opportunities for women on the goldfields. There was also, probably, the prospect of higher wages, at least in the early years. Emily Skinner travelled to Beechworth with several young unmarried women, two of whom had offers of 'situations' at the diggings. At each inn they visited on the way these women were offered jobs, 'but the girls had determined to go to the diggings, where high wages and easy times awaited them'.[53] Whether the easy times eventuated is more doubtful, judging from the number of advertisements in goldfields newspapers calling for 'strong, active girls'.[54] Certainly, the accommodation often left a good deal to be desired, but perhaps the wages made up for it.[55] The large populations of unattached men on the diggings also created an obvious demand for prostitution. Patricia Grimshaw and Charles Fahey have described the hierarchy of brothels that grew up in Castlemaine during the 1850s, and court records in Bendigo suggest a similar variety.[56]

Sly-grog selling represented the other obvious opportunity for women, and there is abundant evidence that they took advantage of it. Gill drew women serving in the grog tents and later as barmaids,[57] while almost every goldfields chronicler mentioned women working in the trade along with men. Most of these accounts were couched in terms of strong disapproval, caricaturing the women concerned as representative of all that was abhorrent,[58] yet there were also more positive portraits. Charles Thatcher's heroine 'Poll the Grogseller' was no repellent hag. On the contrary:

Polly's good-looking, and Polly is young,
And Polly's possessed of a smooth oily tongue,
She's an innocent face and a good head of hair,
And a lot of young fellows will often go there.

Poll was smart too. According to the song, she was 'turning in plenty of tin' (money), and no sly-grog policeman was about to catch her out, as the two city detectives who came visiting found to their cost:

'For if Polly you'd nail, you'd have, without fail,
to get up in the morning early.'[59]

It may be significant that Thatcher's Polly was apparently operating her business without the benefit of masculine assistance.

The gold rushes provided a great stimulus to the entertainment industry. Here was another opportunity for women. As singers and actresses they seem to have made a living touring from town to town, performing in the new theatres and concert halls which sprang up to accommodate the diggers' thirst for entertainment. Women also managed theatres in Bendigo and Melbourne.[60] As Victoria prospered, more eminent entertainers arrived, keen to share in the riches of gold. At one amateur performance in Melbourne, Elizabeth Ramsay-Laye was delighted to catch sight of the infamous Lola Montez in the audience: 'Her beauty was startling. I heard that she had been acting in Melbourne, but no ladies went to see her.'[61] No ladies perhaps, but plenty of gentlemen went (and quite a few women too, to judge from the newspaper reviews) for the actress, dancer and notorious courtesan was perhaps the most celebrated international performer to visit the goldfields. Her suggestive 'spider dance' filled the theatres to capacity wherever she went, the houses no doubt assisted enormously by a spate of disapproving reviews in the press.[62] On the goldfields she was a sensation. But the controversy that dogged her tour highlighted the double moral standard confronting all Victorian women at the time and reminded them that the freedoms of the diggings were largely for men. It was a lesson that another former courtesan, Céleste de Chabrillan, had already learnt to her cost in Melbourne.[63]



Both Lola Montez and Céleste de Chabrillan were extreme examples – women whose notoriety was on a grand scale. More ordinary women in Victoria had to judge the social nuances carefully, seeking the opportunities within the straitjacket that defined respectability. Martha Clendinning was one who saw a window of opportunity in early gold-rush Ballarat and sought to exploit it. Determined not to be left behind in Melbourne but anxious to contribute to the security of their families, Martha and her sister decided they might manage a small shop on the diggings. They had considered and discarded the usual occupations open to women, teaching and needlework, as being irrelevant to goldfields life. Their husbands initially greeted their proposal with 'peals of laughter', predicting that they would be cheated of all their stock. 'We were

**Figure 14.1** Artist unknown, *Lola Montes* [sic] *Dancing the Spider Dance*, pen and ink drawing. (Mortlock Library of South Australiana, State Library of South Australia)

disgusted at the reception our well meant proposal had evoked', she recorded, 'but we kept to our plan'. They sought advice on appropriate stock and set about acquiring stores, but were a little disconcerted by the reaction of the 'gentlemanlike storekeeper' who supplied them in Melbourne, who observed that he hoped they would not long have to pursue

> a business it was quite evident I had not been used to. This speech upset the hope my sister and I entertained, that we should not be distinguished as 'ladies'. We intended to pass as merely respectable women of business; anything more than that would, we felt, expose us to curiosity when we entered on our storekeeping life, and even cause the diggers to be shy of purchasing from us.

In the event, these fears proved groundless and the little store prospered. For two years they sold their groceries and baby clothes, largely to other women, and turned a nice profit, but, as more families moved onto the fields, Martha's ambiguous social position became problematic:

> The time had gone by when, even on the gold fields, a woman unaccustomed to such work could carry on her business without invidious remarks.
>
> I began to fear my husband might be blamed for allowing me to continue at it. After the class of residents on the field had become so superior to those of the working class, whom we had found on our first arrival, to whom all species of employment for women seemed perfectly natural if they could carry it on with success. The doctor had been most anxious and was greatly pleased when I announced my intention [of selling up]. He at once erected a wooden house, for which many of the canvas dwellings had been exchanged. That put an end to all further remarks and allowed us to make our home fit for a lady and her doctor husband to occupy.[64]



Here was the ultimate irony of goldfields prosperity. The very success that had delighted Lucy Hart left Martha Clendinning bereft of an occupation she enjoyed and at which she had displayed some skill. Without her memoirs we might not even have guessed at her experiment, for along with her reminiscences in the archives is an envelope containing the licence issued for her shop – made out in the name of her husband!

It may be the case that the women who decided on emigration at this time, whether accompanying husbands or alone, were among the more adventurous and determined of their generation. Certainly, those letters and diaries which survive suggest resourcefulness, energy and a certain independence of spirit, although this had always to be pursued with care, as Martha Clendinning's story suggests. Woe betide women, like Lola Montez, who challenged convention too openly. This careful judgement of social nuance was probably common to rich and poor alike, however the boundaries themselves may have varied.

**Figure 14.2** Thomas Ham, *Store at the Diggings.* (Gold Museum, Ballarat)

Sarah Davenport's reminiscences included several incidents in which her husband's timidity forced her to act outside the bounds she evidently felt ought to have governed her behaviour. She was at pains each time to justify her actions, but we are left with an impression of a redoubtable woman, continually exasperated by the weakness of her partner but struggling to balance her assertive inclinations against an internalised sense of appropriate womanly behaviour – or, at the very least, concerned to leave her readers with that impression.

One instance must suffice. Sometime after she joined her husband and sons in Ballarat, word reached the field of a new rush to Mount Alexander. Anxious to transport their belongings to the new diggings, Sarah agreed to help another party of diggers to buy a horse and cart in exchange for cartage. Shortly afterwards the men, all 'vandamoinans' as she put it, tried to renege on the deal. They reckoned without Sarah:

> [Two] of the men began to bluster and wanted to frighten us i said 'i paid the money for that what is paid and our things must go if it gose' the party that wanted to go in our place was strong men my husband was quite nervous they saw it and took advantage of it but when they was loading i put some of our things [on] and they put them off i looked at them 'now' i said 'i bought the

> hors and if you do not tak my things i will just brake one of his legs i am a woman of my word you cannot hurt me i bought it' they looked at the receipet and found as i said so they put our things on.

Sarah's bald narrative conjures up a vivid image of the scene – the determined woman in defiance of the ruffians, her nervous husband cowering nearby – but she still felt constrained to excuse her persistence. '[I] should not have been so perseveary', she wrote, 'but my husband seemed so exiteted and i did not wish him to be disapointid'.[65]

Middle-class women like Martha Clendinning confronted the added constraint inherent in notions of 'lady-like' behaviour. Many saw the goldfields as no place for ladies, especially in the early phase of any rush, and some of these ladies had to summon all their stubborn resolution to avoid being left behind in Melbourne. The mere presence of these women on the diggings represented a victory of sorts. Martha herself had made it clear on arrival in Victoria that she did not intend to share her sister's fate in being left behind in Melbourne, a decision which she knew annoyed her brother-in-law:

> I had informed my friends that I had determined to accompany my husband if he went to live on the diggings. To this he willingly agreed, but my brother-in-law was very indignant at such an idea. When he had gone up to Bendigo he had left his wife in Melbourne and still insisted that the diggings was no fit place for any respectable woman. There were, he said, no decent women there, only a few of the 'Vandie's' wives. However, I had made up my mind, and no description he gave of the society I might meet with could turn me away from my decision.



'If I went, he could not prevent my sister going also, and that was what annoyed him', she added shrewdly.[66] Mrs Campbell, a Canadian woman who accompanied her police magistrate husband to Beechworth in 1853, encountered similar opposition from acquaintances in Melbourne. One told her that 'no lady could do it', while another man warned that 'no delicate woman could stand it'. But, marooned with her children in cold, squally Melbourne, Mrs Campbell overrode their concerns: 'I told him my powers of endurance were greater than he gave me credit for and that as I had *made up my mind to go* his kind efforts to dissuade me would be "love's labor lost"'.[67]

Nevertheless, these women were careful to preserve their reputations as ladies, even where they found themselves engaged in (or actively sought out) unusual occupations.[68] Those like Ellen Clacy or Elizabeth Ramsay-Laye who subsequently decided to publish their accounts of goldfields life invariably began their volumes with self-deprecating prefaces justifying the undertaking, although men seldom felt so constrained.[69] Most of these women also went on to address their remarks specifically to their own sex, arguing that many anxious family members left behind in Britain might welcome the insights a woman could offer on goldfields life. While male authors were anxious to invest their memoirs with the twin stamps of accuracy and authenticity,[70] female authors added a palpable veneer of respectability. Sometimes this was too evident to warrant comment – Ramsay-Laye is an obvious example – but others seem to have constructed an elaborate authorial persona to mask a more complex reality.

The chequered history of Mary Fortune, whose journalism was published under the pseudonym of Waif Wanderer, is probably the best known among these women writers, but recent research into the personal history of Ellen Clacy has also suggested some fascinating contradictions. While Mary Fortune only rarely inserted herself into either her journalism or her fiction, a strong moral tone nevertheless pervaded much of her writing. Her tone and perspective was not unlike that of the severe critic of most things Victorian, Louisa Ann Meredith, whose account of a visit to Melbourne and Ballarat was published in 1861.[71] Mary Fortune had little time for weakness of any kind and her perspective was unrelentingly that of the superior, judgemental middle class.[72] None of the compassion of Ellen Clacy or Elizabeth Ramsay-Laye softened her stories. In fact, most of her female characters were very unsympathetically drawn – unfaithful, drunken or complaining wives, neglectful mothers, rude and spoilt young women. Perhaps she imitated the cynical misogyny of the satirical press. Yet what is known of her own life stands in stark contrast to this judgemental voice, encompassing almost certainly both an illegitimate child and a bigamous marriage.[73] Then again, if the high moral tone was camouflage, it probably worked during her lifetime.



Ellen Clacy's history has proved more elusive, so much so that some historians have questioned whether she ever travelled to Australia at all.[74] Recent painstaking research by M. Rosalyn Shennan now confirms that she did indeed visit Victoria briefly, but in circumstances very different from the carefree adventure suggested in her published account. She certainly travelled to Victoria with her brother, and she appears to have gone onto the goldfields, but of Charles Clacy, the husband whom, according to her account, she married just before returning to Britain, or of the marriage itself, there is no trace. Of greater interest, however, is the daughter she apparently bore on board ship on the return voyage. Although the baptismal record for this child nominated a Charles Clacy as the father, there was no evidence in Ellen's subsequent life that he was ever present. She supported herself and her daughter by her writing.[75] Rosalyn Shennan has calculated that the child must have been conceived either just before Ellen left Britain for Australia or actually en route. The latter scenario of course gives added piquancy to one of the stories Ellen told of a shipboard romance which ended tragically in the seduction, then abandonment, of a trusting young female immigrant who then drowned herself clutching her dead baby. Whether Ellen came to Victoria in search of a former lover or met him on board ship, and whether that man was the elusive Charles Clacy, are questions to which we shall probably never know the answers. It may be that the journey was an elaborate ruse designed to conceal an unwed pregnancy and manufacture a colonial marriage, relying on distance to mask the deceit. If so, it was a risky strategy.

Elizabeth Ramsay-Laye included a specific warning in her book to those who sought to reconstruct themselves amidst the anonymity of colonial life:

> What a strange delusion always seems to possess the minds of those who misconduct themselves at home, to fly to Australia, and fancy there, unknown and unsuspected, they may begin a new life! There can be no greater mistake than this . . . Neither is it at all likely that their story will remain unknown; for people of every country and description crowd here, and hourly you meet someone, whom either you have known before, or who is acquainted with your belongings, or with the friends of your friends.[76]

Elizabeth's comments were prompted, so the story went, by her encounter on the Castlemaine field with a former friend who had been persuaded to flee an unhappy marriage in England. By the time Elizabeth saw her she was dying, abandoned by her heartless lover when gossip threatened his new colonial career.[77] Whether this story and that of Ellen Clacy were real or fictional, the message was the same: while a man might put such an incident behind him, a woman could only escape such a lapse through her death.[78] On the goldfields, as in Melbourne or Britain at this time, the double standard of sexual morality held firm. Women like Mary Fortune or Ellen Clacy may have sought anonymity on the diggings, but they could not hope for sexual equality.



Political equality was equally elusive. As opposition to the gold licensing system grew on the goldfields during 1853–54, the rhetoric of the diggers expanded to embrace many of the demands of the Chartist movement, including manhood suffrage. Universal suffrage, however, found few advocates. One champion of women's rights, the radical artist William Dexter, did mount the public platform in Bendigo in August 1853, where he gave an impassioned speech advocating votes for women as well as for men. Unfortunately, the main account of this speech has come down to us through William Howitt, who, although he was later a supporter of women's suffrage, was so outraged by what he saw as Dexter's unpatriotic republicanism that he was principally concerned to disparage everything the artist said. His account does imply, however, both that Dexter argued the case for women's suffrage seriously and that he linked the prevalence of domestic violence on the goldfields to women's political powerlessness.[79] Howitt noted that Dexter's wife 'had figured in London amongst the lecturers on the Bloomer costume',[80] and indeed it does seem likely that Caroline Dexter was a strong influence on her husband. A writer and lifelong friend of George Sand, she followed William to Australia in 1854 and eventually settled in Melbourne.[81]

With this single brief flare, the flame of women's suffrage was extinguished and political rhetoric on the goldfields resumed its characteristically masculine form.[82] 'Is it manliness on our part to allow despotism and such a system to go on?', demanded Henry Holyoake of the diggers at a 'Monster

Meeting' held in Bendigo in December 1854. With resounding cheers the meeting endorsed a resolution for government: 'As all men are born free and equal this meeting demands a voice in making the laws they are called on to obey, and to have a full and fair share in the representation of the country'.[83]

No women appeared on any of these political platforms, nor do they seem to have attended the meetings. If they did, their presence was ignored: the speeches given at these meetings were reported in great detail in the goldfields newspapers and invariably began with the single greeting, 'Gentlemen'. Similarly, none of the reminiscences written by women who were on the diggings at the time gave more than a conventional summary of the licence question. Martha Clendinning had a clear view of the Eureka Stockade from her tent in Ballarat, but she attended none of the meetings and was principally concerned with her personal safety should matters get out of hand.[84] If she yearned for political inclusion she did not say so, although it is difficult to imagine from reading her determined, lively narrative that she would have rejected the opportunity had it been offered to her. Although we suspect that women assisted on the sidelines of the movement, helping to make the banners, sewing the flags,[85] their role was strictly that of 'help-meet'. Indeed, the editor



of the *Bendigo Advertiser* could not have expressed this exclusion of women from the world of politics more clearly than in his outraged defence of Lady Hotham in October 1854. He was responding to some remarks evidently passed at a licence meeting, and his comments underline the common assumption that the hurly-burly of politics was inconsistent with the nature of womanhood:

> We cannot conclude without giving expression to a feeling of disgust at the allusions made to Lady Hotham. In the name of all that's manly, wherefore visit upon her the faults of her husband. We hope to hear no more of this. If we throw down the gauntlet, let it be to men. It must not be said of a people contending for their rights, that their leaders evince a disposition to wage war on women. How on earth any gentleman could for a moment have thought of making use of her ladyship's name in a political question, we must say we are entirely unable to comprehend.[86]

In the final event, Victorian women had to fight long and hard for political equality. For all the early gains of their menfolk, they were the last to be enfranchised in Australia: the women's suffrage bill finally passed the conservative Victorian Upper House only in 1908.[87]

The quest for economic independence was to prove equally fraught. There can be little doubt that some women saw its value – Mary Fortune and Ellen Clacy battled the uncertainties of the literary world; Caroline Dexter attempted to combine art with social reform and writing in Melbourne; while Georgiana McCrae yearned to resume her career in portrait painting, against the firm opposition of her husband. Yet all of these women also knew what a precarious

PLATE 14.1 John Leech, *Alarming Prospect, the Single Ladies Off to the Diggings*, hand-coloured etching, c. 1855. (Rex Nan Kivell Collection, National Library of Australia)

PLATE 14.2 S. T. Gill, *Zealous Gold Diggers, Bendigo, July 1st 1852*, coloured lithograph, 1852. From *Victoria Gold Diggings and Diggers They Are*, Part 2. (Rex Nan Kivell Collection, National Library of Australia)

PLATE 14.3 George Clausen (snr), *Emily Skinner (née Fillan)*, watercolour, c. 1848. (Photo: Susan Duyker, courtesy of Marjorie and Nancy Conquest)

PLATE 14.4 George Lacy, *Digger's Wife in Full Dress*, watercolour, c. 1852. (National Library of Australia)

undertaking it was for a woman at this time.[88] Lola Montez knew the value of economic independence and gloried in the power it gave her. Performing to a full house in Castlemaine, she was hissed by one man in the audience. In the post-performance speech which had become her trademark, she declared defiantly that 'she was the maker of her own fortune . . . she was as rich as any man in the room, that she was worth £70,000 and didn't care a pin for anyone'.[89]

Other goldfields women settled for more moderate success. Martha Clendinning settled down as a prosperous doctor's 'lady wife', Elizabeth Ramsay-Laye accompanied her husband back to Britain, Emily Skinner retreated with relief to a cottage in Coburg,[90] and Lucy Hart counted her modest domestic blessings. Caroline Dexter confounded Melbourne society by founding an Institute of Hygiene and attempting in 1861 to establish a women's monthly magazine, but it ran to only two issues and in the same year Caroline married a prosperous young lawyer, many years her junior. From their home in Brighton she conducted elegant *salons* for Melbourne's literati.[91] Mary Fortune lived a lonely life, excluded from the masculine writers' club founded by Marcus Clarke and making a precarious living from her journalism and crime fiction. She died in poverty, reputedly an alcoholic.[92] And Sarah Davenport? We must hope that she found some respite from the washtub and that her feckless husband did not run through all of their hard-won gold earnings.[93] It seems that dreams of personal independence, economic or otherwise, were not yet for these women in gold-rush Victoria.



But this is not the end of the story. For while these women may not have succeeded in overturning the gendered distinctions of goldfields society, may not, indeed, have sought to overturn them, it would be a mistake to underestimate their personal autonomy. Mere presence on the goldfields represented a victory for some women, while others, like Martha Clendinning or Sarah Davenport, asserted their independence with some success, even where they were later forced to withdraw. All were spirited, intelligent women, determined and resourceful. Their marriages suggest a degree of actual equality at odds with prevailing ideology and with the law. Perhaps, after all, it should not surprise us that theirs was the generation which began the long process of extending equal rights to women.

**otes**

1 The actual (estimated) figures are 584,000 total and 163,800 females. W. D. Borrie, *The European Peopling of Australasia: A Demographic History, 1788–1988*, 1994, Research School of Social Sciences, Australian National University, Canberra, p. 71.

2 D. Goodman, *Gold Seeking: Victoria and California in the 1850s*, 1994, Allen & Unwin, Sydney, pp. 161–2.
3 C. Thatcher, *Thatcher's Colonial Songs: Forming a Complete Comic History of the Early Diggings*, [Coles Book Arcade, Melbourne, 1864], facs. edn 1964, Libraries Board of South Australia, Adelaide, pp. 49–50.
4 H. Anderson, *The Colonial Minstrel*, 1960, Cheshire, Melbourne, p. 10.
5 S. Davenport, 'Sceth of an emegrants Life in australia from Leiving England in the year of our Lord 1841 to 1869', State Library of Victoria, Manuscript Collection, MS 10541. Also published in L. Frost (ed.), *No Place for a Nervous Lady: Voices from the Australian Bush*, 1984, McPhee Gribble, Melbourne, pp. 238–64. For ease of access, page references cited here are from Frost.
6 Her record of these 'tryals' is spare but moving. 'i cold not cry one tear i was stund', she wrote of her baby son's death; and of the 'purmature labour', 'i was almost Dumb with grief'. Davenport, 'Sceth', in Frost, p. 242.
7 L. Hart, letter to her mother from Port Phillip, 3 May 1851, State Library of Victoria, Manuscript Collection, MS 8838. Extracts of the letter have been published in M. Aveling and J. Damousi (eds), *Stepping Out of History: Documents of Women at Work in Australia*, 1991, Allen & Unwin, Sydney, pp. 41–5.

8 'One step forward and two back.'
9 See B. Niall, *Georgiana: A Biography of Georgiana McCrae Painter, Diarist, Pioneer*, 1996, Melbourne University Press, Melbourne, p. 97.
10 E. Ramsay-Laye [alias Isabel Massary], *Social Life and Manners in Australia. Being the Notes of Eight Years' Experience by a Resident*, 1861, Longmans, London, p. 2.
11 Ramsay-Laye, *Social Life and Manners*, pp. v–vi.
12 A phrase coined by William Howitt, or at least repeated by him, in *Land, Labour and Gold or, Two Years in Victoria*, 2 vols, [1855] facs. edn 1972, Australian Historical Reprints Series, Sydney University Press, Sydney.
13 Howitt, for example, although a committed reformer, even, some might have said, a radical, constantly distinguished between gentlemen and others. For all his theoretical support of democratic principles, he found the politics of the goldfields disturbing and close proximity with the uncouth masses distasteful. See, for example, 'You cannot be on the diggings, indeed, without coming in contact with a very low, rude, vulgar, and base population, the scum of the earth', *Land, Labour and Gold*, vol. 1, p. 385; S. T. Gill explored these distinctions in his caricatures *Diggers of High Degree* and *Diggers of Low Degree*, reproduced in M. Cannon (ed.), *The Victorian Gold Fields 1852–3: An Original Album by S .T. Gill*, 1982, John Currey O'Neil, Melbourne, plates 18 and 19.
14 Ramsay-Laye, *Social Life and Manners*, pp. 18–19.
15 Ramsay-Laye's account was full of the networks she accessed, both on the fields and among the more prosperous squatters, but she also made it clear that the networks functioned as efficiently to exclude as they did to include, as Penny Russell has argued for Melbourne at this time. P. Russell, *A Wish of Distinction: Colonial Gentility and Femininity*, 1994, Melbourne University Press, Melbourne, esp. ch. 2.
16 Howitt, *Land, Labour and Gold*, vol. 2, pp. 201–2.

17 M. Clendinning, 'Recollections of Ballarat: A Lady's Life at the Diggings Fifty Years Ago', State Library of Victoria, Manuscript Collection, MS 10102/1, p. 7.
18 A description of such a table, 'made as all tables in a digger's tent were then', appears in Clendinning, 'Recollections', p. 12.
19 A term Elizabeth Ramsay-Laye herself used to describe their arrangements; see *Social Life and Manners*, p. 14. Even on the diggings the Ramsay-Layes managed to find a properly experienced 'housekeeper'/lady's maid.
20 Clendinning, 'Recollections', p. 16. Ellen Clacy described much the same arrangement as Martha Clendinning's; see Mrs Charles Clacy, *A Lady's Visit to the Gold Diggings of Australia in 1852–53*, [1853] 1963, ed. P. Thompson, Lansdowne Press, Melbourne, pp. 55, 63.
21 Ramsay-Laye, *Social Life and Manners*, p. 88.
22 E. Skinner, *A Woman on the Goldfields: Recollections of Emily Skinner 1854–1878*, ed. E. Duyker, 1995, Melbourne University Press, Melbourne, pp. 1–10.
23 Skinner, *A Woman*, pp. 51–2.
24 Skinner, *A Woman*, p. 55.
25 Skinner, *A Woman*, p. 58.
26 C. Woods, *Beechworth: A Titan's Field*, 1985, Shire of Beechworth, p. 28.
27 Skinner, *A Woman*, p. 66.

28 Skinner, *A Woman*, p. 84.
29 'I have remarked in my long experience on the Goldfields that nearly all the men seemed happier at mining than at any other work. No matter what hardships of heat and cold, wet feet, or watching at night, so long as they were working in their own claim with hopes of getting enough, they were right.' Skinner, *A Woman*, p. 54.
30 Clacy, *A Lady's Visit*, 1963, pp. 128, 152.
31 Howitt, *Land, Labour and Gold*, vol. 2, pp. 228–9.
32 Also called Little Adelaide in the early days. See Clacy, *A Lady's Visit*, p. 25.
33 Niall, *Georgiana*, p. 201.
34 Clacy, *A Lady's Visit*, pp. 154–6, adding that, while 'outwardly it seemed a city of gold, yet hundreds were half perishing for want of food, with no place of shelter beneath which to lay their heads'.
35 Among the reasons advanced to explain the large increase in the destitute was the failure of husbands either to return from the diggings or to send funds to their wives and children. See M. Geyer, *Behind the Wall: The Women of the Destitute Asylum Adelaide, 1852–1918*, 1994, Axiom, Adelaide, p. 12.
36 Variants of these stories appear in many gold-rush reminiscences. They probably served a dual purpose, both impressing on readers the wealth to be had from digging and also reminding middle-class readers that 'taste' could not be acquired as easily as gold. See, for example, Ramsay-Laye, *Social Life and Manners*, pp. 126, 166; Clacy, *A Lady's Visit*, pp. 153–4. Colonial artists satirised digger pretensions in caricature, for example Lacy's *Digger's Wife in Full Dress* (see Plate 14.4).
37 An extended discussion of women's occupations, rates of pay and the in-built imperative to marry can be found in M. Anderson, 'Good Strong Girls: Colonial Women and Work', in K. Saunders and R. Evans (eds), *Gender Relations in Australia: Domination and Negotiation*, 1992, Harcourt Brace Jovanovich, Sydney, pp. 225–45.

38 C. Perottet, letter written to her mother, 14 May 1856, State Library of Victoria, Manuscript Collection, MS 10257.
39 Thatcher, *Thatcher's Colonial Songs*, p. 48.
40 *Bendigo Advertiser*, editorial, 22 November 1854, p. 2.
41 Thatcher, *Thatcher's Colonial Songs*, pp. 67–8.
42 Howitt, *Land, Labour and Gold*, vol. 2, pp. 201–2.
43 Clendinning, 'Recollections', p. 19.
44 See, for example, cases reported in *Bendigo Advertiser*, August–September 1854.
45 The inability of a wife to support herself was sometimes used by men convicted of assault to avoid a gaol term. See case of C. Ludwig, heard in the Sandhurst Police Court on 19 August 1854 and reported in *Bendigo Advertiser*, 22 August 1854, p. 3.
46 Reported in *Bendigo Advertiser*, 25 August 1854, p. 3.
47 Although cases of wife-selling have been documented in Britain, there are few known instances in Australia. Nevertheless, on 10 October 1854, one James Hoos was so charged in Bendigo. Specifically, he was charged with 'being idle and disorderly, and assaulting his wife, whom, with his children and two goats, he had sold for the sum of £10'. Perhaps he was drunk or deranged. The *Bendigo Advertiser* noted only that 'as his better half promised to take care of him for the future, he was discharged' (10 October 1854, p. 3). See also S. P. Menefee, *Wives for Sale: An Ethnographic Study of British Popular Divorce*, 1981, Basil Blackwell, Oxford.
48 Mrs Chisholm, on the other hand, approved of low wages for women to ensure that they continued to marry. See P. Grimshaw et al., *Creating a Nation*, 1994, McPhee Gribble, Ringwood, Victoria, p. 88.
49 Quoted in Niall, *Georgiana*, 1996, pp. 321–2.
50 Howitt stressed several times that gentlemen might as well stay at home as attempt gold getting. *Land, Labour and Gold*, vol. 1, pp. 278–9. Most of the goldfields artists had originally intended to be miners but soon abandoned the attempt.
51 M. Kennedy's claim as discoverer was disputed and never formally acknowledged. G. Blainey, *The Rush That Never Ended: A History of Australian Mining*, 1963, Melbourne University Press, Melbourne, pp. 34–5. Davenport, 'Sceth', in Frost, pp. 262–3.
52 In 1854 there was some debate in Bendigo about whether women required licences. *Bendigo Advertiser*, 17 October 1854, p. 2. Examples of pictures showing women at work are Gill's *Zealous Gold Diggers* (see Plate 14.2); C. R. Read, *Thompson's Point, River Turon*, colour lithograph, State Library of Victoria, Picture Collection. Reproduced in A. McCulloch, *Artists of the Australian Gold Rush*, 1977, Lansdowne Editions, Melbourne, p. 129.
53 Skinner, *A Woman*, p. 44.
54 For example, see *Bendigo Advertiser*, 5 September 1854, p. 3.
55 Even in Melbourne, servants' accommodation was often makeshift and miserly, or so recorded Mrs Perry, wife of Archbishop Perry. A. de Q. Robin (ed.), *Australian Sketches: The Journals and Letters of Frances Perry*, 1984, Queensberry Hill Press, Melbourne, p. 72.
56 P. Grimshaw and C. Fahey, 'Family and Community in Nineteenth Century Castlemaine', in P. Grimshaw, C. McConville and E. McEwan (eds), *Families in Colonial Australia*, 1985, George Allen & Unwin, Sydney, p. 94.

57 S. T. Gill, *Sly Grog Shanty*, in Cannon, *The Victorian Gold Fields*, pp. 80–1, and *Hotel Bar*, in McCulloch, *Artists of the Australian Gold Rush*, p. 85.
58 With heavy sarcasm Ellen Clacy and her party dubbed one such woman 'the amiable female'. 'She was a most repulsive looking object. A dirty, gaudy-coloured dress hung unfastened about her shoulders. Coarse black hair unbrushed, uncombed, dangled about her face, over which her evil habits had spread a genuine baccanalian glow, whilst in a loud masculine voice she uttered the most awful words that ever disgraced the mouth of man – ten thousand times more awful when proceeding from a woman's lips.' Clacy, *A Lady's Visit*, pp. 62–3.
59 'Poll the Grogseller', originally published in *Thatcher's Colonial Songster*, in Thatcher, *Thatcher's Colonial Songs*, p. 75.
60 *Bendigo Advertiser*, 25 October 1854, p. 2.
61 Ramsay-Laye, *Social Life and Manners*, p. 152.
62 Although the *Age* critic was prepared to praise where the *Argus* had condemned. See citations in B. Seymour, *Lola Montez: A Life*, 1996, Yale University Press, p. 338.
63 Céleste de Chabrillan, wife of the French Consul, Lionel le Comte de Chabrillan, had hoped to leave her past behind in Paris when she sailed for Melbourne with her new husband in 1854. She might have succeeded better had she not published some particularly frank memoirs just before her departure. Still, Australia seemed a long way away. Unfortunately, for her, the memoirs arrived in Melbourne before she did, and Céleste was to spend a lonely few years in the colony, shunned by officialdom, omitted from the invitations that arrived for her husband, and ignored by the society ladies who might otherwise have competed with each other to entertain a French countess. Lola Montez was prepared to brave a visit, no doubt to the delight of the Melbourne gossips. C. de Chabrillan, *The French Consul's Wife: Memoirs of Céleste de Chabrillan in Gold-Rush Australia*, [1877] 1998, trs. P. Clancy and J. Allen, Miegunyah Press, Melbourne.
64 Clendinning, 'Recollections', p. 29.
65 Davenport, 'Sceth', in Frost, pp. 261–2.
66 Clendinning, 'Recollections', p. 5.
67 Mrs A. Campbell, *Rough and Smooth, or Ho! for an Australian Gold Field*, 1865, Hunter, Rose & Co., Quebec, pp. 57–8.
68 As Martha Clendinning's subsequent history suggests. And for all her determination to be simply a 'respectable woman of business', Martha was not above playing the lady when it suited her. See, for example, Clendinning, 'Recollections', p. 11.
69 See Ramsay Laye, *Social Life and Manners*; Clacy, *A Lady's Visit*, p. 9; Campbell, *Rough and Smooth*, preface.
70 D. Goodman has argued persuasively that historians should read these reminiscences for both their form and their content, in 'Reading Gold-Rush Travellers' Narratives', *Australian Cultural History*, no. 10, 1991, pp. 99–112.
71 L. A. Meredith, *Over the Straits: A Visit to Victoria*, 1861, Chapman & Hall, London.
72 See, for example, her piece entitled 'Twenty-Six Years Ago or the Diggings from '55', in L. Sussex (ed.), *The Fortunes of Mary Fortune*, 1989, Penguin, Harmondsworth, pp. 110–11.
73 Sussex, *The Fortunes*, pp. xv–xviii.

74 J. Molony and T. J. McKenna, 'All that Glisters', in *Labour History*, no. 32, May 1977, p. 39. Goodman, however, argues that the mastery of local detail in *A Lady's Visit* suggests otherwise, see 'Reading Gold-Rush Travellers' Narratives', p. 108.

75 All of this information was uncovered by M. Rosalyn Shennan and is contained in 'A Goldfields Adventurer', unpublished paper, 2001. I am grateful to Rosalyn for allowing me to read and cite her manuscript.

76 Ramsay-Laye, *Social Life and Manners*, p. 42.

77 On her return to Melbourne sometime later, Elizabeth encountered him at a Government House ball and wondered, watching him 'gaily dancing', whether he felt any remorse at her friend's death; see *Social Life and Manners*, p. 94.

78 Mary Fortune's unfaithful wife was also abandoned by her lover, while her vengeful husband took away their child. She became a 'haggard barmaid' on the diggings. Sussex, *The Fortunes*, p. 66.

79 Howitt, *Land, Labour and Gold*, vol. 1, p. 407. Howitt recorded the speech in these terms: 'He was a citizen of the world, and liked all nations – he might have said better than his own. He would vote for women having votes as well as men; and hoped to see the day when every man, woman, and *child* had a vote. I expected he would include the whole household, dog, cat, mice, and all . . . "I have seen", he continued, "men on the diggings thrashing their wives. I don't call that liberty". For my part, it seemed to me taking a very great liberty indeed.' Dexter also claimed to have painted the first digger's banner, complete with the traditional symbolism of labour and trade unionism, as well as the kangaroo and emu to represent Australia (vol. 1, p. 406). Unfortunately, the *Argus* (19 August 1853, p. 3), which reported on the meeting, only summarised that part of Dexter's speech in which he explained the symbolism of the banner.



80 Howitt, *Land, Labour and Gold*, vol. 1, p. 406.

81 Details on Caroline Dexter are slight. See, however, her entry in *Australian Dictionary of Biography*, 1851–90, pp. 64–5, and a brief summary in A. McCulloch, *Encyclopedia of Australian Art*, 1984, vol. 1, Hutchinson, Melbourne, pp. 283–4.

82 Existing accounts of the women's suffrage movement in Victoria seem unaware that the issue was even raised at this point. Audrey Oldfield suggests that universal suffrage was first mentioned in 1857; see *Woman Suffrage in Australia: A Gift or a Struggle?*, 1992, Cambridge University Press, Cambridge, p. 131.

83 Reported in *Bendigo Advertiser*, 6 December 1854, p. 2. The masculinism inherent in early concepts of citizenship is analysed in M. Lake, 'The Politics of Respectability: Identifying the Masculinist Context', *Australian Historical Studies*, vol. 22, no. 86, April 1986, pp. 116–31.

84 Clendinning, 'Recollections', p. 25.

85 Grimshaw et al., *Creating a Nation*, p. 100.

86 Editorial, 'The Anti-License Agitation', *Bendigo Advertiser*, 17 October 1854, p. 2.

87 Miners effectively gained the right to vote through their miner's rights in 1855. Formal manhood suffrage followed in 1857. G. Serle, *The Golden Age: A History of the Colony of Victoria, 1851–1861*, 1963, Melbourne University Press, Melbourne, pp. 177, 264–6. Although Oldfield argues that the Upper House was only one factor in the delayed franchise in Victoria, Serle argued in 1963 (p. 378) that the

conservative influence of the Upper House on Victorian politics was consistently underestimated by historians.

88 Sussex, *The Fortunes*, p. xxi. Ellen Clacy supported herself with her writing until her death at the age of 70 in 1901. Since she left no will, the extent of her estate is unknown. Shennan, 'A Goldfields Adventurer', p. 10.

89 Reported in *Mount Alexander Mail*, 11 April 1856, pp. 2, 5.

90 Where she died after a stroke at the age of 58. Skinner, *A Woman*, p. 4.

91 J. S. Ryan, entry for Caroline Dexter in *Australian Dictionary of Biography*, pp. 64–5. The journal, the *Interpreter*, appeared in January and February of 1861.

92 Sussex, *The Fortunes*, p. xxi.

93 In fact, she lived to a ripe old age and the family seems to have acquired some property, but what this meant for Sarah we can never know. Frost, *No Place*, p. 264.

# After the Gold Rush

## MATERIAL CULTURE AND SETTLEMENT ON VICTORIA'S CENTRAL GOLDFIELDS

**Chapter 1**

*Susan Lawrence*



Gold made the colony of Victoria, giving it wealth, population and a distinctive sense of itself as a society. Almost overnight the population ballooned as tens of thousands of men and women from English-speaking countries around the world crowded in, and Aboriginal people were decisively outnumbered. For the first and only time until after World War II, Australia was also the destination for significant numbers of non-English speakers from a diverse range of European and non-European countries. The new arrivals demanded a new infrastructure of transportation routes, commercial facilities, and institutional services. Forests were felled and sent underground, and farms and market gardens sprang up on the cleared land. Political reform gave all white men the vote and laid a basis for Federation.

One hundred and fifty years after the first discovery of gold in Victoria, modern understandings of the gold rush are telescoped into the first few, frantic years of the early 1850s when the colony seemed in chaos and life was an adventure.[1] But the influence of gold did not end in the 1850s. Gold and goldmining infused Victorian society because of its sustained economic and social influence. While one was bringing prosperity to Ballarat and Bendigo and building 'marvellous Melbourne', gold seeking was also shaping the intimate details of the lives of those who sought it. Small goldfields communities served as homes for thousands of people in nineteenth-century Victoria. The region south of Ballarat is a good example. At the end of the 1800s it was studded with people, in large mining towns like Ballan and Buninyong, in smaller centres like Mount Egerton, and in scores of digging communities with populations in the hundreds.

One of the latter was the Moorabool diggings, about forty miles west of Melbourne. Rushed in 1859 and actively mined into the 1880s by small local companies and the holders of miner's rights, it is typical of the 'poor man's

**Map 15.1** The Moorabool diggings, west of Melbourne. (Drawn by Steve Fieldes and Wei Ming)

diggings' that persisted around Victoria in the wake of the gold rush. Closer examination of this forgotten community suggests what happened when those who came to the colony as transient gold seekers decided to remain. Many chose not to become wage labourers but to retain their independence by working at subsistence mining. Subsistence miners, men who held miner's rights, defined themselves by their search for gold but often worked at a variety of labouring jobs in order to sustain themselves. Subsistence mining entailed the work of women and children too; it was their gardens and flocks which provided both food and income. Evidence derived from archaeological and documentary analysis of the Moorabool can be used to chart something of the transformation from diggings to settled village, and how the change was negotiated in the demographic structure of the settlement, in the homes built and gardens planted, in the consumer goods used and the foods eaten.[2]

One of the most significant aspects of goldmining communities like the Moorabool was the presence of families. While there is no doubt that as late as the 1860s adult men continued to outnumber adult women in the colony, historians have tended to overlook the considerable numbers of women and children in early Victoria. Women actively shaped the ways in which the colonial community operated and defined itself. They influenced the economic base of the settlement, the ways in which work was organised, diet, the appearance of the homes, social activities, and more.

The census of Victoria for 1861 provides the most detailed demographic picture of the people living on the Moorabool. The Moorabool consisted of the three nearby diggings of Dolly's Creek, Tea Tree Creek and the township of Morrisons, and in April 1861 the census-taker recorded 681 people living there, including 345 men, 122 women and 214 children. Most of the adults were young, 70 per cent of them in their twenties and thirties. Only four people in the settlement were over the age of 60. A good many of the adults were married. In the whole of the County of Grant, of which the Moorabool was a part, nearly 70 per cent of the women under 40 were married, as were nearly 50 per cent of the men of that age.



On the Moorabool, one hundred of the adult women told the census-taker that they were married. This figure may not be entirely accurate, as it was not uncommon for women and men to live in de facto relationships while identifying themselves as married. Single women on their own also sometimes had reason to say they were married. Marriage conveyed both protection and respectability. Women engaged in prostitution were particularly likely to disguise their marital status or occupation or both.[3] As there is no means of testing it, the number must be taken as a reasonable, though possibly high,

**Table 15.1** Moorabool population, 1861

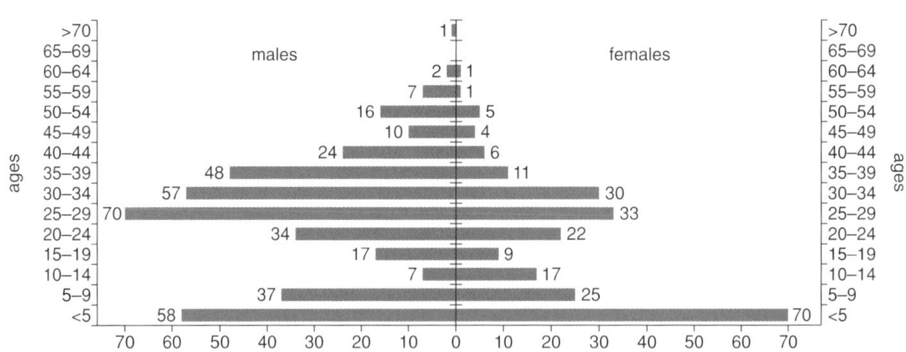


*Source:* Census of Victoria

estimate. However, the number of married women provides a basis for estimating the number of married couples as well. Women in relationships were less likely than men to be on the diggings without their partners and it can be assumed that the number of women identified as widows, or obviously living apart from their partners, was small. Only 2 per cent of the married women in the county were recorded as widows and the percentage on the Moorabool was probably similar.

Assuming that the number of married women was close to the number of married men living with their partners on the goldfield, there were approximately one hundred couples in the settlement. Given their young age, most of those couples probably had children and, factoring in the census-taker's count of the number of children in the settlement, there were two or three children per family. However, given the ages of the children and the average family size at the time, it is likely that the families on the Moorabool were young and still growing. Two-thirds of the children enumerated were less than 5 years of age and only one child in ten was over the age of 10. An indication of the average family size at that period comes from Castlemaine. The average woman giving birth there in 1861 bore seven or eight children over a period of fifteen to twenty years.[4] So a woman with only two or three children in 1861 was probably in the early stages of her reproductive life.



While these figures are speculative, they do provide some indication of the kinds of households that would have been found in the settlement. The emerging picture shows that by 1861 the Victorian goldfields were far from being exclusively or even predominantly male preserves. Half of the population were adult men, but one-third of the population were children. The youth of the population and the high incidence of families would have decisively affected the experience of all those in the settlement. Despite the many unattached men, both women and children are likely to have found numerous companions. Other data suggests that extended family relationships as well as many nuclear families were also a significant part of the community structure. Many of these families on the diggings either were related or became related over time as their children grew up and married.

Historians of the family have demonstrated that, in frontier areas, large kin networks developed quite quickly as early migrants encouraged friends and relatives to migrate and as migrants intermarried.[5] Marriages that linked individuals and families also generated economic and political networks. Business connections between apparently unrelated men in frontier towns were frequently kin alliances centred on the marriages of several women from the same families. Penny Russell has shown that it was the dense network of intermarriage and extended family groupings which supported the structure of genteel society in colonial Melbourne.[6] A similar network of kin ties underlay the otherwise disconnected community structure of the Moorabool diggings, and women's lives in particular were shaped by kin structures that they created and maintained. In such circumstances, neither men nor women necessarily experienced the goldfields as congeries of hostile or indifferent strangers.

Although the grave markers have not survived, cemetery records provide evidence of the webs of family at Dolly's Creek and Morrisons.[7] In 1873 a diphtheria epidemic swept the settlement and among those who died were four little girls, Jessie and Fanny McIntosh and Margaret and Elizabeth McNab. According to their burial records, Jessie and Fanny's father, Robert, was the brother of Margaret and Elizabeth's mother, Grace, making the girls cousins. All the girls were between 4 and 8 years of age and their families had been at Dolly's Creek since at least 1865. The two families formed a close-knit group. All four adults had come from Scotland, and Robert and Alex McNab, his brother-in-law, were both miners, probably working as mates. The two sisters-in-law, Kate and Grace, bore their children at Dolly's Creek and it is easy to imagine each assisting the other during the births. With only four years separating the little girls, they must often have played together.

Families provided companionship and assistance, which the McNabs and McIntoshes would have sorely needed when diphtheria struck. First, 6-year-old Margaret died, followed the next day by Jessie, the oldest, and then a week later by the little ones, Fanny and Elizabeth. The contagion may even have been spread by the loving hands of aunts and uncles who moved between the houses



to do what little they could. Although only one-sixth of the population at Dolly's Creek was adult and female, incidents such as this tragedy corroborate the impression that the settlement was far from being a masculine community. A single woman alone or with only her husband might have felt very isolated in a predominantly male mining settlement but a woman surrounded by sisters, sisters-in-law, cousins or even a mother is likely to have found ready allies in times of need.

The McCormack family provides one example of the rapid creation of a large extended family. Mary McCormack's husband, Thomas Haines, had come to Morrisons as a miner, although he later took up a selection and became a dairy farmer. Mary's father and her three brothers were also miners, one of whom, William, married Sarah Lavery whose father, John, was a farmer at Morrisons.[8] The marriages of these two women made the Lavery, McCormack and Haines households into a clan, providing workmates and companionship for both men and women.

This is not to say that the lives of subsistence mining families were easy. Homes were often little more than shelters of calico and bark, and furnishings were whatever could be packed on a small cart or made on the spot. During archaeological investigations at Dolly's Creek in 1992, the remains of four houses were excavated. The buildings themselves proved typical of the rough, temporary structures generally associated with goldfields architecture, although some of the household goods recovered belied their makeshift surroundings.

Artists and writers have provided many descriptions of homes on the goldfields which suggest what those on the Moorabool might have been like.[9] Transient miners who had to build, transport and rebuild their camps chose

**Figure 15.1** Maddy Atkinson and Greg Jackman excavating at Dolly's Creek. (Photo: Susan Lawrence)

expedient methods and locally available materials such as calico, blankets, timber and bark. This resulted in buildings that were simple in design but which varied considerably in form and material. At Bendigo in 1852, the Clacy family lived initially in canvas tents purchased in Melbourne. Later they built a shelter using branches from a nearby tree as tent poles, covered with blankets carried in their swags. Antoine Fauchery described his tent of canvas and branch poles as a 'white forage cap'. Pictorial evidence of similar tents is featured in the work of other goldfields artists, including that of S. T. Gill and Edward Snell (see Griffiths and Platt's chapter).

Some impression of the range of buildings on the Moorabool is conveyed in the census of 1861. The census-taker categorised houses by type of building material and by number of rooms and also recorded how many people were living in each type of house. He recorded 215 buildings in which 619 people lived. For the remaining sixty-two people, no building type was noted. Of the recorded homes, 54 were of wood or stone but the majority, 161, were of bark or calico. The census defined tents as any building with a calico roof, so it is possible that some 'tents' had bark or timber walls and glass windows. There was archaeological evidence of this at three of the houses excavated. Most people recorded in the census lived in tents; only 200 occupied wood or stone buildings.

All of the houses excavated were canvas structures, although some stone was used in the lower walls of one of the buildings. The buildings are typical of the kind of domestic accommodation found on the diggings. Collapsed mounds of mud and stone that once formed chimneys are still visible at the house sites. They stand at one end of levelled terraces cut into the hillside, and the terraces are partly defined by drainage ditches still faintly visible. The size of the houses can be estimated from the area of the terraces and ditches. These terraces are approximately 16 × 12 feet, and the tents would have been slightly smaller, the size of a single small room. To modern eyes the one-roomed tents appear impossibly small, but the artisans and unskilled labourers who found their way to the Victorian goldfields were familiar with such spaces. In scale, the houses were similar to the workers' homes they had left behind elsewhere in Australia and the United Kingdom. In Adelaide, Melbourne and Sydney during this period, working families lived in one-room row cottages only twelve foot square, which were modelled on typical workers' housing in England.[10]



Hardware used in the buildings provides some idea of their construction. Square-cut iron nails and round wire nails were used to hold the timber frames together, while brass tacks were used to fasten the calico covers in place. Window glass was also found at the houses, though, so there must have been sufficient framing in the walls to hold at least one window at each house. Ditches were dug to channel rainwater away from the interior of the buildings. Diarists such as Edward Snell and Seweryn Korzelinski describe having to do this during the rainy winter months. It was a custom evidently improved by trial and error, as experiments by those who lived at one Moorabool house testify. Archaeological excavation showed that the first attempt was a shallow scooped ditch dug along the uphill side of the tent. This soon proved inadequate, however, and a second, longer ditch was constructed which lay further from the tent and which carried the water into a gully nearby. This second ditch was deeper than the first and, in order to improve drainage, had straight sides, a flat, level base lined with rubble and sheets of iron and tin.

Homes were frequently dominated by fireplaces that occupied most of one end of the tents. The fireplaces were built of roughly shaped quartzite blocks held together by mud mortar. Once the houses were abandoned, the mud washed out from between the stones and the fireplaces collapsed, leaving mounds approximately three feet in height. When they were still standing, the fireplaces were probably about chest height, with chimneys made of wood, corrugated iron, or even green bullock hides stretched around branches. All of these varieties have been described or drawn by gold seekers.[11] Sometimes barrels were added to give extra height, but fire was a constant threat, as diarists frequently complained. Many tents on the Victorian goldfields had no fireplace at all. If people only expected to stay a short time, it was easier simply to cook outdoors over an open fire in front of the tent. However, when people stayed longer on a field, or if the weather turned cold and wet, fireplaces would be built and the tents made more comfortable and convenient. The prevalence of fireplaces at Dolly's Creek thus suggests that the diggings were occupied in the winter months and that the settlement was relatively permanent.

Someone who lived at one of the houses had taken the trouble to finish the hearth with a render of kaolin, or pipeclay. This fine white clay is frequently found in the shallow mine shafts on the Moorabool and evidently had been brought from the mines to the house. Applying the pipeclay render was a job usually done by women and girls. This chore was vividly remembered by Elsie (Argent) Bayard, who grew up at Morrisons early in the twentieth century, the youngest child in a large mining family. When Elsie was a child her mother regularly carried hessian sacks of pipeclay back from the family mine. Every day Mrs Argent would mix the pipeclay with water to form a whitewash and carefully apply a fresh coat to the fireplace in the kitchen.[12] Given the size of the fireplaces and the way they physically dominated the small tents, whitening them in this way would have made the interiors much brighter and lighter.

One of the houses excavated was markedly different from the others and also unlike those drawn by artists on the goldfields. There were at least two buildings in the complex and both were built partly of stone. Only a corner of one remained standing, so it cannot be described, but the other was still  remarkably intact. It was a one-room structure, approximately the same size as the buildings at the other two houses excavated: 12 × 7 feet. The lower part of the walls, up to about waist height, was built of roughly shaped blocks of stone held together by mud mortar. Nails, tacks and iron spikes suggest that above the stonework stood a timber-framed structure that was originally calico, probably a tent, but which over time had been made more solid. A fragment of the building material survived under a section of collapsed wall, revealing a complex story of home improvements.

The fragment comprised a multi-layered sandwich of paper, fabric and tin. In the centre was a layer of calico, the original tent. That had been strengthened by the addition of several layers of newspaper and a layer of heavy, embossed cardboard. Then, on the exterior, sheets of tin had been nailed, making the structure snug and watertight. The interior was not left as plain calico, but had been covered in wallpaper with a delicate foliage pattern that completed the transformation from temporary tent to permanent home. Other building debris confirms the substantial nature of the structure. It boasted at least two glass windows, one in the front wall of the house beside the door. The door itself was a solid affair, with iron hinges, a strong lock and a brass doorknob. The door and window were placed in the long wall of the house, and a brick fireplace, rather haphazardly built in comparison to the rest of the structure, occupied the house's gable end.

The sturdy construction of the buildings at the Stone House requires some explanation. There must have been good reason to spend the time and energy in building stone walls. If the building was a store or pub, as the artefacts seem to indicate, the added security of stone would make sense. Tent stores on the goldfields were frequently robbed by thieves who used knives to cut through the canvas walls at the back of the store and then pulled goods through the holes.[13] A storekeeper-cum-publican determined to prevent robberies would see the benefit of a building more substantial than those of ordinary householders. Identifying the building as a store or pub does not, however, preclude it from having been used as a residence, because it was not uncommon for storekeepers to live in their shops, just as they did in inner-city neighbourhoods.[14] The artefacts found at the Stone House – nails stored in an old pickle jar, a stack of shovels and a cask, maybe full of grog, maybe of water – echo the contents of a shop, while the fireplace for warmth and cooking and the pretty wallpaper suggest a home as well.

These houses built by subsistence miners typify the delicate process of accommodation between freedom of movement and a modicum of comfort. The structures were small and simple in style, made of readily transportable materials, such as calico, or readily available ones such as timber, bark and local stone. At the same time, the houses had substantial fireplaces and even wallpaper. The household goods carried by subsistence miners were similarly both expedient and fashionable. Furniture was made on the spot by bush carpenters, using odds and ends of branches and posts. Other household goods tended to be minimal, not necessarily because of poverty but in order to reduce the quantity to be packed and carried. At the same time ceramic dishes were often chosen instead of more utilitarian tin, and their varied and multi-coloured patterns did double duty as ornament. Further ornamentation was provided by fabric – chintz wall-coverings, sacking carpets – and by small, portable items such as brass mantel clocks.



The artefacts from Dolly's Creek not only tell much about the homes that people made there, but also what people ate, drank and wore. Through them we glimpse who the inhabitants were and how they constructed their social worlds. The overall impression conveyed is of people using what fitted most conveniently into a mobile lifestyle, choosing selectively to create comfort and signify decent respectability. On the whitened mantelpiece in one house a clock stood in its ornamental brass case. Perhaps the pressed glass dishes found during the excavation stood there as well, or some of the transfer-printed plates, cups and saucers. The multitude of patterns and colours represented would have created a brilliant display within the overall austerity of the home.

Decoration was also achieved through the use of floral-patterned wallpaper. A teapot stood with mixing bowls on a table or a shelf, together with bottles of condiments, medicine and beverages. An iron hook held a pot or a kettle over the hearth, ready for preparing meals. Slate pencils, clay pipes, ink bottles and a thimble and needle, all recovered during the dig, suggest a rounded life of study, work and relaxation. At night the rooms were lit by the fires, supplemented by candles stuck in bottles or on saucers and by kerosene lamps: parts of brass lamp bases and glass chimneys were found at two of the houses. In the 1860s kerosene was still a recent innovation which was in the process of displacing whale oil, so these households were very up-to-date.

For the most part furnishings were minimal, consisting of beds, tables and cupboards often made on the spot with available materials.[15] Little archaeological evidence of furniture has survived at Dolly's Creek. However, there were two iron drawer-pulls, of different sizes and styles, so probably (but not necessarily) from different pieces of furniture. A brass mantel clockcase was the only other item of homeware excavated. An elaborate piece, it stands seven inches high and rests on two curved feet. The space for the central dial is surrounded by floral motif moulding and the original surface was probably gilded to produce a French ormolu effect. The mechanism itself was of a standard size and form – a two-inch dial, probably with a white enamel face – of a type available for separate purchase. Similar cases and clocks were mass-produced and commonplace. While only the fairly well-to-do could afford to carry large pieces of furniture with them, smaller household goods like these could be transported more easily. Portable consumer items surviving within the archaeological record reveal otherwise lost or hidden dimensions of goldfields homes.



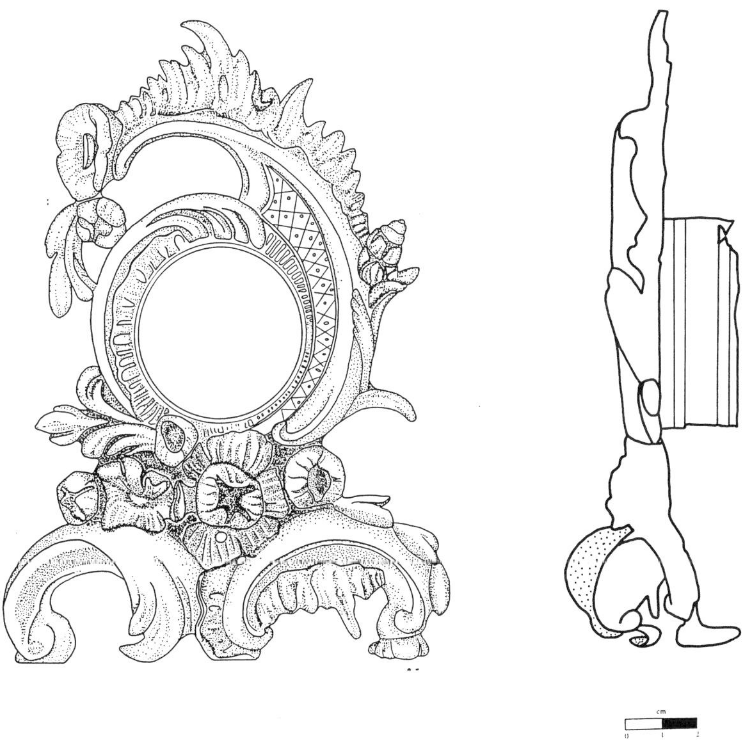

**Figure 15.2** Brass frame for a mantel clock found at Dolly's Creek. (Drawn by Michelle Denny)

Like the tents and the furnishings, kitchen equipment described in the written documents was for the most part basic and expedient, reflecting the need to be easily transportable. In addition to mining tools, the goods mentioned most frequently were a kettle or billy, a frying pan, a dutch oven, a tin plate and mug, and a knife, fork and spoon. Written evidence emphasises the ubiquitous use of tin dishes.[16] Here the archaeological evidence provides a direct and dramatic contrast. Large quantities of broken crockery were recovered from all four houses excavated, which suggests that ceramic plates and cups were far more popular than contemporary participants would suggest. Had gold seekers been using only tin dishes, there would have been no crockery to leave behind.

At least seventy-six dishes were represented at Dolly's Creek, all in fragments. Moreover, these are only the dishes that were broken and thrown away. Each household would have used many more dishes than this each day. All of these dishes were made from common, inexpensive earthenwares but were decorated with diverse patterns: at most, only two dishes matched at any one house. More than forty patterns were found, a few being popular enough to be used on several dishes. Asiatic Pheasant and Blue Willow were the most popular: five dishes of each were found, and every house had at least one dish in each of these designs. Thirteen other patterns were also found at two or more houses, but all others appeared only once. No household attempted to display matched sets or even to set a table with all dishes of uniform colour. Most designs had been transfer-printed in monochrome, primarily in the colours of blue, mulberry, sepia and green, but a few were in red or black.



Matching or not, the fact that the dinner plates, teacups and saucers of these goldfields households were ceramic, not tin, comes as something of a surprise. Documentary evidence suggests that tin was preferred because it was versatile and portable; tinwares were light, they did not break and they could be put on the fire. Pottery, on the other hand, was fragile, heavy and entirely impractical. Itinerant gold seekers choosing crockery to use in their bush camps were clearly aiming to create permanent settlements and to signal their membership of respectable society. Blue Willow teacups and Asiatic Pheasant plates were markers of gentility, used because of their implied social qualities

**Table 15.2** Distribution of excavated fragments of ceramics at Dolly's Creek

| | Garden House | Fireplace House | Clock House | Stone House |
|---|---|---|---|---|
| dinner plates | – | 5 | 5 | 5 |
| teacups | 1 | 6 | 16 | 3 |
| saucers | 1 | 3 | 7 | 3 |
| other | 1 | 3 | 15 | 2 |
| Total | 3 | 17 | 43 | 13 |

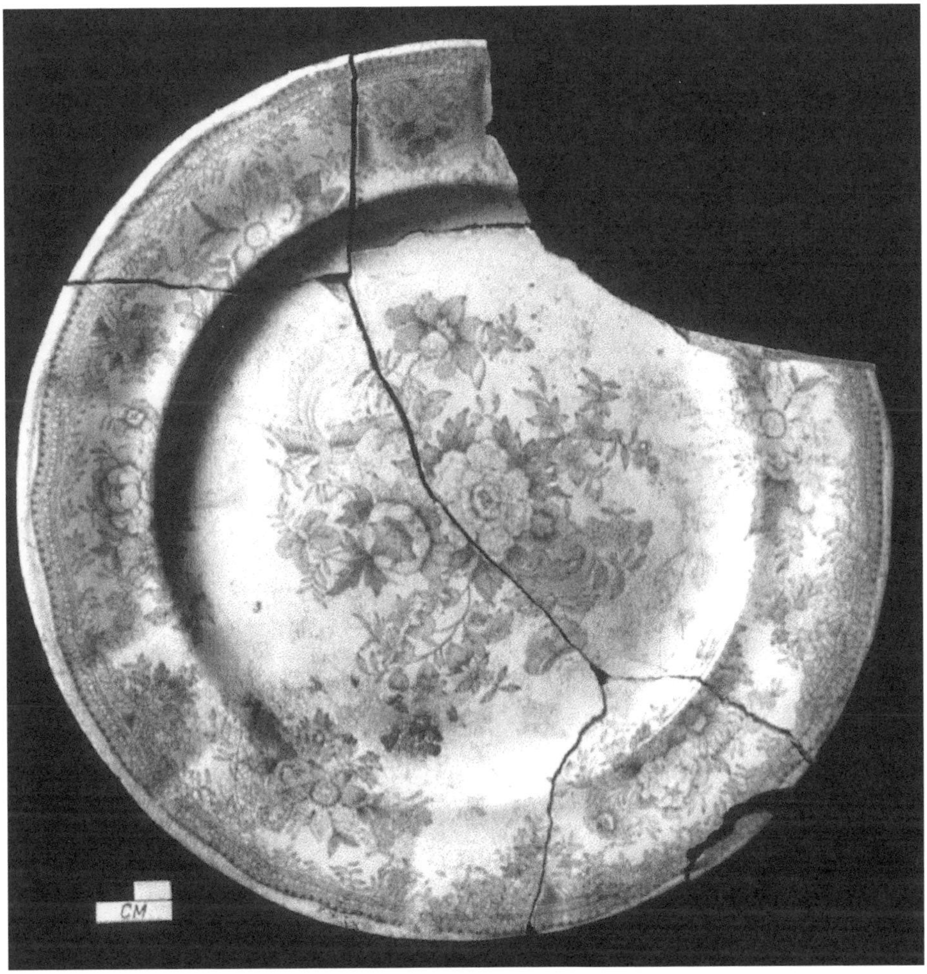




**Figure 15.3** A dinner plate in the popular Asiatic Pheasant pattern.

and distinctions. The choice of other household goods, including a stemmed serving dish in pressed glass, a ceramic teapot, the wallpaper and the mantel clock, all reinforce this impression. These were the types of objects that denoted membership in decent and self-respecting society.

The pervasive presence of crockery at Dolly's Creek is indicative of its widespread adoption among all those living on Victorian diggings. Despite what written evidence attests, physical traces left by the people themselves demonstrate that everyone, male and female, settled and on-the-move, preferred to eat their meals and drink their tea from ceramic plates and cups whenever possible. Perhaps the diarists and letter writers, in emphasising the exotic and romantic nature of their experiences, neglected these more mundane and ordinary aspects of their lives. Goldfields rubbish, carelessly or at least unconsciously discarded, tells another story. Symbolically embedded in their crockery is the deeply felt longing for order and domestic predictability, for a secure base upon which it was possible to build, accumulate and improve. Perhaps crockery was a tangible way of reaching for what David Goodman has called 'quietness': a settled peace-of-mind many outside observers felt to be lacking on the goldfields.[17]

Archaeological evidence of gardens and food remains hints also at other ways in which the goldfields were changing by the 1860s. By then there was often space and opportunity to keep domestic animals and to grow crops and vegetables, made possible through the provisions of goldfields commons and the terms of miner's rights. The land on most diggings was set aside as goldfields commons, which reserved the area for the public interest. Land within the boundaries of the common could not be sold and miners worked that land by applying for miner's rights and leasing claims.[18] Commons set the land aside for the purposes of mining and also provided miners and their families with a place to live. By purchasing a miner's right, miners were entitled to build and maintain a dwelling on a block of land (20 perches in the Ballarat District) on the goldfields common, as well as having the right to work a claim, and they were also permitted to graze livestock on common land.



From the physical remains excavated, it is evident that these mining families were often creative and enterprising in working their land. At three of the houses, fist-sized quartz cobbles had been arranged in deliberate patterns of lines and shapes to form pathways and borders around garden beds. Unfortunately, the topsoil in this region is very thin and poor, and no archaeological evidence survived of the plants or enriched garden soils. The stone arrangements were the only direct evidence of gardening found. However, indirect evidence of related sorts of domestic enterprise was recovered. The faunal remains – that is, the scraps of bone left from cooking and eating meat – came from cattle, pigs, sheep, goats, rabbits and chickens. Eggshell also survived. If people had cattle and goats, they likely had access to milk, butter and possibly cheese, although none of these would show up in archaeological evidence. Popular representations of the Australian diet in the nineteenth century, and particularly on the goldfields, consist of little more than mutton chops, damper and tea. The bones of Dolly's Creek tell a different story.

As Margaret Anderson has shown us, the recollections of Emily Skinner, who lived on the goldfields in north-eastern Victoria in the 1860s, provide further insight into household gardens and food production.[19] Her first garden grew 'peas, beans and all kinds of salad stuffs', including radishes, as well as potatoes, cabbages and onions. Later, on another goldfield, she again had a 'very good garden, filled with fruit trees and vegetables. Nearly every kind of English fruit grew well with us and others, such as Cape gooseberries, pomegranates and melons of all kinds.'

Their house had a wide verandah, covered with creepers, and a flower garden in front in which Emily tried without success to grow some of the local native plants. Behind the house stood a large garden and a yard fenced off for the goats and fowls. The Skinners had a large flock of goats, which wandered on their own during the day but returned in the evening to be milked, and Emily made butter from the milk. She wrote that 'nearly all the people kept cows or goats, pigs and poultry of all descriptions and grew their own vegetables'.

Contemporary accounts of the goldfields associate gardens with women. Emily Skinner certainly invested the gardens she saw and created with a particular set of symbolic meanings associated with women and domestic comfort. One such example was provided by the prospective arrival at the diggings of a new bride. When a neighbour and long-time friend of the Skinners was about to be joined by his new wife, recently arrived from England, Emily and others participated in the preparations: 'We had all helped to "fix up" the house and had then got a few things growing in the small garden which bye and bye was to be a wonder of beauty and neatness'.[20] Establishing a garden was likewise one  of the first tasks she herself had undertaken when a new bride on the goldfields. In both instances, the husbands had lived on the goldfields for some time, yet only when they were joined by their wives were 'homes' established. Gardens, as Suzanne Hunt's chapter shows us, were an important part of this process. The creation of a garden was a public indication that women were there and that domestic comfort was complete.

In other cases, women extended gardening activity into managing the small farms which economically underpinned subsistence goldmining. Joseph Jenkins, an itinerant labourer on the Victorian goldfields in the 1870s, wrote in his diary about the circumstances of his friends, the Evans and Lewis families. In both cases, the men mined and prospected while the women did much of the work involved in managing the farms.[21] Mrs Lewis' work was particularly critical for her family because, in addition to running the farm single-handedly while the men were away prospecting, she kept a cow and goats. Money made from selling milk and butter kept the family going when, as was frequently the case, prospecting was unsuccessful. Louisa Lawson, feminist and mother of author Henry Lawson, similarly spent her childhood and young adulthood on the goldfields of New South Wales during the 1860s and 1870s. Her husband, Peter, was away mining more frequently than he was home, and she and her young children did all the work of ploughing the land and planting and harvesting crops, raising cattle, keeping a dairy herd and growing vegetables. In addition, Louisa managed the home and raised five children, ran a shop and was the local postmistress.[22]

Women on the Moorabool also made good use of their miner's right entitlements. Animals were grazed communally on common land. By 1863 the Morrisons goldfields common was home to 200 cattle, 40 horses and 'many goats and pigs'. All were kept there for a small fee and a herdsman was employed to watch over them.[23] Dairy cows and chickens could be kept close to the homes, while goats did so well that their feral descendants are a continuing problem in the area today.

Legislative changes in the 1860s and 1870s encouraged the selection of small farms in goldfields regions.[24] By the 1880s, aided by a downturn caused by lack of capital and technology, subsistence mining had largely been pushed to the margins of colonial life. As it became unviable as a primary means of support, farming activities, which underpinned many mining families, assumed greater importance. Many of those who took up the newly available land around the Moorabool were mining families already settled in the area. Small-scale farming on these blocks preserved some of the independence that the gold seekers had treasured, even if it lacked the romantic allure of gold. More importantly, it could be considered an extension of an existing subsistence strategy rather than a radical departure from it. Often, too, part-time mining could be continued.



Archaeology has been a crucial means of gaining a more nuanced understanding of the settlement on the Moorabool. The landscape of houses and gardens, the domestic rubbish, and the memorials of the population have all provided rich sources of information. Details of daily life shed light on the better-known social changes under way in the colony's larger towns. The large-scale economic shift away from gold was facilitated by smaller changes at the individual and domestic level, in which women and family became integral to the survival and adaptation of Victoria's diggers.

## Notes

1 G. Blainey, *The Rush That Never Ended*, 1963, Melbourne University Press, Melbourne; G. Serle, *The Golden Age: A History of the Colony of Victoria, 1851–1861*, 1963, Melbourne University Press, Melbourne; W. Bate, *Lucky City: The First Generation at Ballarat*, 1978, Melbourne University Press, Melbourne; D. Goodman, *Gold Seeking: Victoria and California in the 1850s*, 1994, Allen & Unwin, Sydney; R. Annear, *Nothing but Gold: The Diggers of 1852*, 1999, Text Publishing, Melbourne.

2 For a more detailed analysis of the investigation, see S. Lawrence, *Dolly's Creek: An Archaeology of a Victorian Goldfields Community*, 2000, Melbourne University Press, Melbourne.

3 For an example of contemporary attitudes to the accuracy of the census, see the humorous song 'Taking the Census' by the popular goldfields entertainer Charles

Thatcher, quoted in L. Murray, 'The Narrow-Columned Middle Ground: 19th Century Australian Poetry', in L. Murray (ed.), *Blocks and Tackles: Articles and Essays*, 1990, Angus & Robertson, Sydney, pp. 1–35.

4 P. Grimshaw and C. Fahey, 'Family and Community in Nineteenth-Century Castlemaine', *Australia 1988*, Bulletin no. 9, 1982, p. 106.

5 L. Derry, 'Fathers and Daughters: Land Ownership, Kinship Structure and Social Space in Old Cahawba', in A. Yentsch and M. Beaudry (eds), *The Art and Mystery of Historical Archaeology: Essays in Honor of James Deetz*, 1992, CRC Press, Ann Arbor; P. Grimshaw and G. Willet, 'Family Structure in Colonial Australia: An Exploration in Family History', *Australia 1988*, Bulletin no. 4, 1980; E. McEwen, 'Family, Kin and Neighbours: The Newcastle Coalmining District, 1860–1900', *Australia 1988*, Bulletin no. 4, 1980; J. Beer, C. Fahey, P. Grimshaw and M. Raymond, *Colonial Frontiers and Family Fortunes: Two Studies of Rural and Urban Victoria*, 1989, Melbourne University History Monograph Series, Melbourne.

6 P. Russell, *A Wish of Distinction: Colonial Gentility and Femininity*, 1994, Melbourne University Press, Melbourne.

7 Ballan Historical Society, Morrisons Cemetery Records.

8 Morrisons Cemetery Records.

9 Mrs Charles Clacy, *A Lady's Visit to the Gold Diggings of Australia in 1852–53*, 1963, Lansdowne Press, Melbourne, p. 85; A. Fauchery, *Letters from a Miner in Australia*, 1965, Georgian House, Melbourne, p. 60; S. T. Gill, *The Victorian Goldfields 1852–53: An Original Album*, 1982, Currey O'Neil for the Library Council of Victoria, Melbourne; T. Griffiths (ed.), *The Life and Adventures of Edward Snell*, 1988, Angus & Robertson, Sydney.



10 Austral Archaeology, Archaeological Report on the Rookery Site, vol. 1: Excavation Report, p. 14; Godden Mackay Heritage Consultants and Karskens, Cumberland/Gloucester Streets Site Archaeological Investigation: Volume 1, Executive Summary; J. McCarthy, Archaeological Investigation: Commonwealth Offices and Telecom Corporate Building Sites, The Commonwealth Block, Melbourne; L. Caffyn, *Workers' Housing in West Yorkshire 1750–1920*, 1986, West Yorkshire Metropolitan County Council and the Royal Commission on the Historical Monuments of England.

11 M. Tipping (ed.), *An Artist on the Goldfields: The Diary of Eugène von Guérard*, 1982, Currey O'Neil, Melbourne, pp. 21, 51, Plate XII; Clacy, *A Lady's Visit*, p. 63; Griffiths, *The Life and Adventures of Edward Snell*, pp. 60, 291, 305, 315; L. Sussex (ed.), *The Fortunes of Mary Fortune*, 1989, Penguin Books, Melbourne, p. 24.

12 There is a similar story from the Lisle goldfield in Tasmania, where a woman also remembered the chore in the early twentieth century: 'The whitest of pipeclay was used as a wash for the fireplaces and housekeepers would search for the best and carefully remove any yellow pieces from their whiting buckets so that the final result would be most immaculately snowy'; C. Coroneos, *A Poor Man's Diggings: An Archaeological Survey of the Lisle Denison Goldfields*, 1993, Tasmanian Parks and Wildlife Service, Hobart, p. 166.

13 S. Korzelinski, *Memoirs of Gold Digging in Australia*, 1979, University of Queensland Press, Brisbane, p. 137.

14 Seweryn Korzelinski lived in his shops on the Alma and Maryborough diggings, and Mary Fortune lived in her father's shops at Kangaroo Flat (Castlemaine), Buninyong and Inkerman. Korzelinski, *Memoirs*; Sussex, *The Fortunes of Mary Fortune.*

15 Korzelinski, *Memoirs*, p. 59; Fauchery, *Letters from a Miner*, p. 60; T. Lane and J. Serle, *Australians at Home*, 1990, Oxford University Press, Melbourne, p. 26. Emily Skinner recalled that on the Woolshed diggings: 'Our men were first rate bush carpenters and had made several articles of furniture: cupboard, safe, tables and some really comfortable chairs. They looked quite nice when cushioned and covered with neat chintz and we spread bags and matting over the earthen floor'; Edward Duyker [ed.], *A Woman on the Goldfields: Recollections of Emily Skinner, 1854–1878*, 1995, Melbourne University Press, Melbourne, p. 75.

16 Clacy, *A Lady's Visit*, pp. 55–6; Korzelinski, *Memoirs*, pp. 60, 71.

17 Goodman, *Gold Seeking.*

18 J. Powell, *The Public Lands of Australia Felix: Settlement and Land Appraisal in Victoria 1834–91 with special Reference to the Western Plains*, 1970, Oxford University Press, Melbourne, pp. 81–3; *Victorian Government Gazette*, 1861, p. 260; local mining boards could and did pass by-laws that altered provisions of the Mines Act in order to suit local conditions.



19 Duyker, *A Woman on the Goldfields*, pp. 58–9, 103.

20 Duyker, *A Woman on the Goldfields*, p. 123.

21 W. Evans, *Diary of a Welsh Swagman 1869–1894*, 1978, Macmillan, Melbourne, pp. 45, 107.

22 B. Matthews, *Louisa*, 1987, McPhee Gribble/Penguin Books, Melbourne, pp. 113–14.

23 *Victorian Government Gazette*, 1858, p. 1127; Minute Book, Morrisons Gold Fields Common, 3 November 1863.

24 Powell, *The Public Lands.*

Chapter 16

# Vegetable Plots and Pleasure Gardens of the Victorian Goldfields

*Suzanne R. Hunt*



Between 1858 and 1900, the cultivation of public and domestic 'pleasure gardens' in the central goldfields towns of Ballarat, Castlemaine and Bendigo became an integral part of the ethos that defined these young cities.[1] Not only did they help to beautify 'the howling wilderness',[2] but they also sought to convey to the rest of the world an image of civilised living and to the individual, a sense of place.

The collective heritage and legacy of these pleasure gardens constitute one of Victoria's most prized cultural resources. Municipal botanic gardens hold significant collections of exotic and native plants and ensure the continuation of horticultural study and research. Featuring mature and rare species of trees, shrubs and nineteenth-century garden structures such as rockeries, pavilions and landscape design, they are now major tourist attractions. These gardens are also the focus of state and regional festivals and are popular venues for recreational activities. As residues of nineteenth-century scientific and aesthetic endeavour, the history of these pleasure gardens encompasses the sociology and philosophy of gardening in colonial Victoria. Collectively, they also embody the liberal philosophy that informed social and land reform during the period.

The cultivation of gardens on soil churned up by goldmining was a remarkable achievement. Although the gold rushes created enormous wealth, the methods used to extract the precious metal left the landscape in a deplorable condition. In 1855, a writer in the *Mount Alexander Mail* expressed grave concern about the drastic effect of increased quartz mining and steam power on the environment. He predicted that 'in a few years the wood crowned summits of our hills will look as desolate and barren as the flats and gullies do already'.[3] His prediction proved to be correct. The demand for timber to shore up mine tunnels and to feed steam boilers continued unchecked until the 1860s, when the Victorian government was forced to control the industry.[4] Other forms of environmental damage were inflicted when the sludge from puddling dams, used in alluvial mining, drained into creeks and streams, clogging up fresh-water supplies. Adding to these problems were hydraulic sluices that played powerful jets of precious water onto a working face, blasting away the soil and permanently erasing the original contours of the land.

The *Mount Alexander Mail* commentator was not the only person alarmed. James Sinclair, a prominent garden journalist, appalled by the extent of the damage, pleaded with miners and members of the government 'to bring under cultivation the wastelands of the colony'. He argued that if a 'lucky digger gets a few acres he may still reclaim his future' and 'may raise up around him a scene of beauty and richness that few European monarchs could ever accomplish'.[5] Despite their reputation for recklessly exploiting the environment in search of material gain, it would seem that many of the diggers who settled on the goldfields shared his optimism. As early as the 1850s, it is possible to discover the beginnings of domestic vegetable and flower gardens in photographs and sketches of residential areas in Castlemaine and Ballarat.[6] A model of a cottage and garden by James Cowling, now held in the Castlemaine Art Gallery and Historical Museum, provides a good illustration of the simple layout and geometric design of garden beds that were commonly used. His model won an award at the prestigious Intercolonial Exhibition held in Melbourne in 1866–67, and it was suggested that the exhibition committee should enter it in the Paris exhibition to show the people of Europe the sorts of houses that middle-class Australians lived in.

**Figure 16.1** Cuthbert Clarke, *South-West Corner of Hargraves and Campbell Streets, Castlemaine*, watercolour, 1858. (Castlemaine Art Gallery and Historical Museum)

Today, it seems grandiose to imagine that Europeans might have been interested in viewing such a design. But immigrants who had fled from the overcrowded industrial towns of England in search of personal wealth and land ownership were driven by a keen desire to impress upon the world the physical evidence of material success and civilised living. The visiting historian James Froude was clearly impressed by the number of established gardens that he saw during his travels in 1885 and, in his book *Oceana*, made a point of comparing them favourably with those in the United States. He found that, in America, 'the ordinary suburban house rises bare in the midst of indifferently kept grass, and even the palaces of the millionaires stand in ground poorly laid out'. Whereas in Melbourne, and in these colonies universally, owners seemed to wish to surround themselves with graceful objects, and especially with the familiar features of their old home – oaks, maples, elms, firs, planes, and apple-trees.[7]

Froude's comparison reveals a neglected dimension of life on the goldfields: why did Australian immigrants, supposedly unlike their American counterparts, cultivate gardens amidst the frenzy of settlement and mining activity? This chapter seeks both to address this question and to argue for the importance of garden cultivation as a form of material culture that articulated the aspirations of a colonial society nurtured during the gold boom in Victoria. 

James Sinclair's call for government intervention reflected the public's growing dissatisfaction over the amount of land held by pastoralists and squatters. During the 1850s, an emergent liberal philosophy of land reform based upon the 'yeoman' farmer ideal was actively canvassed by Chartists and other radicals who came to Victoria to mine for gold.[8] Inspired by their vision, diggers, settlers, merchants and manufacturers held rallies and conventions, agitating the government to release the land. They contended that the development of smaller farm allotments clustered around an administrative centre would create a new social class of settler. Opening up the land to support agricultural communities around these proposed rural townships was, as Tony Dingle commented, 'the road to true nationhood for, as the yeoman built his house and tended his fields . . . his intentions are gradually and insensibly transferred to his land of adoption'.[9] Land ownership symbolised prosperity and provided a means to achieve it.

By the 1870s, many small farming communities had grown up adjacent to the goldfields (although not on the scale intended) and the population in nearby towns had begun to stabilise. Prominent citizens eager to promote their communities, and themselves, combined with municipal councils to transform what was essentially a raw frontier into something of which they could be proud. This growth in civic pride is best illustrated in the formation of libraries, art galleries, schools, mechanics institutes and, importantly, in the creation of public and domestic pleasure gardens. Other key elements that help to explain a phenomenal growth in gardening activity during this era include the presence of a temperate climate in central Victoria for growing crops and plants, and the high number of professional nurserymen and gardeners who settled in provincial towns.

The principles of gardening in vogue rested upon the nineteenth-century British notion that gardening was a moral and serious pursuit. Gardens, like painting and literature, could be constructed as a form of visual art.[10] In simple terms, man did not have to rely upon nature to create a beautiful scene or vista, he had the freedom and ability to self-impose original and imaginative ideas onto a landscape. Trees, for example, could be strategically placed to create imposing views, and cottages and suburban villas could be tastefully decorated with flowerbeds to reflect the late eighteenth-century English aesthetic, the Picturesque. Popular garden magazines, such as the *Gardener's Magazine* and *Gardener's Chronicle*, spread the message of this new garden ideology, but it was left to the professional gardeners and nurserymen to translate these ideas into the work that they undertook.

The greatest advocate of such theory was John Claudius Loudon (1783–1843), a prolific writer and landscape designer.[11] He formulated a landscape style that modified the English painterly vogue for 'the Picturesque' with an idea that he termed 'Gardenesque'. In essence, he proposed that by correctly clumping trees and shrubs, allowing for different heights and growth habits (as opposed to dotting them about in rows or singly), one enriched the appreciation of a specimen without diminishing a sense of the whole design. His ideas were widely acknowledged in Australia, and by the 1860s elements of his style were being freely utilised in both public and domestic gardens.



Like painting and literature, gardens in the mid-to-late Victorian period in Britain reflected changing tastes in fashion and style. It was believed that gardens had the ability to help educate the public about 'beauty' and to impose social order over chaos.[12] Technical innovations were also rapidly adopted. Labour-saving machines, such as the lawn-mower, permitted larger areas to be landscaped with grass, while improved technology in the construction of the glasshouse enabled the general public to cultivate new plant species. Perhaps the most significant change was the shift in status of the head gardener, from semi-skilled worker to respected professional. His specialist knowledge of botany and scientific principles in relation to horticulture was an important factor in the successful propagation and introduction of exotic plant material into England from around the world.[13]

Such ideas found their way to Victoria's goldfields towns by way of local newspaper articles, garden books and journals, and professional gardeners and nurserymen, and through the formation of horticultural and acclimatisation societies. A report in the *Mount Alexander Mail* in 1861 on the benefits of a horticultural society for Castlemaine citizens clearly articulates the perceived public role that such societies represented as arbitrators of 'taste' in frontier towns:

> It is needless to advocate the claims of an institution of this kind to public support. The advantages which it confers can fortunately be appreciated by all. A considerable portion of the community find a pure and refined pleasure in

> the cultivation of flowers from a keen appreciation of the beautiful, but there are others whose sympathies can be excited only by a mere material stimulus. A society which has for its object at once the development of taste and skill in a humanising pursuit, and the enlargement of our gastronomic resources, deserves the most liberal support that can be accorded it.[14]

The article further contends that the moral and social health of the populace would be promoted through gardening: it gives citizens a 'strong attachment to the soil on which they live, [and] strengthen[s] their patriotism'.

The aims of the Horticultural Society were to promote horticulture in all its phases, particularly in the area of cultivation, and to exhibit plants, fruit, flowers and vegetables. Members of the society – an all-male affair – included professional nurserymen, gardeners and enthusiastic amateurs. Women were not excluded, however, from exhibiting their produce at the annual show and were frequent prize-winners in the floral 'Bouquets for the Hand' section. The inaugural Horticultural Exhibition held at Ballarat in 1859 offered seventy-two sections and featured a dizzying display of flower varieties – from roses, dahlias, gladioli, pansies, carnations, picotees, stock, verbena and potted fuchsias to balsam. There were also vegetables – cucumbers, carrots, parsnips, cabbages and cauliflowers – as well as soft fruits, such as raspberries, cherries and currants. At the Castlemaine Show in 1861, attractions which reportedly caught the public's fancy included a cage of Persian cats, Indian and Colonial birds in cages, a collection of stuffed animals, mud turtles and 'two small specimens of the Brazilian aloe in bloom'.[15] Only a few examples of indigenous flora featured among all this exotica; most fruits and vegetables exhibited at horticultural shows were grown from imported seeds.



A reliable supply of fresh produce was essential to the well-being of people living in these pioneering communities. This partly explains the important function of horticultural societies and the prominence of their members, many of whom were involved in other civic projects as well. Several of the members of the Ballarat Horticultural Society, for example, were equally actively involved in the management of the botanic garden and on the first Street-Tree Planning Committee.[16] These energetic goldfields men took their civic responsibilities seriously. A number of them were also leading nurserymen whose skills and entrepreneurship ensured the success of the boom in gardening.

Ninety-seven plant nursery catalogues published in Victoria between 1860 and 1880 were located by Rosemary Polya during her research.[17] While these are only a small sample of the number actually published, they nevertheless indicate the extent of the industry and the vast range of plants propagated. Many of the nurserymen came from Scotland, bringing with them sophisticated grafting techniques and the ability to experiment with new varieties of plants from around the world. Massive plantings of *Cupressus macrocarpus* and Monterey pine testify to their skill. These trees were used for shelter and shade, as were the remnants of gorse and hawthorn hedges, once planted in their thousands, that still stalk fence lines across the paddocks in Victoria. Thomas

Lang was one such enterprising nurseryman. He imported seeds direct from the Himalayas, and other seeds (possibly vegetable) were brought in from Shanghai. The latter seeds were displayed in his window, complete with the Chinese labels.[18] He designed and planted the first Ballarat Maze, and his discovery that woolly aphids did not infest the roots of the northern spy apple saved the ailing apple-growing industry in Australia.[19]

George Smith of Ballarat was another leading nurseryman. He came there to dig for gold in 1854 but decided that there was more future in horticulture. His catalogue, published in 1862, proudly listed his credentials as 'late of her majesty's Royal Fruit and Flower Gardens, Windsor Castle'. It contained practical suggestions for collecting composting manure from road scrapings and listed the types of tools required to maintain a domestic garden. Ideally, he proposed, garden beds should be formed in squares with two cross walks and edged with soft plants like 'thyme, thrift, strawberry, camomile' or herbs. He discouraged the use of quartz edging, describing it as 'ungardenesque', and he positively shuddered over the popular use of 'bricks, bottles or bones'.[20] As well as his horticultural interests, Smith developed a distinguished career as a member of the City Council, the Ballarat Water Commission, Ballarat Hospital, Benevolent Asylum, the Orphan Asylum, and the Agricultural and Pastoral Societies. In honour of Smith's horticultural work, Baron Ferdinand von Mueller, the celebrated Government Botanist, nominated him as a fellow of the Royal Horticultural Society in England.[21]



As members of the Horticultural Society, men of the calibre of Lang and Smith worked closely with like-minded members of the Acclimatisation Society in provincial towns to introduce exotic and rare plants, and they called upon gentlemen of 'taste' to assist with the re-creation of the landscape in the image of the 'parks and pleasure grounds' of the 'old country'.[22] The role that these groups performed in the development of public and domestic pleasure gardens lay in their ability to act both as arbiters of taste, and as practical agents of change who supplied seeds, plant material and, importantly, professional skills and advice.

Prior to the establishment of the nursery trade around 1855, settlers and miners who arrived in gold-rush towns had to improvise. Many brought seeds and cuttings with them but some, like Owen Jones who supposedly planted the first vegetable plot in Castlemaine on the southern bank of Forest Creek in 1852, relied on their wits:

> When the rains had set in for winter, me and my boys cultivated a piece of ground, about a quarter of an acre. We had to act the Robinson Crusoe, in picking up onions and other things that the store keeper threw away, and in a few months it grew, and was a green patch in a wild desert.[23]

Jones, like many others, also sent away to Melbourne or Adelaide for seed and cuttings, but often with heart-breaking results. 'We are sorry to hear', wrote John Bailey and Son about a consignment of fruit trees, 'that the rats took a fancy to the trees, but shall be glad to hear of their final success with you'.[24] 'Rats' of the two-legged variety were still more of a pest. Diggers subsisting on mutton chops and damper could not resist the temptation to taste luxuries like fresh peaches and apples. Complaints were even registered in the *Mount Alexander Mail*, where an angry citizen castigated Jones for tantalising the local population with forbidden fruit grown on public grounds![25]

Fresh vegetables and fruit carted from Geelong and Melbourne sold at vastly inflated prices. Local newspapers seized upon the absurdity of the situation and encouraged residents to take up gardening to produce food for themselves.[26] This problem was partly addressed by the Victorian government when a Select Committee in 1853 advocated changes to the mining licence fee to 'allow the miner to cultivate a garden near his claim'.[27] This was a landmark decision because it was the first step towards granting miners a residence licence. By 1857 the holder of a miner's right (as it was now called) was entitled to a quarter-acre block on Crown land for residential purposes, and by 1865 improvements such as a house or business could be sold. Miners now had tenure over their place of residence, although the land still belonged to the Crown.[28]

Thousands of cottages were eventually constructed on designated areas under this scheme. Tute's Cottage, the last in Castlemaine to be occupied under a miner's right, still stands at Ten-Foot Hill, near the property formerly occupied by Owen Jones, and was recently placed on the Victorian Heritage Register. The cottage is unique because its original architectural character remains intact.[29] The layout of the garden, however, is sadly missing. We do know that cottages similar to Tute's usually featured both a front and back garden area, and an article on 'Our Cottage Gardens' in the Adelaide journal *Garden and Field* gives a good description as to how individual and eclectic they were:



> A majority of them, no doubt, were intended to contribute to the necessities of the occupiers, by supplying them with vegetable food: but after a time came fruit trees, then shrubs and flowers, until at last it would puzzle many people how to describe the garden, because something of everything almost is contained in it.[30]

The owners of these gardens were not concerned with the dictates of fashion, preferring instead to create their own sense of place. They seem to have made do by exchanging cuttings with their neighbours or by purchasing the popular seed varieties that were sold in nurseries and hardware stores. For the more sophisticated gardener wanting to design and lay out a garden, help was readily available in the form of specialised books, journals and catalogues (Plate 16.2). The earliest published horticultural information was sourced from well-known English writers such as Loudon and Edward Kemp. Victor Crittenden's *A History and Bibliography of Australian Gardening Books, 1806–1900* lists more than seventy books published in Australia,[31] with several of them, such as William Adamson's *The Australian Gardener*, running to many editions.[32] While these books may have been beyond the means of most people, many were available in public libraries.[33]

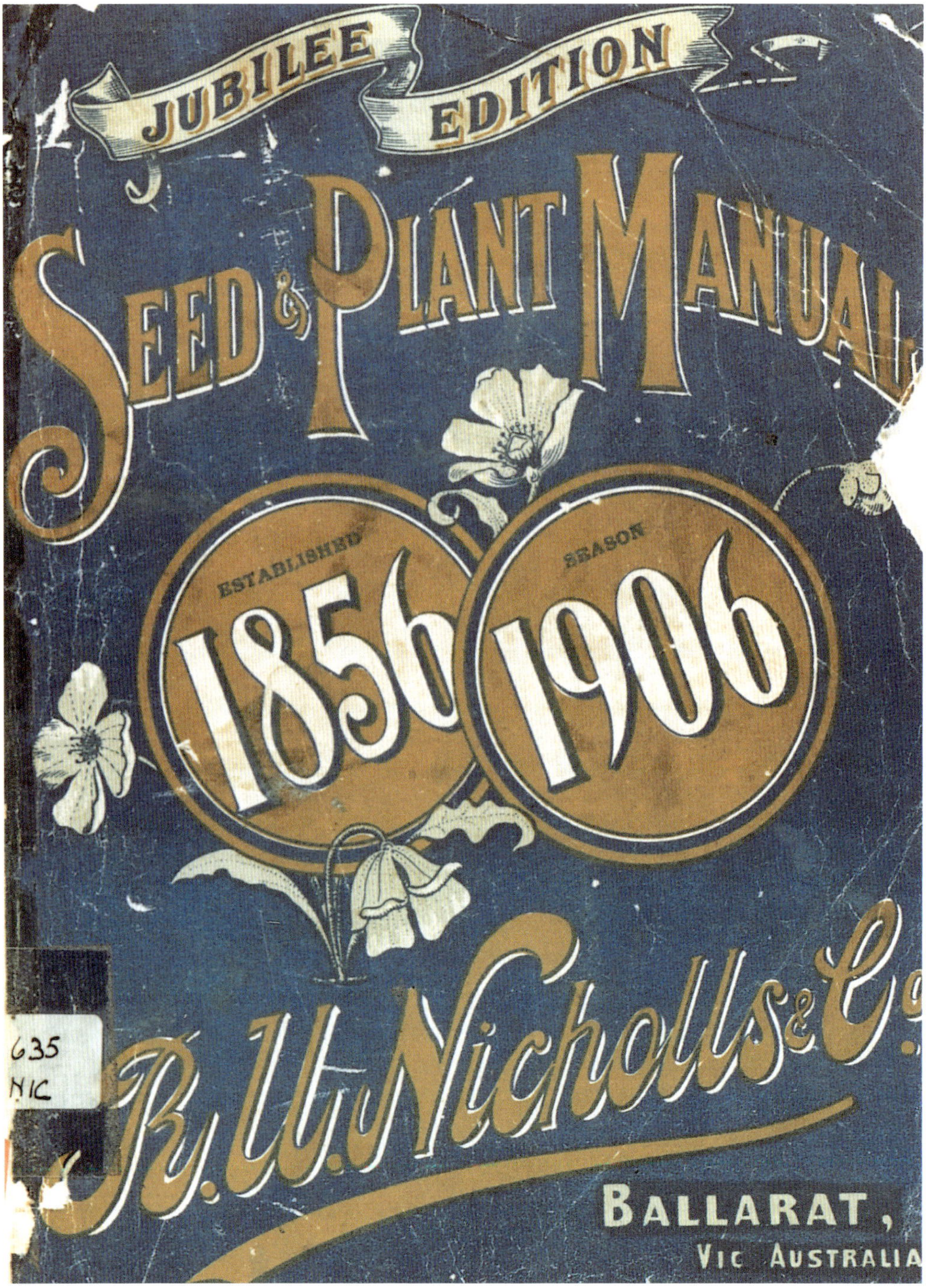


PLATE 16.2 Front cover of jubilee edition of R. U. Nicholls & Co.'s catalogue. (Ballarat Historical Society Inc.)

features included an Italian water garden and a fountain after the style of a water fountain at Pompeii. It is possible that the Italian style in the garden derives from his holiday abroad in 1875 and a subsequent longer stay in England during the 1880s. During this same period, an international movement favouring a return to the Italian Renaissance began to dominate the garden literature.[40] In contrast to the Victorian preference for colour and a mixed palette of plant material, the revived 'Italian garden style' recalled the dramatic use of lakes, lawns, simple planting and walks studded with classical antiquities. Elements of Lansell's garden and mansion certainly echo the mood of this movement.

Another little-known area of goldfields garden history and provincial social life is revealed through the ambitious activities of commercial pleasure gardens. These were usually attached to nurseries or hotels, or were developed as an extension of a private garden. In 1862 when the English cricket team suffered their first defeat at Wattle Flat against a country Victorian team, Chewton celebrated in great style at Merediths Pleasure Gardens, near Castlemaine. Planks were thoughtfully placed in the branches of the oak trees to make seats so that all the visitors could get a view of the dancing and entertainments. In November 1864, the *Mount Alexander Mail* recorded in detail another such event, held at nearby Hirschi's Nursery and Pleasure Garden:



> A numerous assemblage of pleasure seekers met at Hirschi's gardens last evening, to witness the opening of the Casino, and Victoria Bridge. Some of the members of the volunteer band were present, and discoursed some enlivening music. The numerous garden walks were brilliantly lighted up with coloured lanterns, which with the bonfires of blue and other fires gave the whole scene quite an air of 'fairyland'. Some of the visitors enjoyed themselves by dancing; others walked up and down; others again took to the 'Mountains of strawberries and lakes of cream', while some, fatigued after a day's labor, rested themselves in quiet, sipping nectar and enjoying pleasurable conversation.[41]

Belmont at Beaufort remains an outstanding example of an extant private pleasure garden. The water races and puddling dams on this former goldmining site were fashioned into ornamental ponds, turning a relatively dry location into an 'oasis in the wilderness'.[42] Photographs record the extensive grounds and the large crowds of people who flocked there on weekends to enjoy the flowers and ferns, to row on the lake and to sample the fruits and wines.

Although domestic gardens, from the humble cottage style to the grand development of Fortuna Villa, were linked to current fashions in gardening, they were essentially personal definitions of space. Public gardens, on the other hand, reflected a combination of community needs and government initiatives. The Victorian government had always been aware of the need to provide space for public facilities. From 1853 to 1858, 3000 acres of Crown land were earmarked for recreational use as cricket grounds, racecourses, parks, gardens

**Figure 16.2** Open Day Visitors at Belmont, 1864. (Photograph courtesy of present owner, Mrs J. Reid)

and public utilities. As a consequence of the gold rushes, Victoria's booming population and rising wealth afforded the impetus and finances needed to develop these facilities. Between September 1857 and June 1874, more than 1770 sites (1,311,532 acres) were set aside.[43]

The designation of public space fell essentially under the control of Clement Hodgkinson, the Deputy Commissioner for Crown Lands. Requests to his office took the form of either a direct plea from a municipal council or a prominent citizen, or came through the mediation of organisations like the acclimatisation societies.[44] The justification for botanic gardens (as opposed to parks and other recreational spaces) in provincial gold towns reflected the importance assigned to horticulture by a developing country, a broader Imperial interest in plant introduction and exchange, and the collective local desire to transform mining wasteland into something that demonstrated the progress and achievement of its inhabitants. So intense was the demand from the public to establish these gardens that legislation was gazetted in December 1854 to make it possible for municipal and borough councils to make by-laws to set

**Figure 16.3** R. U. Nicholls & Co.'s Doveton Nursery, Doveton Street North, Ballarat, c. 1900. (National Library of Australia)

aside sites for their development.[45] As a result, thirty such gardens had been established in Victoria by the end of the century.

Perhaps the greatest champion for the establishment of provincial botanic gardens was the Director of the Melbourne Botanic Gardens and Government Botanist, Dr Ferdinand von Mueller (1857–73). He introduced a variety of exotic trees and flora into Victoria and an exchange of plant material from around the world. Firmly believing in the scientific purpose of botanic gardens, he regularly supplied them with seeds and plants. A scientific purist, his regimented style of planting and labelling of specimens in serried rows did not countenance the indulgent use of lawned areas, curved gentle walks or scenic vistas. By the 1860s, public support for his planting style had waned and he was criticised in the press for 'failing to provide a beautiful place of resort'.[46] Anthony Trollope's travelogue sums up the public's annoyance that the gardens were 'a perfect paradise of science for those who are given to Botany rather than to beauty . . . But the gardens though spacious are not charming, and the lessons which they teach are out of the reach of ninety-nine in a hundred.'[47]

A board of inquiry, set up to investigate the state of the gardens and the position of the director, resulted in von Mueller's replacement in 1873 by William Guilfoyle under whose curatorship the Melbourne Botanic Gardens underwent a complete transformation.[48] Guilfoyle's design successfully embraced a combination of the Picturesque and Gardenesque styles made popular by

**Figure 16.4** The Ballarat Gardens in 1881, Showing the 'Walks' and 'Views'. (Ballarat Library, Historical Society Collection)



Loudon and other English designers, without sacrificing the integrity of the scientific purpose of a botanic garden.[49] This trend, possibly copied from Kew Gardens in London,[50] and certainly used by contemporary landscape designers in Europe and Australia, swept across Victoria, changing the way goldfields provincial botanic gardens were landscaped and used by the public in the next three decades.

A comparison of a description of the Ballarat Botanic Gardens from the *Ballarat Miner* in 1859 with one made by Froude in 1886 illustrates the transition from wilderness to paradise. In 1859 they were said to be

> a place of tall grass and some flowers blushing unseen, as two miles from Lombard street is too far to travel on shanks pony, so the Botanical Gardens is merely a tale that is told to the majority of Ballarat people. Plenty of gum trees and graceful lightwoods adorn the reserve. The Curator has planted some trees he brought from Home. Some day these might be worth the long journey to see. About seven acres are enclosed by a post and rail fence, but the rest of the 60 acres is a wilderness.[51]

By the late 1880s the botanic garden beside Lake Wendouree had become a showpiece of the city. Froude remarked that 'it was as if the town council had decided to show what gold and science could do with such soil and climate'.[52] He was overwhelmed by the magnificence of the gardens:

> The roses which bloom ill on the hotter lowlands were here, owing to the height above the sea, abundant and beautiful as in Veitch's nurseries at midsummer. Besides roses, every flower was there which was fair to look upon or precious for its fragrance. There were glass houses to protect the delicate plants in the winter; but oranges and camellias, which we know only in conservatories, grow without fear in the open air, and survive the worst cold which Ballarat experiences . . . A broad gravel-walk led up the middle of the grounds, with lateral paths all daintily kept. Dark shadowy labyrinths conducted us into cool grottoes overhung by tree-ferns, where young lovers could whisper undisturbed, and those who were not could read novels. Such variety, such splendour of colour, such sweetness, such grace in the distribution of the treasures collected there, I have never found combined before, and never shall find again.[53]

Facilities in the gardens were continually improved and expanded. A tramline and steamers now conveyed people to the gardens and a merry-go-round, model railway, fishing (courtesy of the experimental hatchery set up by the Ballarat Fish Acclimatisation Society), a maze, and animal enclosures drew crowds and delighted children. Sailing and rowing on Lake Wendouree flourished. Picnics were held on the lawns and bands played in the rotundas.

**Figure 16.5** Buda House. (Castlemaine Art Gallery and Historical Museum)

Unfortunately, saline ground-water at White Hills, Bendigo's botanic garden, destroyed the vegetation and it was left to Rosalind Park, the former police camp for the Sandhurst goldfield, to compete with those in Ballarat. A contemporary observer was fascinated that such 'seclusion and quietude of a fern-tree gully' could be found, 'only a short remove from the scene of some extensive mining operations'.[54] The gardens were the pride of the city and, as far as one writer was concerned, unequalled in any European metropolis:

> Imagine such a place within a few minutes walk of the Royal Exchange in Cornhill, of the Place de la Bourse in Paris, the Burghstrasse in Berlin, or of the Borse in Hamburg! Hereafter, when Bendigo becomes a city of 100,000 inhabitants, they will prize as one of their precious possessions this 'beautiful 'rus in urbe'.[55]

By the turn of the century, goldfields botanic gardens were major tourist attractions and the atmosphere of a 'pleasure park' had superseded the original scientific intent proposed in the 1850s. A significant legacy of the gold-rush era, the collective heritage of public and private pleasure gardens has been largely overlooked. As a form of material culture, pleasure gardens evoke the private aspirations of their creators and reveal a population intent on leaving its vision and a permanent mark upon one of the most feverish periods in Australian history.



## Notes

1 The concept of a 'pleasure garden' originated in the seventeenth century and is defined in Andre Mollet's *Jardin de Plaisir* (see P. Hobhouse, *Gardening through the Ages*, 1992, Simon & Schuster, New York, p. 169, for a fuller explanation). In simple terms, it meant that gardens were designed to create rooms, vistas and walks. Elements such as statues, rockeries, pavilions and choice specimen trees were added as special features to be enjoyed beyond the 'growing' components of the garden. It was adapted into the late eighteenth-century English landscape style (Picturesque) and incorporated into the nineteenth-century Gardenesque style (a form of which was used extensively in Australia).

2 W. Bate, *Lucky City: The First Generation at Ballarat 1851–1901*, 1978, Melbourne University Press, Melbourne, p. 43.

3 Heritage report by D. Bannear, 'A Study of Historic Activity Sites in the Box-Ironbark and Midland Areas of Victoria', February 1997, CRA, Victoria, p. 4.

4 R. Wright, *The Bureaucrat's Domain: Space and the Public Interest in Victoria 1836–84*, 1989, Oxford University Press, Melbourne, pp. 117–18.
5 J. Sinclair, *The Gardeners Magazine and Journal of Rural Economy*, 1855, Melbourne, vol. 1, p. 4.
6 C. Clarke, *South-West Corner of Hargraves and Campbell Streets, Castlemaine*, watercolour painting, 1858, Castlemaine Art Gallery and Historical Museum.
7 J. A. Froude, *Oceana or England and Her Colonies*, 1886, Longmans Green & Co., London, p. 94.
8 T. Dingle, *The Victorians: Settling*, 1984, Fairfax Syme & Weldon Associates, Sydney, p. 60.
9 Dingle, *The Victorians*, p. 60.
10 B. Elliot, *Victorian Gardens*, 1996, Timber Press, Portland, Oregon, p. 10.
11 M. L. Simo, *Loudon and the Landscape: From Country Seat to Metropolis 1783–1843*, 1988, Yale University Press, New Haven.
12 Elliot, *Victorian Gardens*, ch. 1.
13 Elliot, *Victorian Gardens*.
14 *Mount Alexander Mail*, 22 November 1861, p. 2.
15 *Mount Alexander Mail*, 22 November 1861, p. 2.
16 *Ballarat Historic Landscapes Trees and Gardens*, 1983, report published by Royal Botanic Gardens, Melbourne, pp. 28–9. 
17 R. Polya, *Nineteenth Century Plant Nursery Catalogues of South East Australia: A Bibliography*, 1981, La Trobe University Library, Bundoora.
18 H. Hutton, 'Thomas Lang and Co.', report located in the Sovereign Hill archives on gardening. It is tantalising to speculate that these seeds may have been brought in to supply the many Chinese market gardeners with familiar plants and herbs to cater for the large Chinese population who lived in the Ballarat area.
19 *Ballarat Historic Landscapes Trees and Gardens*, p. 31.
20 G. Smith, 'The Cottage Gardener', in Victorian Pamphlets, vol. 3, 1862.
21 *Ballarat Historic Landscapes Trees and Gardens*, p. 30.
22 Royal Horticultural Society of Victoria, Minute Books, 19 November 1859 – 10 March 1873, Rare Books Section, State Library of Victoria.
23 R. A. Bradfield, *Owen Jones: Labour Pioneer on the Castlemaine Goldfield*, 1978, Castlemaine Mail, Castlemaine, p. 14.
24 See Bradfield, *Owen Jones*.
25 Bradfield, *Owen Jones*, p. 15.
26 *Ballarat Times*, 30 September 1854.
27 R. W. Birrell, *Staking a Claim: Gold and the Development of Victorian Mining Law*, 1998, Melbourne University Press, Melbourne, p. 26.
28 Birrell, *Staking a Claim*, p. 91.
29 Victorian Heritage Register, Tute's Cottage (formerly Hannan's), file no. 608448.
30 *Garden and Field*, 21 September 1875.
31 V. Crittenden, *A History and Bibliography of Australian Gardening Books 1806–1900*, 1986, Canberra College of Advanced Education Library, Canberra.
32 W. Adamson, *The Australian Gardener: An Epitome of Horticulture for the Colony of Victoria*, 1863, Mason & Firth, Melbourne.

33 P. Fox, 'What's in a Historic Garden?: The Villa Garden and the Landscape', in *Papers presented at the Historic Gardens Conference, Rippon Lea*, 16 April 1986, Melbourne, p. 9.
34 Ballarat Horticultural Society, *First Show Schedule*, 25 November 1859.
35 *Ballarat Evening Mail*, 25 March 1870.
36 'Trees and Gardens from the Goldmining Era: A Study of the Maldon Landscape', report prepared by Royal Botanic Gardens Melbourne, Department of Crown Lands, 1980–81.
37 P. Cuffley, 'Buda Garden', *Journal of the Australian Garden History Society*, vol. 1, no. 4, January/February 2000, pp. 9–21.
38 P. Watts, *Historic Gardens of Victoria: A Reconnaissance*, 1983, Oxford University Press, Melbourne, p. 200.
39 'Fortuna, Bendigo: Garden, Conservation Management Plan 1996', report prepared by Francine Gilfedder & Associates for Department of Defence.
40 Elliot, *Victorian Gardens*, pp. 238–9.
41 *Mount Alexander Mail*, 29 November 1864, p. 2.
42 'Belmont', National Trust of Australia (Victoria) Classification Report, prepared by R. Aitken, 1997.

43 Wright, *The Bureaucrat's Domain*, pp. 112–13.
44 Wright, *The Bureaucrat's Domain.* Land was reserved by these groups ostensibly for experimental programs for 'public purpose'. See p. 116.
45 F. Gilfedder, 'The Provincial Botanic Gardens in Victoria and Their Relationship with the Royal Botanic Gardens, Melbourne', *Victorian Historical Journal*, vol. 67, no. 1, April 1996, p. 145.
46 J. M. Powell, 'A Baron under Siege: von Mueller and the Press in the 1870s', *Victorian Historical Magazine*, L, 1, February 1979, pp. 18–35.
47 Watts, *Historic Gardens of Victoria*, p. 60.
48 R. T. M. Pescott, *W. R. Guilfoyle 1840–1912: The Master of Landscaping*, 1974, Oxford University Press, Melbourne, pp. 63–78.
49 R. Aitken, 'Picturesque but Instructive: William Guilfoyle's First Decade at the Melbourne Botanic Gardens', *Journal of the Australian Garden History Society*, vol. 7, no. 5, March/April 1996.
50 R. Desmond, *The History of the Royal Botanic Gardens Kew*, 1995, Hartwell Press, London.
51 *Ballarat Miner*, 29 November 1859.
52 Froude, *Oceana*, pp. 130–1.
53 Froude, *Oceana*, p. 131.
54 J. Smith (ed.), *The Cyclopedia of Victoria 1904*, 1904, Cyclopedia Co., Melbourne, p. 332.
55 Smith, *The Cyclopedia of Victoria*, p. 332.

# Part V  Art, Visuality and Material Culture

# Edward Snell

Chapter 17

SKETCHING A FORTUNE

*Tom Griffiths and Alan Platt*



Here is an emigrant gold digger's dream: 'It was a bitter cold night tonight and caused me to dream that my mother was dead and that I was digging a grave for her at the North Pole and finding gold in the bottom of the hole'.[1] The winter of 1852 was closing in on the Bendigo diggings and this homesick English miner, Edward Snell, was calculating the emotional costs and contours of fortune hunting. He had travelled to the Antipodes for adventure and fortune and was digging his way back: 'Oh for a slice of luck and then hurrah for England!'[2] When could he justifiably return home (for return he must, but only with a fortune)? How much gold constituted a fortune? What would he lose in gaining it?

In 1849, writing in his diary on the boat to Australia on his twenty-ninth birthday, Snell reflected: 'When I was 21 I calculated on making a small fortune by the time I was 30 but I have made very little headway in that line as yet'.[3] In the next decade he was to fulfil his wish. He became a spectacular beneficiary of the Australian gold rushes, making his money away from the fields even more than on them.

Edward Snell (1820–80) was born at Barnstaple in Devon, and it was his mother, Elizabeth, who had most influence on the shape of his life, restraining him from a career at sea and pulling the strings for his engineering apprenticeship. He became an engineer, surveyor, artist, adventurer – and, for five months in 1852, a gold digger. But, above all, he was a diarist. He kept two diaries: one in England from 1842 to 1849 and one in Australia from 1849 to 1859 (Plate 17.1).[4] They are both literary and artistic delights, the repositories of humour, wry observations and the tales of two hemispheres. His Australian diary presents a lively and irreverent picture of colonial society, before and after the discovery of gold, and reveals a man who loved life – and often loved to make fun of it.

It was not gold that drew Snell to Australia. His destination just before he boarded ship at London's East India Docks in late July 1849 could easily have been the United States of America or New Zealand. Since childhood he had felt a yearning for the sea; he was after escape and opportunity, motivated by what he called 'a love of variety'. After a seven-year apprenticeship to Henry Stothert at the Newark Foundry in Bath, his career as an engineer had flourished in the English railway mania of the 1840s, and by 1845 Snell had become head draughtsman at the Swindon works of Daniel Gooch's Great Western Railway Company, earning two guineas per week. In August a year later, he was appointed superintendent of the new extensions to the factory and shortly afterwards was made assistant works manager. But, in 1848–49, industrial economies meant that Snell's wages were cut, more severely than those of some of his colleagues, and he gave in his notice, not for the first time in his working life. Within a few weeks, this man of impulse and initiative was bound for South Australia on board the *Bolton*, sketching the trim of the sails in the second volume of his journal and breathing deeply the sea air that he loved.

Snell's first encounter with Australia, then, was with a pastoral society, and a small and faltering corner of it. He arrived in South Australia at a time of growing unemployment. A few years earlier there had been a scarcity of labour, but by late 1849 local groups of unemployed were declaring themselves to be in a worse state than in England and were complaining about 'the delusive representations given at home'.[5] The land of opportunity left many feeling lonely and frustrated.



Two months after his arrival in Adelaide at the end of November 1849, Snell confided to his diary: '[I] am not making money as fast as I expected to do'. During his first year in the colony, he tried his hand at sketching and painting by commission, trading, carpentering, surveying, wharf work and land speculation. He undertook surveying and sketching excursions to the Yorke Peninsula and Lake Alexandrina and struck up a liaison with an Aboriginal woman, Tanne Arrito: 'As she seemed to take a fancy to me I took her down to the place where the sheep were dipped, filled one of the coppers with water and lighted a fire under it, when the water was warm I made her ladyship get in and scrub herself all over from top to toe, after which I took her into my service and she stuck to me like a brick'.[6] By early 1851, he again had thoughts of 'clearing the colony', feeling tired of the uncertainty of his life and of his lack of money and bad debts. The hot summer and insects wore him down: 'May the devil seize this country, the colonists say there is only a sheet of brown paper between Hell and South Australia and I think they are about right'.[7]

From June 1851, nuggets and news began to find their way back from freshly discovered goldfields in the eastern colonies. The walls of Adelaide buildings sported advertisements of ships 'about to sail to the diggings' in New South Wales and, soon, also in Victoria. By December Snell recorded: 'The Gold fever very bad in Adelaide – every body selling off to go to the diggings and all business at a stand still'. South Australia was almost denuded of its workforce as local papers advertised the gold finds and advised travellers on favourable overland routes. Such was the exodus of people that, by early 1852, Snell thought there was 'every prospect of the Colony being totally abandoned'. He unsuccessfully fossicked for gold in the Adelaide hills and panned the mud of the Torrens River, finding 'a few small pieces the size of pins' points'. At the end of January 1852, he and several friends pooled their money, drew up a partnership contract and joined the line of drays laden with stores that were winding out of Adelaide towards the Victorian diggings.

They spent five weeks and three days walking from Adelaide to the Mount Alexander (Castlemaine) diggings, encountering Aborigines, catching and drawing spiders, discovering a lone settler's grave, traversing swamps, spending sleepless nights attacked by mosquitoes, and passing public houses with placards announcing 'No credit till the Gold fever is over'. They tried to live off the land when they could. They shot and ate a carpet snake and skinned a wombat brought to them by some Aborigines, but the meat was stolen during the night by wild dogs. One day they 'shot 3 ducks, 2 Cockatoos, a magpie and a crow and as there were 7 of us we drew lots for them'. Snell scored a cockatoo.

On 4 March 1852 they had their first sight of the diggings at Mount Alexander, 'a scene which beggars all description', but Snell nevertheless had a go:



> Thousands upon thousands of tents extending through the gullies for about 10 miles in every direction, lots of stores distinguished by flags, and slaughter houses which might be nosed a mile off, enough to breed a fever in the place – the ground full of immense holes, many of them 30 feet deep, and the surface cut up by carts and midleg deep in dust. Ruffianly unshaved vagabonds strolling about with gallows plainly written in their countenances and the creek thronged with cradles and tin pans, and fellows washing in every direction.[8]

Here he began almost five months of life as a gold digger, and his words and sketches capture the drudgery as well as the danger of the goldfields. He worked with a small, changing band of mates, sharing the gold they found as well as the duties of cooking and camping life. Their lives were typical of this early phase of cooperative surface mining. They sank holes of up to ten feet, occasionally fossicked out small nuggets 'dry' from the soil, and constantly dug out loads of earth to wash in tub, cradle and tin dish by the creek. Neighbours crowded in around them, violence and theft were frequently reported at nearby tents ('This seems a precious country for robbery and murder'), books and news were swapped, and old friends from England or the ship or Adelaide were regularly encountered. On Bullock Creek near Bendigo, theirs was the first tent pitched and the grass was knee deep. Within a week or two, they were engulfed by the opportunistic society of the goldfields, the grass was eaten quite bare and the water ran like treacle. Every creek and waterhole became so crowded that there was 'no chance of shoving in even a pannikin'.[9] 'The diggins', recorded Snell, 'look very pretty at night, thousands of fires in all directions, the flash of a gun or pistol every few seconds, two or three rows always going on, and every here and there the noise of a flute or fiddle playing Nancy Dawson, Jack's the lad, Paddy will you now, and similar tunes, make the place quite lively'.[10]

It was a physically demanding life: as journalist Alfred Clarke put it, if you had 'spirit enough to attempt it, and endurance enough to carry it on for three months, why there is gold and rheumatism in store for you'.[11] Each night, after a mutton pie and damper (but never vegetables), Snell wrapped himself in his possum rug and turned in exhausted, nursing sore legs and eyes, listening to the endless firing of guns in the dark (he estimated about 1500 shots a night), and calculating the arithmetic of their bizarre goldfields economy. A licence cost half an ounce of gold a month, and water sold at a shilling a bucket. In March, flour was £3 per bag and by July £18, at a time when an ounce of gold was worth little over £2 on the goldfields. In winter, a brief supply of damper could thereby swallow half a pound's weight of hard-won gold.

A month after they had arrived on the goldfields, Snell and his partners had only four and a half ounces of gold to show for their efforts and they 'felt rather down upon the diggin . . . it doesn't keep us in victuals and the horse is half starved and getting as poor as can be'.[12] The winter closed in on them and even iced up their dreams. On good days they washed out a few ounces of gold to share, sometimes a little more, often less or nothing. More than an ounce of gold a day was not an unusual average for a digger in the first year or two of the rushes.[13] In June, their daily combined tally jumped briefly and excitingly to eight ounces and the next day, fourteen, and the next, fifteen. Word got around and diggers crowded in on their claim. Their weekly individual earnings of a few ounces of gold soared for a week or two to over a pound's weight, giving them about £35 each (in goldfields prices) for six days of digging. Their success was dogged and steady, and climaxed with this one brief burst of luck.



Snell supplemented his income by selling water (which kept them in 'butcher's meat'), by judiciously trading animals and belongings, and by cooking up any stray sheep or kangaroo that happened into his hole. By the end of July, he had accumulated more than 100 ounces (buried nearby, secreted in a flour bag or kept close to his skin), which he calculated would realise about £400 in England and which he eventually cashed in Melbourne for £341. Taking into account his expenses of over £100, he made something over £200 clear in five months. But goldfields prices could so easily and quickly erode profits – the dray to Melbourne cost £5, being pulled out of a bog took a further £2, and the security of a gold escort to Melbourne for his takings cost another £4 8s.[14]

Snell, with his usual strategic luck, had been on the Victorian diggings during what Geoffrey Serle has described as 'the period of greatest opportunity', when 'the individual had, for the first and only time, a better than even chance of winning much more than his working expenses'.[15] From late 1851, the rich alluvial fields of Mount Alexander and Bendigo were opened up. Yet it was not until August 1852, the month that Snell arrived in Melbourne from the diggings to cash his gold, that the rush of migrants from Britain began to reach

**Figure 17.1** E. Snell, *Portrait of a Successful Digger.* (La Trobe Australian Manuscripts Collection, State Library of Victoria)

**Figure 17.2** E. Snell, *Raining Hard – Spell Oh!* (La Trobe Australian Manuscripts Collection, State Library of Victoria)

the colony. Serle calculates that, from mid-1852, 'eight out of ten diggers made no more money than the equivalent of reasonable wages, paid their way, or lost money'. Only 10 per cent of all diggers made more than £100 and the other 10 per cent fell somewhere in between.[16] Snell had picked the golden moment. He concluded with satisfaction: 'I think on the whole gold digging, though a dirty business, pays better than engineering'.[17] He was wrong.

Snell's English and Australian diaries are unique in their coverage of two early railway booms: one in England in the 1840s and the other in Victoria in the 1850s. Writing to his old boss Daniel Gooch in 1856, he commented: 'We are railway mad now out here. I have seen nothing like it since the mania in 1845.'[18] The gold rushes brought to Victoria the money and population that made railways both possible and necessary, although they had long been mooted. Men who had come to the colonies with railway experience suddenly found

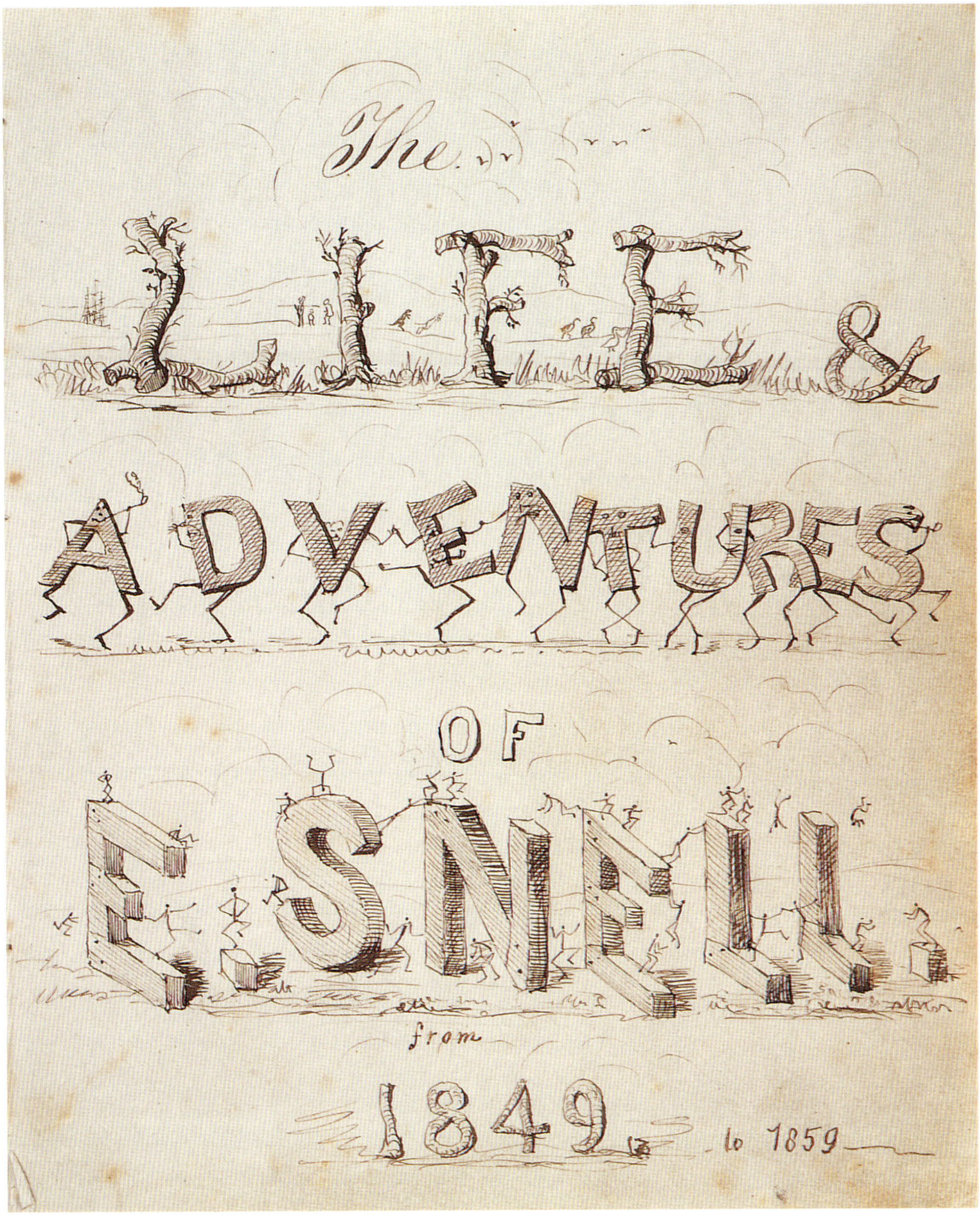


PLATE 17.1 Edward Snell's Australian diary, 1849–59. Angus and Robertson edition. (La Trobe Australian Manuscripts Collection, State Library of Victoria)

Plate 17.2 E. Snell, *Interior of Our Tent on the Bullock Creek.*
(La Trobe Australian Manuscripts Collection, State Library of Victoria)

Plate 17.3 E. Snell, *Bird's Eye View of the Golden Gully – Bendigo.*
(La Trobe Australian Manuscripts Collection, State Library of Victoria)

273

The foregoing sketch shews a section of the cradle. A is the first hopper with a bottom made of hoop iron woven together the stuff is put into this and rocked down into the second hopper B, the bottom of this is made of sheet iron perforated with 3/8 holes. the stones that are too large to pass through this fall over one end of it on to the upper part of the dash plate C, and is carried with the earth that falls through the hopper to the head of the cradle D, it is then gradually washed forward by the action of the water (depositing the gold on the bottom of the cradle in its passage) over the bridge E, and finally out over the tail of the cradle F, G,H, are plugs for letting the water off when the cradle is cleaned out I is the handle for rocking it, J K, the rockers & L, M, the blocks on which the cradle rocks. Very little gold passes over the first bridge E and the stuff that remains in the head of the cradle is taken out & sifted into a tin dish shaped thus it is then shaken up with plenty of water and carefully washed by hand until all the earth is washed away and nothing but the gold remains at the bottom of the dish this last process requires great care & some skill or the gold is liable to be washed away with the earth

Washing with the Cradle & tin dish
our tent on Barkers Creek Mt Alexander

PLATE 17.4 E. Snell, *Washing with the Cradle and Tin Dish . . . Our Tent on Barker's Creek, Mount Alexander.* (La Trobe Australian Manuscripts Collection, State Library of Victoria)

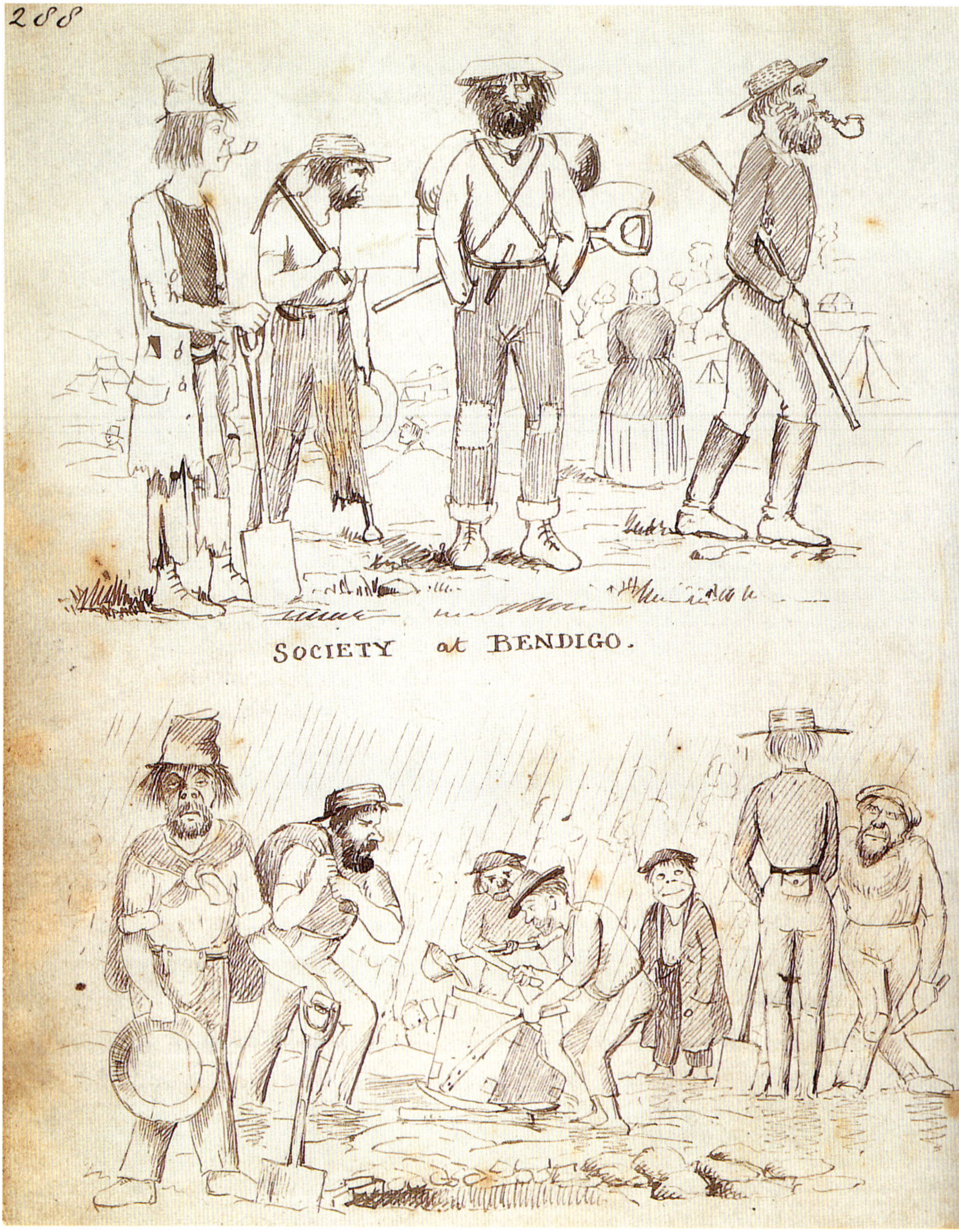


PLATE 17.5 E. Snell, *Society at Bendigo.* (La Trobe Australian Manuscripts Collection, State Library of Victoria)

themselves in demand. It is not surprising, then, that within a month of leaving the goldfields Snell got a job with the newly formed Geelong and Melbourne Railway Company. He saw his advantage clearly: 'The fact is nearly all the Engineers out here are only Surveyors and don't understand mechanics', he remarked to Gooch, and he put the same argument to a parliamentary inquiry to suggest that his kind should be well paid for their services.[19]

The Geelong company was promoted in the belief that it would make Geelong more competitive with its rival port, Melbourne. Many believed that the railway would supplant the heavy sea traffic between the towns and even that overseas ships would unload at Geelong and despatch their cargoes to Melbourne by the new railway. Snell's personality and style, his confidence and bravado, must have appealed strongly to those launching a competitive enterprise like the Geelong railway. The scheme's promoters were mostly local merchants whose business fortunes were tied closely to those of their town. Snell's enthusiasm and self-assurance – the qualities of a salesman – were exactly what they wanted. People found relief in his straight, no-nonsense talk, his goldfields swagger. A story told in later years by his son Charles Scott Snell, whether true or not, captures this appeal:



> Mr Snell was interrogated by the Government agent as to his fitness for the work. In reply, Mr Snell said, 'The questions you have been asking me convince me you know nothing whatever about railway construction and if you were in my employ instead of the Government's, I should give you the sack, and if I had the Government purse to pull upon, as you have, I would put the statues of the twelve Apostles on my bridges.'
>
> 'Bravo', called a voice, 'how refreshing after so much about curves and gradients'.[20]

Snell made it his business to get the job done. He had negotiated an extraordinarily generous contract – he stood to earn £17,500, that being 5 per cent of the £350,000 he estimated the railway would cost. These were the terms he had suggested, pointing out that a salary would encourage him to spin out the work, whereas the terms he proposed ensured that 'it would be in my interest to finish the work as quickly and as cheaply as I could. In addition to this, I am a shareholder in the line . . .'[21] But Snell's concern for ends over means, and his impatience with the finer points of 'curves and gradients', were to undermine his engineering standards and later to draw criticism.

The Geelong–Melbourne line opened on 25 June 1857, the first (and worst?) country railway in Australia. It had been completed a year or two later than had been hoped and at a cost far above the original estimate, of which Snell gained 5 per cent. Much of the company's capital had been sought in England, and the company was forced to raise an additional £262,500 on mortgages. Yet even these circumstances were not extraordinary or disastrous. It was on more general economic, strategic and safety grounds that the new line was widely considered a failure. By the time of the opening, it was clear that it would not be the boon to Geelong that people had hoped it would be and that a line to the goldfields would have been more profitable and useful. In the eyes of some, it became an example of what happens when 'the competition of private interest has full sway'.[22] In this way the line became embroiled in the wider debate about the perils of private control of railways, and this dimmed the limited success it had. Captain Andrew Clarke, reporting on the affairs of the company in 1856, concluded: 'This company has apparently been successful . . . [but] the line is now generally perceived to be a misdirected enterprise, not only as regards the interests of Geelong, but those of the country generally'.[23] He believed that there was not a big enough monied class in the colonies to support an investment with such delayed returns. A parliamentary Select Committee, reporting in August 1857, supported the recommendation that railways should henceforth be the responsibility of the state.[24]

A far more serious and particular criticism of the Geelong–Melbourne Railway was made on the grounds of safety. Even the railway's opening day was tragically marred by a fatal accident, in which locomotive engineer Henry Walter struck his head on a bridge while looking from the train. One parliamentarian declared in October 1857, after trains had been running on the line



for four months, that 'it was notorious all through Melbourne that no person was safe in going on this line'.[25] In November 1856, Messrs Mungo Park Smith and Thomas Oldham, engineers, investigated the state of works on the railway and submitted a comprehensive and detailed report condemning its engineering standards. The course of the railway, the permanent way, embankments, cuttings, culverts, bridges, fencing, crossings and signals were all found wanting. It was estimated that £130,000 would be needed to bring the line up to standard and make it safe. In 1858 government commissioners again seriously criticised its condition. Replying in 1857 to questions regarding the strength of his bridges, Snell declared, 'I do not think it is worth while for us to build for posterity with wages at 15s. to 20s. a day for mechanics; our children and grandchildren would not thank us for it, especially if they have to repay the money borrowed by us to carry out the works'.[26] One senses that even Snell himself may have been sceptical of the product of his labour, preferring in early 1858 to take the steamer to Geelong rather than the train that he had worked to set on its rails. In 1860 the government took over the Geelong–Melbourne Railway at a cost of £800,000, and it was estimated that a further £250,000 would be needed for repairs.

The railway work was not Snell's only occupation during his time in Geelong. He made drawings for a quartz crushing machine for a Geelong company and, most importantly, did the drawings for James Harrison's invention of the world's first large-scale, commercial ice-making machine, a development of great significance, especially for the colonies.[27] At the beginning of 1853, he entered into partnership with Frederick Ferdinand Kawerau and established a very active surveying and architectural business. During the seventeen months of their partnership, they benefited from the new capital produced by the gold rushes. They called for at least forty-four tenders, only a few of which related to the railway. In early 1853 Snell noted in his diary: 'Trade flourishing to no end. £80,000 worth of work in the office not reckoning the Rwy', and four months later he wrote, '£119,000 worth of work in hand besides the railway'. But the boom was short-lived. In 1854 and 1855 Geelong's population and prosperity were fluctuating, with much speculative importing and rapid financial gains and losses. As the recession loomed, there was a scramble for survival. Snell's railway contract helped him to survive these uncertainties. In extracting success from instability, Snell's advantage was his timing and versatility.

Snell's journals measured the ebb and flow of his life. When time was on his hands they burgeoned, much of his energy being poured into creating them. But when he was busy, the daily jottings tailed off and the diary was neglected for weeks or months. Although his years in Geelong amounted to more than half of his time in Australia, this period occupies just over a tenth of his diary. It was an exciting, breathless time in his life – work flooded his office in the early days of his private business, the railway appointment offered him prospects of money and influence, and his personal life, too, showed a new sense of commitment. On 23 June 1853 Snell married Charlotte Elizabeth Bayley, after a lightning courtship – a courtship so fast, in fact, that his first confidence to his diary about this new attachment was on his wedding day. Charlotte, a few years his junior and originally from London, thereafter rates a mention in the diary virtually only when producing offspring. In 1853–54 Snell built a house in Geelong, and by the end of 1856 he had had three children, Emily Charlotte, George Stothert and Edward Lisle, the first of whom died within a few weeks of birth.



The diary was treasured. It caught the overflow of Edward Snell's creative energies and still sparkles with his spirit. It had been his almost constant companion. Sometimes he would write up several days at one sitting or, as he became busier, he would summarise whole months in a single entry. Occasionally, he left the journal behind on his excursions and wrote it up later, doing his sketches from memory or possibly from rough notes made on the spot. The remarkable thing is how often the diary went with him, ready for a note to be penned or a scene sketched. Relaxing by the fireside on the goldfields one night in 1852, he wrote: 'Our tent is in a precious litter and here's a sketch of the interior of it and by Jove while I've been sketching I've forgotten the damper and it's burnt as black as coal, there it is on the right hand side of the sketch'.[28]

The diary was not the only book in Snell's swag. The range and volume of his reading were striking, and Snell is a good example of the sophisticated reader common among the surprisingly literate goldfields community of Victoria in the 1850s. He joined mechanics institutes but also borrowed books from 'Scotch Jock a shepherd' and 'some Irishmen in a tent adjoining ours'. His reading covered a wide range, including histories and documentaries (*Wars of Europe, China Illustrated*), books of local interest (Leichhardt's *Overland*

*Journey from Moreton Bay to Port Essington*), Dickens, Byron and Shakespeare, poetry and songs, 'milk and watery novels' (when little else was available), technical books (perhaps on morse, navigation or mining), the Bible and religious tracts, British newspapers and journals, mechanics magazines, and local papers, to which he occasionally wrote letters.[29] And, surely, he must often have browsed through the pages of his own journal.

Snell was seen by his friends as 'the writer' or 'artist' in their midst. He painted their portraits, composed or wrote out the legal deeds of their agreements, edited the ship's newspaper and tattooed designs on the arms of his love-struck mates. They must have known of his diary and perhaps looked over his shoulder at his sketches, although Snell obviously disliked this, commenting unfavourably at one point about ship's passengers who 'make no scruple of looking over one's shoulder while writing'.[30] He refers to a 'writing desk' that he had with him on his travels, and his sketching pens were almost as important to him as his pipes.

Snell was a skilled artist. He used his talent in his professional draughtsmanship, and in the colonies he supplemented his income by painting scenes and portraits. Although he claimed that he did not 'deal much in caricatures',

his diary sketches show a brevity of line, a round simplicity and a comic form that invite this description. Yet Snell's drawing also served his scientific curiosity and shows considerable technical precision, particularly in his sketches of ships or engineering works, two of his passions. He could draw accurate panoramic views, detailed insect sketches and cameos of Aboriginal culture, but it is the spice of humour and sense of movement that give his drawings life and distinction (Plates 17.2–17.5).

Snell brought to the vast frontier of Australia the jolly air and intimacy of an English village. He found a face-to-face world of familiarity, even in a strange land. The society he portrayed was rumbustious, intimate, sometimes violent. Avid reader and writer as he was, Snell had also absorbed the strong oral culture of his roots. He was alert to superstition and ritual, embraced spiritualism, quickly grasped and used the local vernacular, loved singing, and was sensitive to rhythm and poetry. He could recall songs with ease and was delighted to discover new ones, which he often transcribed in his diary, and he was challenged by the strangeness of Aboriginal words, some of which he tried to record and translate. Snell's diaries are fascinating for the glimpses they give us of this mixture of oral and literary traditions, of superstition and rationality, of village intimacy and worldly adventurousness.

At the end of March 1858 Edward Snell and his family boarded the *Norfolk*, bound for England. He was going home for good. It is perhaps surprising that he should leave Australia just when he was apparently reaching fulfilment in both his personal and professional life. There seemed many reasons to stay. Snell had met his wife in Geelong, and they had married and been happy there. His children were born in the colony. In 1857 his sister Elizabeth and her

293

Our Tent near the Eagle Hawk Gully

the horse came back of himself. After dinner wrote up this log & made the above sketch of our tent.

Monday June 6th 1852

No rain yet & every body very badly off for water. Tom & I finished our first hole in this gully & got 2 ozs out dry. Finished reading the Insurgent chief to night.

7th

Tom & I sunk a new hole & got on the bottom just before dark cleaned up a foot or two of it but didn't see much gold. Bob fossicked 1¾ ozs out of his hole to day. Saw Crany again at dinner time & had a long yarn with him.

8th

Hurrah! rain at last & plenty of it. obliged to cut drains both inside & outside the tent to carry off the water, and as working in the wet was out of the question I amused myself by silver mounting a clay pipe the greater part of the day. Picked up a few nuggets on our heap of washing stuff but got no gold, to day worth mentioning. Here's a sketch of the pipe which is the best flavoured one I ever had, & beautifully coloured. I made the ring round the top out of a sixpence & the ferrule round the stem from the end of a silver pen holder given me as a keepsake by J.W. Cockle in England – the pen holder is none the worse though for being shortened a little and I am writing with it now.

My silver mounted pipe

9th

Rained hard & blew a stiff gale from N.E. all day & night. We went to work in the rain & most of the holes in the gully being filled with water washed about 2½ loads of stuff out of which we got 8 ozs of gold being our best days work on the diggings yet. Bob bought half a sheep to day & made a jolly stew for dinner with

**Figure 17.3** Page 293 of Snell's Australian diary. (La Trobe Australian Manuscripts Collection, State Library of Victoria)

family had come out to live in Geelong; his brother-in-law lived in Melbourne; and his close friend Edward Prowse, who first had persuaded Snell to come to Australia and accompanied him on the *Bolton*, had also established himself in Victoria and was not to return. In Snell's work, opportunities abounded. In 1856 he had boasted to Gooch that he was expecting to handle more railway work: 'As I stand A1 here at present I do not think I shall return until I have made enough to retire on'.[31] Why, then, did he return to England?

Snell surely left partly because of his declining reputation as an engineer. Although he may have stood 'A1' in 1856, by late 1857 and early 1858 government reports on the railway and community opinion had tarnished that image. Writing in 1857, one commentator on Victoria's railway system blamed the power of engineers for much that had gone wrong.[32] Another asserted that the main reason why the Geelong company had failed was the number of technical mistakes that had been made.[33] Although he would have considered such criticism unfair, Snell was sensible enough to catch its drift and to act accordingly.

Yet there were other, perhaps stronger, reasons why he returned home. Snell had always been restless. Throughout his eight and a half years in the colonies, he had moved often and many times had planned to clear out – back to England, or to Sydney or New Zealand or the United States. He had kept in close contact with home through his letters and hankered after the English landscape. He had worked very hard for five years and felt that he needed a holiday. And if one believes that a diary does not just reflect life but also helps to focus it, then Snell must have noticed that the pages of volume two, the record of his Australian adventure, were drawing to a close. And an adventure it had been: he had always intended to return, and to return a success. He had, after all, made his money, and at the age of 37 he 'had enough to retire on'.



To make his fortune had been an oft-repeated ambition, and Snell had left England partly because he did not feel that his homeland had properly provided for him. He was later to say that, in the colonies, he had made 'two piles' – one from gold digging and the other from the railway.[34] His experience exemplifies the judgement that those who prospered most from the diggings were gold-buyers, storekeepers, property-owners, merchants and publicans – and the occasional, opportunistic engineer.[35] Snell's £341 cache from gold digging was certainly a significant reward, but it was the railway that paid him handsomely. On 27 November 1857, a few months before departure, he noted: 'I received the balance of my account with the Company in debentures to the tune of £4500 – the interest being 7 per cent, principal redeemable at any time within ten years, so that I may reckon my income rather over £300 per an', and on 24 December he recorded that 'the total amount received by me for the Engineering on the Geelong and Melbourne Railway, including the preliminary survey and all extras, was [£18,676 1s 1d]'. This was at a time when a building mechanic in the colony might have earned less than a tenth as much during the same five years. In addition, Snell must have made some money from his private practice, despite the debts that followed the recession in Geelong and the ending of his partnership. The boom years brought a phenomenal amount of trade to his office, and he later summarised this period as a time of 'making lots of money'.[36] He was certainly able to send generous gifts of money to his family back home. He had found gold in the bottom of his hole at the end of the globe, and he was free to return.

## Notes

1 E. Snell, *The Life and Adventures of Edward Snell*, ed. Tom Griffiths with assistance from Alan Platt, 1988, Angus & Robertson and Library Council of Victoria, Sydney, p. 310. Sections of this chapter are drawn from our introduction (pp. xi–xxxvi) to this published edition of Snell's entire Australian diary. Further information about Snell's life and more detailed references can be found there.
2 Snell, *The Life*, p. 295.
3 Snell, *The Life*, p. 42.
4 Snell's English diary was entitled 'The Life and Adventures of E. Snell, 1842' and remains unpublished and in private hands.
5 Quoted in D. Pike, *Paradise of Dissent: South Australia, 1829–1857*, 1967, Melbourne University Press, Melbourne, p. 321.
6 Snell, 27 July 1850, *The Life*, p. 131.
7 Snell, 20 January 1851, *The Life*, p. 204.
8 Snell, 4 March 1852, *The Life*, p. 278.
9 Snell, 12 March 1852, *The Life*, p. 282.
10 Snell, 8 April 1852, *The Life*, p. 290.
11 *Geelong Advertiser*, 19 September 1851, quoted in G. Serle, *The Golden Age: A History of the Colony of Victoria, 1851–1861*, 1963, Melbourne University Press, Melbourne, p. 22.
12 Snell, 4 and 15 April 1852, *The Life*, pp. 287, 292.
13 Serle, *The Golden Age*, p. 11.
14 Snell, 30 July, 9 August 1852, *The Life*, pp. 315, 321.
15 Serle, *The Golden Age*, p. 23.
16 Serle, *The Golden Age*, p. 85.
17 Snell, 5 July 1852, *The Life*, p. 311.
18 Snell to Gooch, 22 November 1856, Geelong and Melbourne Railway papers, RAIL 1008/28, Public Record Office, England. For more on Gooch and his relationship with Snell, see A. Platt, *The Life and Times of Daniel Gooch*, 1987, Alan Sutton Publishing Co., Gloucester.
19 'Report of the Commissioners appointed to enquire into the best mode of providing for the internal communication of the colony', *Votes and Proceedings* (hereafter *V & P*), Victorian Legislative Council, 1854–55, vol. 1, A-No. 4, p. 62.
20 C. S. Snell, 'Memo re Edward Snell', 2 February 1950, MS (H16745), Box 35/4, La Trobe Collection, State Library of Victoria.

21 *V & P*, Victorian Legislative Council, 1854–55, p. 62.
22 R. Woolley, *Railways: By Whom Shall They Be Made and Managed?*, 1855, Argus, Melbourne, p. 13.
23 'Railways: Report of Andrew Clarke R.E., M.P., Surveyor-General, 15 November 1856', *V & P*, Victorian Legislative Assembly, 1856–57, vol. 4, no. 33, p. 13.
24 'Report of Select Committee of Legislative Assembly on Railways', printed 11 August 1857, *V & P*, Victorian Legislative Assembly, 1856–57, vol. 2, D-No. 37a.
25 Snodgrass, *Victorian Parliamentary Debates*, 1856–57, vol. 2, p. 1299.
26 'Report of Select Committee of Legislative Assembly on Railways', p. 129.
27 Snell, 21 October 1855, *The Life*, p. 357. James Harrison applied for his patent in Victoria on 13 November 1855, only three weeks after Snell recorded his own involvement.
28 Snell, 9 April 1852, *The Life*, p. 291.
29 An index of Snell's reading can be found at the back of Snell, *The Life*.
30 Snell, 6 September 1849, *The Life*, p. 14.
31 Snell to Gooch, 22 November 1856.
32 F. A. Corbett, *Railway Economy in Victoria*, 1857, W. Fairfax, Melbourne, pp. 9–10.
33 Mr Guthridge, quoted in R. J. Beveridge, 'Victorian Railway Policy, 1850–1883', MA thesis, University of Melbourne, 1952, p. 38.
34 *Harbinger of Light*, 1 November 1887, pp. 3566–8.
35 Serle, *The Golden Age*, p. 86.
36 English diary, chronology in final pages.

# Antoine Fauchery

hapter 18

## A FRENCH ARTIST'S VIEW OF THE GOLDFIELDS

*Dianne Reilly*

First-hand reportage of life in Victoria in the 1850s is rare, thus making Frenchman Antoine Fauchery's reflections on his experiences both fascinating and valuable. The great interest in his work is due largely to the non-Anglo-Saxon viewpoint he expressed in his writings and photographic images (Plate 18.1). He was one of the most perceptive observers of Victorian life in the nineteenth century, and his written and pictorial depictions of life in the infant colony during the gold-rush era are among the earliest published. 

Fauchery gained contemporary fame and more recent acclaim with his volume *Lettres d'un Mineur en Australie,* published in Paris in 1857, which documents his first visit to Victoria in 1852–56. This remains his most notable literary work. Until the 1965 publication of A. R. Chisholm's translation of the letters, neither the author nor his book was widely known in Australia. The work is in the same genre as James Bonwick's *Notes of a Gold Digger and Gold Digger's Guide* (1852), Mrs Charles Clacy's *A Lady's Visit to the Gold Diggings of Australia in 1852–53* (1853), and C. Rudston Read's *What I Heard, Saw, and Did at the Australian Goldfields* (1853). The difference is that Fauchery presented a particularly European view of the new country, not unlike that conveyed by Bill Bunbury's woodliners (in this volume), or in Céleste de Chabrillan's *Les Voleurs d'Or* (1857), Hubert de Castella's *Les Squatters Australiens* (1861) and *Notes d'un Vigneron Australien* (1882), and C. Brout's *Guide des Émigrants aux Mines d'or en Australie* (1855). A legacy from Fauchery's second period of residence in Victoria from 1857 to 1859 is a series of photographs produced under the title of *Sun Pictures of Victoria.* These graphically depict life at the diggings and in the developing city of Melbourne, and the eminent inhabitants of the new city.

Antoine Julien Nicolas Fauchery was born in Paris on 15 November 1823. Little is known of his early years other than details, including the fact that 'he was

**Figure 18.1** Nadar, *Antoine Fauchery*. (Bibliothèque Nationale, Paris)

born without a patrimony'.[1] In his introduction to *Lettres d'un Mineur en Australie*, the poet Théodore de Banville gave much personal data about the author. While prone to descriptive hyperbole, as in his physical description of a man 'with sparkling eyes, Olympian hair in frizzy disorder, a well-shaped arched nose, heroic eyebrows, thick, red, luminous, ironical, sensual lips',[2] he expressed affectionate admiration for the author and keen appreciation of the sincerity and simplicity of 'these gently ironical letters'.[3] For the poet, Fauchery represented eternal youth and he found this youthful quality in his writings and in his personal characteristics:



> I don't know anyone who has been, and is, as young as Fauchery. Naively gay and sad . . . driven crazy by a sunbeam and wiping away, between one smile and the next, some furtive tear, his tales and short stories had on them, as he himself had, that charming peach-down that nothing can imitate, the amorous, confident audacity of adolescence.[4]

At the time of their meeting in a café in the Latin Quarter, Fauchery had already attempted several careers: in architecture – 'he would have liked to continue at Cologne the gigantic, unfinished cathedral';[5] as a wood-engraver, a field where he lacked the patience required for such exacting work; and in painting as a pupil of Léon Cogniet, the popular historical painter and portraitist.

At about the same time as his meeting with de Banville, Fauchery also made the acquaintance of numerous other writers prominent in bohemian circles. These included Henry Mürger, Charles Baudelaire and Gérard de Nerval, all of whom began their literary careers contributing to the popular *Le Corsaire-Satan*. This was to be his next career change and, from 1845 to 1848, he wrote many articles for the journal. Fauchery would later be immortalised as the painter Marcel in Murger's *Scènes de la Vie de Bohème*, on which Puccini's opera *La Bohème* is based.

It was about this time that Fauchery formed a close friendship with Nadar, the greatest of the early French photographers. Nadar photographed most of the celebrated people of his time for his famous *Panthéon*, published in 1854, and his caricatures of prominent writers and actors, including Fauchery, Alexandre Dumas and Baudelaire among many others, appeared in the *Journal Pour Rire* in 1852. From a balloon, Nadar was the first photographer to take aerial photographic views. His influence would prove crucial in the subsequent development of Fauchery's career.

After a short creative burst as a writer, and as a would-be champion of the international campaign for Polish liberation from Prussia, Antoine Fauchery submitted to the call of adventure. On 23 July 1852, he sailed from London on board the *Emily*, bound for Port Phillip. Fauchery spent nearly four years in Australia on this occasion, recounting his observations of life in the colony in his series of eight letters dedicated 'To my friend' – possibly de Banville – but this may have been just a device to intrigue a wider readership. As with those many thousands of others who made their way to the goldfields in 1852, Fauchery no doubt hoped to 'strike it rich'. His account of life on the diggings 
revealed his interest in the wealth acquired by a few, but, in general, his *Letters* suggest a stronger interest in places and people than in the elusive metal.

The first of the eight *Letters* describes the practicalities of departing Europe for the goldfields, and the early days of the non-stop voyage are noted in diary-form, recording Fauchery's shrewd and ironic observations of his fellow-passengers. He juxtaposes the excitement of the passengers with the indifference of the ship's crew, whose gloomy attitude should have warned them that all would not be as expected:

> The Englishman shouts frantic hurrahs, the English women wave their handkerchiefs, all the passengers are delighted; it is a veritable pleasure excursion, a fairy-like trip . . . only the sailors remain unmoved . . . they smile only after giving us all that look that is so searching and . . . so peculiar to the navigator.[6]

It is not long before Fauchery reveals the reason for the sailors' reluctance to participate in embarkation festivities. Four days later, he commences his diary entry with a droll: 'Things are not so good . . . A general uneasiness seems to have come upon everybody, and the sea casts a green reflection on some very pitiful faces.'[7]

Fauchery saw his English fellow-travellers as lacking in sensibility but eminently suited for successful emigration. He believed that the essential difference between the English and the French was one involving a degree of sensitivity:

> They are, of all races, those who possess the most serious qualities for putting into practice and radically applying the most advanced theories concerning life in common. Their coldness, their impassibility, the indifference that they profess

> for the *small detail* . . . will serve them powerfully whenever they might have the idea of going in for one of those new systems that we have invented, but in which our temperament will never allow us to hold out.[8]

He described the boredom of the long days at sea, broken only by the call when hot water was to be distributed, the crowded state of the dark sleeping boxes which made him think of 'paupers' coffins',[9] the long alley where most of the passengers spent all their time, and the Englishwomen, careless of all appearances and still wearing the same dresses, in spite of the ravages of the voyage: 'What can I say, indeed, about a dress of thin material with three or four flounces, a Sunday-go-to-meetings dress, which has suffered the effects of seasickness for more than three weeks!'[10]

The reader shares the relief of the passengers when at last land is sighted. Yet the vessel had to wait for three days outside Port Phillip Heads, so great was the number of ships bringing gold seekers. The ships that lay at anchor in the bay, deserted by their crews, were numerous too:



> From the thick of this forest of masts no face, merry or serious, greets us as we go by. Most of the crews have deserted, they tell me. About all that is left on board is some octogenarian sail-maker, with a denuded forehead, whose infirmities chain him to his bench.[11]

Of his arrival in the embryonic city of Melbourne, after frustrating and irritating delays, he wrote: 'The enormous number of newcomers to this little capital makes it an impossible place to stay in. Neither gold nor silver would get you a room there.'[12] Nevertheless, he admired the future city with its well laid-out streets, while deploring the high cost of all the necessities of life, including a room shared with five others and giving a full view of the gallows: 'All this for three pounds a week (75 francs)!'[13] His chagrin with those guidebooks which had misled prospective miners by describing Australia in idyllic terms may have inspired him to write his own letters, as a guide for future visitors to the goldfields:

> Oh, where are you, green shores, purple corals, mysterious forests, flowers and butterflies with the colours of the sun and moon, as in those little green books? Oh, half-virgin nature, where one can still listen, in cool shades, to the murmuring of that sweet melody that the leaves sing to the branches, where are you? Further on perhaps? – No.[14]

Fauchery concentrated on the enormous cost of everything, from an entrance-tax for the town to vastly overpriced food and accommodation. As he passed an inn in Bourke Street, it amused him that he could have earned, had he had any culinary skills, a huge salary 'at the Black Bull hotel, where they are offering 500 francs a week for a French cook!'[15]

The reader senses Fauchery's cultural shock after the sophistication of Paris. He was alternately fascinated and appalled by life in Melbourne and stayed there only a few days before setting out for Heidelberg where he had heard of a French priest, known only as 'Father B'. Father B was able to give him much background information, not all of it accurate, on political and social events in Australia's history in their brief time together. Because of his limited funds, Fauchery then lost no time in setting out on foot for the Ballarat goldfields, having taken a steamer as far as Geelong.

His spirit of adventure had brought him far from the stimulation of his bohemian friends, yet, shortly after reaching Ballarat, he joined the camp of the artist Eugène von Guérard and his party of seven Frenchmen. Von Guérard's diary of 23 January 1853 recorded, none too accurately, that 'a Frenchman, named Fougery, with his wife, and several others has arrived from California – they have rigged up their tent close to ours, and make a pleasant addition to our party'.[16] There is no doubt that Fougery and Fauchery were one and the same person. Von Guérard noted an occasion when Fauchery reaped the rewards of his hard labour:

> 20 September [1853]. Another rush to a gully in line with Canadian, Prince Regent and Sailors, from which glowing accounts of the richness of the surface diggings are reaching us. My old friend Fougery is said to have got several pounds of gold after a few days of digging to a depth of only a couple of feet.[17]



Fauchery was thus an accepted member of this European enclave where teams were formed to work more than one mine at a time:

> 2 October [1853]. Our little company of nine had a meeting today, at which we arranged a scheme of work. We are to work in three sections of three at a time, doing a six hours' shift, two hours underground, and four at the windlass. This will leave us time for resting and progressing with the preparation of planks for the shaft, of which we need at least a thousand.[18]

However, this arrangement lasted no longer than a month, after which the group disbanded and Fauchery went his own way, recording what he saw for an audience at home in France. Concerning the lust for gold, he wrote:

> The mine! that is the one centre of attraction, the goal of all hopes, the dreamland where the sun rises! . . . It's the gold-fever; the fever for pure gold . . . driving all those who are stricken with it to throw up suddenly the most lucrative positions to run away and look for the uncertain.[19]

He described, for the benefit of future prospectors, every aspect of the daily routine of the digger:

> The miner's equipment is the simplest you can imagine: an iron shovel with a short handle, shaped like a spade, a pick of the smallest size, a fairly strong rope, a washing-dish, a bucket and a tub, are all he needs to do all the digging and seeking known in the district, a few yards of canvas or calico for making a tent, an axe, a frying pan, a saucepan and a tin mug: that meets the needs of domestic life.[20]

Although he did not make any great fortune himself, Fauchery was actually in Ballarat when 'a piece of gold weighing 132 pounds!'[21] was excavated by some 'new chums' just arrived from the steamship the *Great Britain*, which was on its maiden voyage. Fauchery wrote of the gold discovery by two of the ship's passengers who had formed a company with sailors who deserted from the vessel. They had only just arrived in Canadian Gully and everyone laughed at the first of the novices to go down the mine 'like a beginner at billiards'.[22] Fauchery saw the man an hour later: 'He was a negro . . . his face still bore traces of the livid pallor peculiar to coloured people that make the white lips stand out strangely from a grey mask. He was speaking very softly like a man who has just committed a crime.'[23] These singularly fortunate partners sold their claim the next morning and went back to England on the return voyage. The Union Bank at Geelong exhibited the nugget and charged an admission fee in aid of the Hospital and Benevolent Asylum.



There was not much diversion for the miners after a hard day's work at the diggings, and one of their few pleasures was to visit the 'lemonade tents' set up in the camps to disguise the illegal selling of alcohol. The successful digger was frequently served some concoction which knocked him unconscious and made him easy prey for thieves. It took some courage, Fauchery wrote, to accept the offer of a drink in a 'sly-grog' shop: 'Good Lord! this Bourchardy stage-setting,[24] these muffled tremolos from the orchestra, these uneasy glances into the wings, these strange gestures, you might almost say these trap-doors – all this for a glass? – Yes indeed, and what awful brandy at that!'[25]

After working for twenty-two months at Ballarat, Fauchery became despondent, believing that most surface gold had been taken and that finding any at all was becoming increasingly elusive:

> For months at a time, day and night, you have to pump . . . Humid, airless depths, heavy timbering, fetid streams, quicksands which, if you are not careful, pull you in by the feet and submerge you, all this is a mere joke, a mere, tough matter of patience, elbow-grease and agility plus a few nasty colds. But what is really bitter, what breaks your heart and discourages you . . . is after six or eight months of trying . . . to find at the bottom of a sterile shaft the ruin of all your hopes.[26]

He sold his half-share in a non-productive mine to a speculator, the payment not even covering his expenses – 'all my savings, eaten up by seven months' useless work!' – and turned his back on Ballarat.[27]

On his return to Melbourne, Fauchery reflected on the social changes he observed in the developing city:

> I found Melbourne modified, practically transformed. Canvas Town and the bazaar on the banks of the Yara-Yara [*sic*] no longer existed. The government had replaced the canvas city with vast, spacious board huts for the use of newcomers with no resources. Philanthropic establishments such as '*pawn-brokers*' and '*auction rooms*', in great numbers . . . took under their roof the new and second-hand merchandise formerly displayed in the open. The town had developed, the shops had grown bigger, the takings in the hotels and public houses had increased tenfold. Two theatres, elegantly constructed, one for dramas, the other for equestrian feats, had replaced the old 'circus', that modest tent in which I had heard the Ethiopian harmonists from the Gaîté theatre, delighting the colonial audience. In the streets, the blue or red shirts, the muddy trousers and the big gendarme-boots had become rare, and the consideration once accorded to this costume now seemed to be given entirely to black coats and white collars. The miners themselves, coming down from the mines, hastened instinctively and unanimously to shed the dusty livery of labour, to rank as citizens by putting on the garb of the '*regular gentleman*'.[28]

Fauchery had come back with only £60 in his pocket, and he saw his future in the Café Estaminet Français which he opened at 76 Little Bourke Street East  as a place of retreat for non-British emigrants who 'lacked any sort of social milieu and sought in vain a little corner where they could sit in comfort'.[29] He rapidly spent his money on the essentials for the café – a French billiard table, a coffee pot, and some glasses and plates – and opened for business:

> My house had become the rendezvous for all foreigners . . . the best part of my clientele did not come exclusively from the consulates. A nucleus of compatriots and many foreigners, such as Swiss, Italians, Jersey people, Canadians, Mauritians and other such folk . . . flocked to the one and only establishment that differed in kind and manner from the public house.[30]

Despite his prospectuses advertising 'BILLIARDS, SOFAS, LUNCHES, COLD SUPPERS',[31] the venture was not a success, in part because Melbourne was in the throes of a serious financial crisis. There was little activity on the goldfields, with bitter protests against the licence fee imposed on miners, and martial law prevailed. Fauchery had no sympathy for those responsible:

> This deplorable conflict had a fatal effect on public opinion. It was thought for a long time that the insurgents' ulterior motive was political revolution, aiming at independence for the district of Victoria and the establishment of a republic. This latter apprehension was not likely to calm worried minds in town. I, who as a foreigner had no concern with it all, could only contemplate sadly my customers' desertion.[32]

In mid-1855, Fauchery returned to the diggings, this time to the Jim Crow goldfield near Daylesford, to try his luck not as a miner but as a storekeeper.

Once again he failed, leaving Daylesford after six months in business. He returned to Melbourne on foot, recording in his final letter his sensitive impressions of the scenery, the Aboriginal people, whom he pitied, and the squatters, whom he considered 'Australia's real savages . . . [who] live like bears amid their sheep and cattle'.[33] Fauchery commented on the large number of Chinese who had found their way to the goldfields. He admired their industry and honesty and expected that they would play a major part in developing agriculture in the new land. However, he noted the general concern at the influx of 30,000 Chinese emigrants to Australia within three weeks, only two of whom were female, two girls of 7 and 12. In an attempt to give both sides of the argument, Fauchery quoted at length the eloquent plea of one Quang-Chew, put forward to Governor Sir Charles Hotham by the Chinese, against the threat of deportation.

Anticipating his departure, Fauchery obviously looked forward to his return to Paris: 'I shall perhaps look back with regret at life on the diggings. But I shall not regret Melbourne,' which, despite shops, gas lighting and five new theatres, still lacked sophistication, so much so that at one particular theatre 'the company, made up of people from all countries, sings the same



opera in French, English, German and Italian, each man singing in his own language! – I pity the conductor if he is a musician – which is not highly probable.'[34]

On 5 March 1856 Fauchery sailed from Melbourne for London on the *Roxburgh Castle*. While homeward bound, he missed the successful staging, on 12 March at the Vaudeville Theatre in Paris, of *Calino*, the play he had written in collaboration with Théodore Barrière. Another play by Fauchery, one co-written with Henry Mürger, was published while he was in Paris in 1856. It was *La Résurrection de Lazare*, a drama in letter-form.

From 9 January to 8 February 1857, Fauchery's *Lettres d'un Mineur en Australie* were serialised in fifteen instalments in the Parisian newspaper *Le Moniteur Universel*, official journal of the French Empire under Napoleon III. He may have been commissioned by the newspaper to send the letters home by mail for publication; the final two delayed their appearance until after his return to Paris. He remarked in a footnote to the eighth and final letter that 'this letter and the preceding one had been written by me while I was still in Australia. I found them again at the London post office where they had remained for lack of a suitable address.'[35]

It may be safely hypothesised that, while in Paris, Fauchery contracted with Poulet Malassis et de Broise – the publishers of Baudelaire's then-notorious work *Les Fleurs du Mal* – to bring out a collected edition of the *Lettres*, since the single volume appeared in 1857. It was reviewed in the *Argus* in November 1857, possibly by James Smith, the prominent Melbourne journalist:

> More entertaining [than Céleste de Chabrillan's *Les Voleurs d'Or*] is a thin volume of *Lettres d'un Mineur en Australie*: they are written by a Parisian Australian gold-digger, M. Fauchery. Now a Parisian once fairly out of Paris, if he does not

> commit suicide at once from the severity of the *Nostalgie des Boulevards*, becomes one of the most amusing and least instructive of all possible travellers. I say least instructive – i.e. he gives you little correct knowledge of the foreign country he is in: on the other hand, he furnishes the most truthful glimpses of Paris from the antipodes, and teaches the practical philosophy of cheerfulness in every page.[36]

In Paris, Théodore de Banville reviewed the book, summing up the appeal of these reminiscences so well:

> If I were a critic, I should feel no hesitation in praising the book. I would congratulate the author for having succeeded in being, always, naif, intimate, interesting, with a thorough grasp of his subject, without ceasing for a moment to be clear and logical, in accordance with the old French method . . . For my part, I am a lover of poetry . . . and if I can speak to you about the *Lettres d'un Mineur*, I can do so only as a poet and as a good friend of the author's.[37]

On 15 January 1857, Fauchery married Louise-Joséphine Gatineau at the church of St Pierre de Montmartre. Soon after, he approached the Ministry for Public Instruction and Religion for official accreditation and funding to enable him to set out on another expedition and to send back his written impressions and photographs of various monuments and geological features seen in Australia, India and China. He was granted the sum of 500 francs for this mission, and he left France and Europe forever when he sailed from London with his wife on board the *Sydenham* on 20 July 1857.



Photography was still very much in its infancy, and a subject of great public interest, when Fauchery joined forces with the English geologist Richard Daintree to produce *Sun Pictures of Victoria.* They created this collection of photographs between late 1857 and early 1859, using the collodion wet-plate process, and this is the only such collection known to have been produced in the 1850s. The subjects reflect the popular interests of the day and echo those already featured in folio collections of engravings and watercolours, and in the illustrated newspapers of the era.

Little is known of the circumstances surrounding the association of Fauchery and Daintree, but there are a number of coincidences in the patterns of their lives. They had both tried their luck on the Victorian goldfields in the early 1850s, and both returned to Melbourne to take up different occupations in 1854 – Fauchery as a restaurateur, and Daintree as an assistant mineralogical surveyor. Each went back to his homeland in 1856 and, as chance would have it, they independently became involved in the new science of photography, both returning to Melbourne in 1857.

On his arrival, on 4 November 1857, Fauchery established himself as a photographer at 132 Collins Street East. James Smith recorded in his 'Recollections of an Octogenarian' a vivid portrait of Fauchery as he was at that time:

> In a small four-roomed cottage, standing a little back from the south side of Collins-street East, on a block of land now covered by the Austral buildings, I found a French gentleman domiciled as a photographer, who, accompanied by his wife, was cheerfully reconciling himself to the discomforts of colonial life at the time, and exhibiting a vivacity and buoyancy of spirits all the more remarkable as he had occupied a prominent position in Paris as a man of letters. This was M. Antoine Fauchery.[38]

In January and February 1858, in four advertisements in the *Illustrated Melbourne News*, Fauchery advertised for sale photographs he had brought back with him from Paris and so made his new profession of photographer known to interested Melburnians:

> Mr. A. Fauchery, just returned from Paris, begs to inform his friends and the public that he has brought with him a splendid collection of PHOTOGRAPHIC PORTRAITS AND ENGRAVINGS executed by himself during his sojourn there, and for which he solicits their early inspection at his rooms, No.132 Collins-street East. Mr. A. F. is ready to take portraits, landscapes, and views of public buildings or private residences, in large sizes, by quite a new process; and also to operate the reproduction of Portraits, Engravings, or Pictures into as many copies as desired. Likenesses taken after death.[39]



Since Fauchery was an intimate friend and longstanding associate of Nadar, it is reasonable to assume that he had learned from him his craft as a photographer and had updated his knowledge during this last visit to Paris. In March 1858, Fauchery was awarded the gold medal for photography at the Victoria Industrial Society's Eighth Annual Exhibition of Manufactures, Produce, Machinery and Fine Arts, for 'various portraits on paper, from collodion negatives'.[40]

It is not known whether Fauchery and Daintree had met before 1858 when their collaboration as photographers was first noted in the Melbourne press. Since one of Fauchery's aims in Australia was to photograph items of geological interest, it is not surprising that he should have sought the scientific knowledge of Daintree which was lacking in himself. Their collaboration on this project was fortuitous. The result of their joint work was a collection of fine photographs which otherwise might never have been produced at this time, so far from the centres of the cultured world. They brought to their photography a variety of experience from their very different backgrounds. Fauchery was the artistic member of the partnership, his vision informed by the French cultural tradition. Daintree, pioneer geologist and later explorer of some of the most difficult terrain in Australia, had studied at the Royal School of Mines in London under Dr John Percy and Robert Hunt, two early experimenters in the science of photography. Fauchery and Daintree planned to produce a series of photographs, in ten monthly instalments, consisting of views representing life in Victoria at the time. These were, in fact, the first photographs ever produced commercially as a series in Australia.

**Figure 18.2** A. Fauchery, *Comte Lionel de Chabrillan, Consul for France at Melbourne, 1852–1858*, 1858. (La Trobe Picture Collection, State Library of Victoria)

**Figure 18.3** A. Fauchery, *Bourke Street, Melbourne – Looking West from Spring Street.* (La Trobe Picture Collection, State Library of Victoria)

The project was favourably reviewed, probably by James Smith, on 13 August 1858 in the Melbourne *Argus*, as *Sun Pictures of Victoria*:

> Under this title Messrs. A. Fauchery and R. Daintree have just published the first monthly part of a work which it is proposed to extend to ten numbers, and which is to comprise 50 large photographs, in illustration of our colonial celebrities, our landscape and marine scenery, and our private and public architecture. The invention of the stereomonoscope, by means of which the object exhibited in a sun picture, of any size, assumes solidity and relief to the eye of the spectator, gives an additional value to photographic transcripts of nature. The collection under notice are admirable specimens of this branch of art, for art it is; as, irrespective of the skill requisite to manipulate successfully, the manipulators must also possess the artistic faculty in the choice of subjects, in the selection of the most picturesque point of view, and in discerning the most favourable aspects or accidental dispositions of light and shade. With this faculty Messrs. Fauchery and Daintree appear to be endowed, and it appears to advantage in more than one of the pictures now before us. These are five

PLATE 18.1 Antoine Fauchery, *Self-Portrait*.
(La Trobe Picture Collection, State Library of Victoria)

PLATE 19.1 C. H. May (att.), *Swans and Swags* brooch, 1895–1905, 18ct gold and amethyst, 5.0 × 1.8 cm. (Photo: Douglas Elford, courtesy of the Western Australian Museum collection)

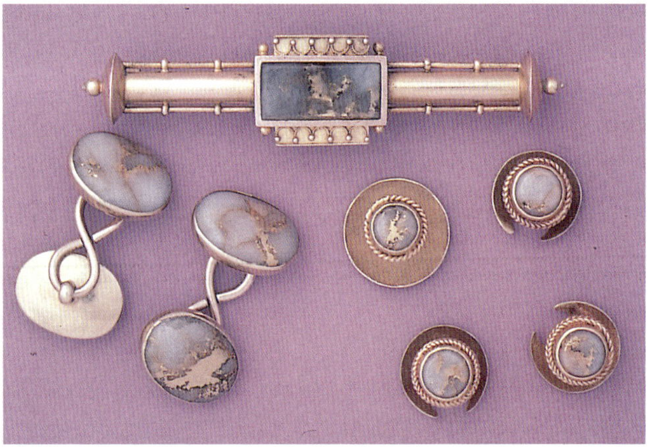

PLATE 19.2 Jonathon Ferry (att.), cut and polished quartz, set in 18ct gold and amethyst, 5.0 × 1.7 cm, private collection, Perth. (Photo: Douglas Elford, courtesy of the Western Australian Museum collection)

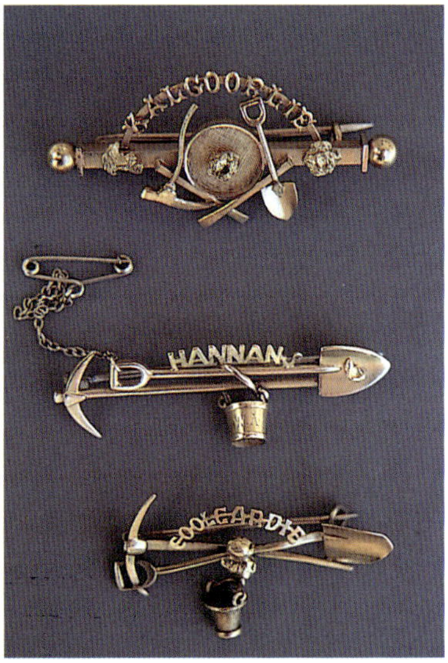


PLATE 19.3 G. R. Addis, *Kalgoorlie* brooch, 1897–1907, 18ct gold, 5.5 × 2.5 cm, House of Riches, Cottesloe; Piaggo & McKinlay, *Hannan's* brooch, 1895–1902, marked 'PIAGGIO & CO', 'PERTH', a swan, '15 ct', 'Rd No 9', 5.3 × 1.8 cm; Larard Bros, Coolgardie brooch, 15ct gold, a golden fleece and a star over heraldic bar, marked '29/5/94 REG No 6, 15'. (Photo: Dorothy Erickson)

PLATE 19.4 Joseph Pearl (att.), *Abbotts Stamper* brooch, marked '18ct', 5.0 × 2.6 cm, Rumsey Collection, Sydney. (Photo: Dorothy Erickson)

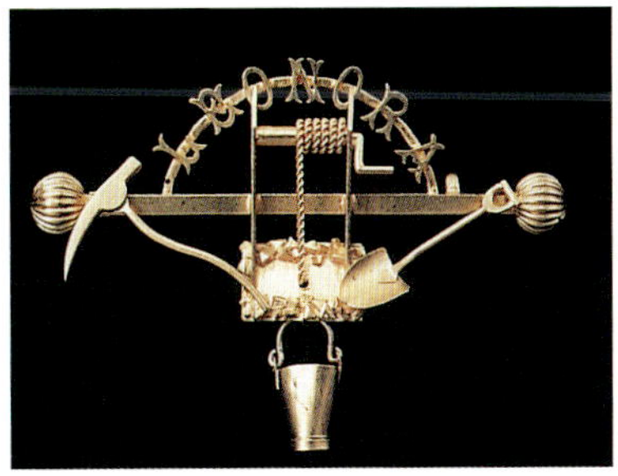


Plate 19.5 *Poseidon* nugget-brooch, gold and pearls, 1895–1910, 5.5 × 2.0 cm. [and] Donovan & Overland, *Leonora* brooch, marked '18ct', 1898–1907, 5.7 × 4.0 cm, formerly in Whiteman Collection, Guildford. (Photos: Douglas Elford, courtesy of the Western Australian Museum collection)

Plate 19.6 C. E. Band and Donovan & Overland, *Kurrawang* brooch, 18ct gold and opal, marked 'BAND', and a key, c. 5.0 × 2.5 cm, private collection, Sydney. (Photo: Dorothy Erickson)

Plate 19.7 Mazzucchelli & Downes, bar brooch, c. 1906, 15ct gold, marked 'M&D', 5.5 × 1.8 cm, Erickson family collection, Perth. (Photo: Douglas Elford, courtesy of the Western Australian Museum collection)

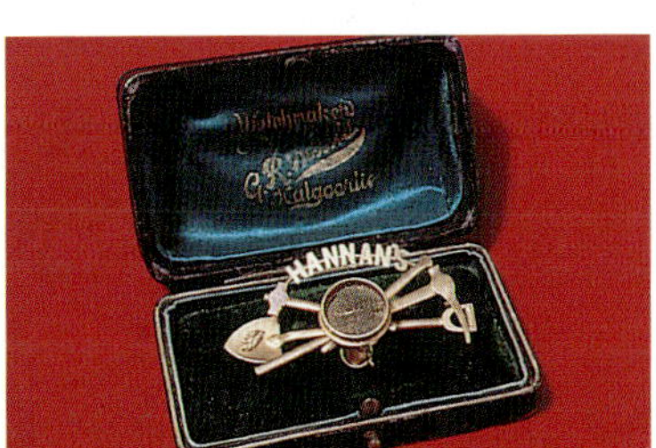

Plate 19.8 G. R. Addis, *Hannan's* brooch, c. 1894, marked '18ct', 5.6 × 3.5 cm (Photo: Douglas Elford, courtesy of the Western Australian Museum collection); [and] *Rosenthal Aronson* brooch, registered design 'WA 10', 5.0 × 1.7 × 1.0 cm, Erickson family collection, Perth. (Photo: Dorothy Erickson)

PLATE 20.1 Edwin Stocqueler, *Australian Gold Diggings*, oil on canvas, c. 1855. (Rex Nan Kivell Collection, National Library of Australia)

PLATE 20.2 G. Pearce (*sic*: Augustus Baker Pearce), *The Myers Creek Rush near Sandhurst, Victoria*, watercolour, 1867. (Rex Nan Kivell Collection, National Library of Australia)

> in number, and merit separate notice . . . We may add that the pictures are mounted on cardboard, and will constitute, when the tenth number has been issued, a work of permanent value and interest.[41]

It was not until the 1860s that many other photographers took advantage of the possibilities of the collodion process for publication. A flourishing industry for the sale of photographs, individually and in albums, was then begun, continuing throughout the nineteenth century. An infinite number of prints could be made from the wet-plate negatives, and for this, as in *Sun Pictures*, albumen paper was most commonly used.

While Fauchery and Daintree appear to have produced multiple copies of photographs for sale, only three sets of their joint work are known to exist today: one in the La Trobe Picture Collection, State Library of Victoria, and two slightly different sets in the John Oxley Library, State Library of Queensland. The most intriguing question about their collaboration is related to the contribution each made to these collections. Their different backgrounds and personalities introduced a variety of emphasis to the photographs they produced. There is a great body of Daintree's landscape and geological photography extant, mainly related to his work with the Victorian Geological Survey and in north Queensland, but it is not known whether Daintree was ever in business alone as a photographer. On the other hand, Fauchery advertised as a portrait photographer and it may be presumed that the portraits in the album are his. The majority of the mining and landscape images have been identified clearly as Daintree's, but, given the physical difficulties inherent in the collodion wet-plate process, joint photographic expeditions with both partners participating cannot be ruled out. Whatever the attribution, it is remarkable that the Fauchery–Daintree collaboration produced such high-quality work so early in the history of photography. They provide a rare and invaluable record of life in Victoria in the 1850s.



Two images in the collection dramatically reveal Fauchery's deliberate re-creation of his own earlier experience as a gold seeker. By 1858, large-scale mining companies employing waged labour had virtually replaced small cooperative operations. Fauchery posed groups of miners as in earlier years, and indeed, these particular miners may also have been fossickers who panned for gold when their work for the mining company was over.

In *A Group of Diggers* (Figure 18.4), the pose is calculated to present an image of the camaraderie which existed on the goldfields:

> I must say that in general, at the mines, those that show proof of a willingness and energy often fall in with people ready to help them, and that these acts of friendliness, inspired by feelings of loyalty and rustic honesty, are free from any sort of ulterior motive based on self-interest or greed.[42]

The excitement of the earlier years of surface alluvial gold discoveries by individuals is well conveyed by the men, gathered here in companionable fashion to inspect the contents of the pan. Given that an exposure time of up to two minutes was necessary to take a photograph such as this, Fauchery succeeded in conveying great spontaneity, and the comradeship of the men who had 'staked their all on the great lottery of the Ballarat leads'.[43]

*Washing Out a Good Prospect* is the second of the groups of miners photographed by Fauchery. Many of the same faces come alive again before the camera. These men were possibly attached to the Argus Flat Company at Chewton, and yet, an impression of the fever of speculation of earlier days is successfully conjured up in this image. These men, many of educated background, toiled at the exhausting work of mining in the hope of success. Fauchery described the suffering:

> It's only the first blister that hurts. That blister, I can tell you, is terrible, immense; it is as big as the two hands that it invades, swells and deforms. It really hurts. The pain starts at the finger tips, goes under the nails and moves up as far as the shoulders. I am not sure whether it is the same pain that extends from the shoulders along the spinal column, sticking needles in between all the vertebrae, torturing you to such an extent that it is sometimes impossible to stand up.[44]



Attire was, of course, informal, but headgear of one sort or another appears to have been obligatory. Perhaps the most popular adornment was the broad-brimmed 'cabbage-tree hat', which was originally woven from the leaves of the cabbage palm to provide serviceable and durable shelter for those constantly working in the open.

In his agreement with the French government, which had subsidised his return to Australia, Fauchery had undertaken to send home photographic documentation of life in the Antipodes. He had as his sitters some of the most eminent Victorian leaders of the time, including the Governor, Sir Henry Barkly, the Premier, William Clark Haines, and the first Bishop of Melbourne, Charles Perry. His portrait of the dashing French Consul, the Comte de Chabrillan, first official representative of France in Melbourne and husband of the notorious courtesan, Céleste 'Mogador' de Chabrillan, superbly shows the man's character and style (see Figure 18.2).

The photographs of Aboriginal people in the album reveal a great understanding and sensitivity on the part of Fauchery towards the indigenous population. The dramatic *Wurundjeri Tribesman Ornamented for a Corroboree*, perhaps his best-known image, conveys the fascination of the European observer for the rituals of Aboriginal people and is evidence of the rapport the photographer had established with the tribesman, something which, as Ian Coates shows us, was by no means common.

It is not known when or under what circumstances the partnership of Fauchery and Daintree was dissolved. Daintree had already returned to the Victorian Geological Survey under Alfred Selwyn by 17 January 1859, and he was destined

**Figure 18.4** A. Fauchery, *A Group of Diggers.* (La Trobe Picture Collection, State Library of Victoria)

to make an even greater name for himself as an explorer of the Queensland hinterland and, later, as Agent-General for Queensland in London.

Early in 1859, Fauchery departed Melbourne for further adventures in the Philippines. He was next heard of in China. He joined the French Expeditionary Forces as war correspondent and official photographer and took part in the Arrow War which opened China to the West. He sent reports in the form of letters to *Le Moniteur* in Paris from October 1860 to February 1861. These were serialised in the newspaper as 'Lettres de Chine', and record such events as the surrender of the Taku forts and the taking of the Summer Palace in Peking by French and English forces. Unfortunately, the photographs taken by Fauchery in China have not been found. On 12 January 1861, he embarked at Shanghai for a voyage to Japan on board the P&O steamer *Cadiz.* On 27 April 1861, only three months after his arrival in Japan, Fauchery died in Yokohama at the age of 37 from the combined effects of gastritis and dysentery.[45]

Antoine Fauchery, although a minor figure in the bohemian circles of Paris, was an artist, writer and traveller who contributed much of great historical value during his short life. From his experiences in Victoria, he published a lively account of his own life on the goldfields, and his interest in the magical new science of photography contributed to the production of an important record of the young colony. His work in Victoria provides a fascinating insight into the perspective of a nineteenth-century continental adventurer who responded to the lure of gold.

## Notes



1 A. Fauchery, *Letters from a Miner in Australia*, trs. A. R. Chisholm, 1965, Georgian House, Melbourne, p. xiv.
2 Fauchery, *Letters from a Miner*, p. xii.
3 Fauchery, *Letters from a Miner*, p. xix.
4 Fauchery, *Letters from a Miner*, p. xiii.
5 Fauchery, *Letters from a Miner*, p. xiv.
6 Fauchery, *Letters from a Miner*, p. 5.
7 Fauchery, *Letters from a Miner*, p. 6.
8 Fauchery, *Letters from a Miner*, p. 10.
9 Fauchery, *Letters from a Miner*, p. 9.
10 Fauchery, *Letters from a Miner*, p. 12.
11 Fauchery, *Letters from a Miner*, p. 19.
12 Fauchery, *Letters from a Miner*, p. 21.
13 Fauchery, *Letters from a Miner*, p. 23.
14 Fauchery, *Letters from a Miner*, p. 22.
15 Fauchery, *Letters from a Miner*, p. 32.
16 M. Tipping, *An Artist on the Goldfields: The Diary of Eugène von Guérard*, 1982, Currey O'Neil, Melbourne, p. 40.
17 Tipping, *An Artist on the Goldfields*, p. 56.
18 Tipping, *An Artist on the Goldfields*, p. 58.
19 Fauchery, *Letters from a Miner*, p. 33.
20 Fauchery, *Letters from a Miner*, p. 55.
21 Fauchery, *Letters from a Miner*, p. 77.
22 Fauchery, *Letters from a Miner*, p. 76.
23 Fauchery, *Letters from a Miner*, p. 77.
24 Joseph Bourchardy, writer of many French cloak-and-dagger plays.
25 Fauchery, *Letters from a Miner*, pp. 56–7.
26 Fauchery, *Letters from a Miner*, pp. 78–9.
27 Fauchery, *Letters from a Miner*, p. 81.

28 Fauchery, *Letters from a Miner*, pp. 83–4.
29 Fauchery, *Letters from a Miner*, p. 85.
30 Fauchery, *Letters from a Miner*, pp. 87–8.
31 Fauchery, *Letters from a Miner*, p. 86.
32 Fauchery, *Letters from a Miner*, p. 89.
33 Fauchery, *Letters from a Miner*, p. 99.
34 Fauchery, *Letters from a Miner*, p. 105.
35 Fauchery, *Letters from a Miner*, p. 82.
36 *Argus*, 11 November 1857.
37 Fauchery, *Letters from a Miner*, pp. xviii–xix.
38 J. Smith, 'Recollections of an Octogenarian, V, Bygone Celebrities', *Leader*, 27 July 1857, p. 43.
39 *Illustrated Melbourne News*, 16 January 1858, p. 38; 23 January 1858, p. 50; 30 January 1858, p. 60; 6 February 1858, p. 67.
40 *Argus*, 4 March 1858, p. 5.
41 *Argus*, 13 August 1858, p. 6.
42 Fauchery, *Letters from a Miner*, p. 68.
43 Fauchery, *Letters from a Miner*, p. 79.
44 Fauchery, *Letters from a Miner*, p. 64.
45 *Australian Dictionary of Biography*, 3, 1851–90, pp. 158–9.

# Cinderella's Jewellery

THE GOLD-RUSH BROOCHES OF WESTERN AUSTRALIA **Chapter 1**

*Dorothy Erickson*

316 The electrifying news of rich gold rushes in Victoria and New South Wales in 1851 attracted goldsmiths as well as miners from around the world. Jeweller Edward Fischer arrived from Vienna, William Edwards from London and Sylla Denis from France, all settling in the booming city of Melbourne. P. J. Wendt and Julius Schomburgk migrated to Adelaide from Germany, while Julius Hogarth and Conrad Erichsen (Erickson) came from Scandinavia to set up in Sydney. All of these artists made memorable work from eastern Australian gold. Perhaps from an instinctive desire to shape a new sense of identity in their new land, the most distinctive and enduring tendency of such transplanted Australian jewellers was to use local motifs.[1] These included wondrous great brooches adorned with native flora, and a series of miners brooches featuring implements used to extract the gold. Notable examples were to be seen in Bathurst, Ballarat and Bendigo during the 1850s, 1860s and 1870s. Most of these ebullient works seem designed to celebrate the good fortune of both miner and goldsmith alike.

In the 1890s, the geography of Australian gold fortunes shifted abruptly. Just as the east coast began to suffer the effects of a worldwide depression, fabulous gold discoveries were made in the west. The Cinderella colony of Western Australia saw her population multiply nearly seven times in only twenty years. From 1901 to 1910, about 50 per cent of Australia's gold production came from the 'golden west'. Men from all walks of life came to try their luck on the West Australian goldfields. For a while, it was literally possible to pick up a fortune off the ground, particularly after rain had washed the red dust from the precious metal.[2] Within five years of the discovery of gold in Western Australia, mining was taking place in four widely separated areas, ranging from the tropical north of the colony to the remote eastern deserts.

Jewellers already in the colony were, of course, well positioned to take advantage of these golden opportunities. The northern Kimberley, Pilbara and

Murchison fields were serviced by goldsmiths in the port towns to the south. Alfred Jackson in Geraldton advertised to make up Kimberley gold, as did Vincent Nesbit, Anthony Fouchard and John Wellby in Perth. Fouchard, Nesbit, Charles May, Edward Hooper, and Wheeler & Bennett all became substantial firms. The fourth area, serviced by the ports of Albany and then Fremantle, was 300 miles into the dry inhospitable interior. It was destined to become the richest area of all – the fabled Yilgarn.[3]

Notwithstanding the hardships entailed in reaching the Yilgarn, goldsmiths were soon in residence. The first town established was Southern Cross, a railhead which launched the further push to Coolgardie, Kalgoorlie, Boulder and elsewhere. The Coolgardie field was discovered in 1892 by Arthur Bayley and William Ford, 125 miles east of Southern Cross across an almost waterless, windswept, red sand plain dotted with gum trees. The following year 'The Old Camp', as Coolgardie was called in its first decade, was declared a field. Jewellery followed hard on the heels of such discoveries, not least because women lived and worked on the Yilgarn goldfields even in the gruelling early days. At first, miners were forced to live in makeshift huts and tents, enduring continual high temperatures, shortages of water and food, choking red dust and little or no sanitation. Inevitably, they paid a high cost in dysentery and typhoid. Along  with some hardy miners' wives, nurses came to these fields to help with the typhoid epidemics, and grateful diggers often gave them nuggets as presents.

In 1894 Paddy Hannan gave one such nugget to Clara Saunders, later to be the bride at the first wedding celebrated in Coolgardie. Clara, who worked in one of the hostelries, had the previous year given up her room to Paddy when he was sick. For her kindness she was rewarded with what he claimed was the first piece of gold he had collected at a new find named 'Hannan's'. The Golden Mile of Kalgoorlie–Boulder was destined to be one of the richest goldmines in the world throughout the 1890s. Clara celebrated her good fortune by having the nugget mounted on a bar brooch of a style fashionable at the time: a twenty-two carat red-gold arrow.

Throughout the goldfields small nuggets were used as currency, while larger ones often became mementoes. Many were attractively shaped, inspiring the successful prospector to have them fashioned into jewellery. It was a simple matter to solder a nugget to a gold pin. Moreover, gold was an ideal metal for a peripatetic goldsmith working in primitive circumstances in remote locations. The metal is one of the few found in its native state that is both malleable (can be beaten) and ductile (can be drawn into a wire). With its low melting point and lack of oxidation, gold is easy to anneal (soften) in the crudest of fires and can then be fashioned into objects which neither rust nor tarnish.[4]

Goldsmiths, unlike silversmiths, can work with very few tools, and it is not difficult for them to move from field to field. Many may well have originally come to prospect as a part-time occupation, only to find richer rewards through practising their art. By 1900 at least fifty goldsmiths had set up premises to service the needs of the miners and residents of the 'Eastern Goldfields', as the area came to be known. A still greater number were itinerants, spending a few years moving restlessly from place to place among the gold towns, then disappearing when the alluvial gold ran out. Roving goldsmiths were noted at such places as Mount Magnet, Leonora, Malcolm, Laverton, Day Dawn, Sandstone, Cue and Norseman. Many of these settlements are now ghost towns, commemorated only in the intriguing little mining brooches bearing their names.

The West Australian gold rushes had an equally dramatic impact on Perth, which was still a charming country town of about 15,000 people in 1895. Yet the town was well positioned for expansion: a harbour was constructed to allow deep-sea ships to enter, and railways soon pushed out to settlements in the interior. Seemingly overnight, Perth changed from a sleepy hollow to a bustling metropolis, temporarily a focus of world interest. Like Graeme Davison's 'Marvellous Melbourne', it had become an 'instant city'. Few other cities or towns in Australia could offer such good work prospects during these generally depressed years, and a cosmopolitan migrant society came rapidly into being. Engineers, financiers and even smiths commuted between America, South Africa and Western Australia. Young men out for adventure added colour and verve to the goldfields. Celebrities such as Lord Percy Douglas (later the Marquess of Queensberry) and David Carnegie became potential clientele for the local goldsmiths, ensuring the repatriation elsewhere of much that they made.



As a result of shifts in fashion, the jewellery that they purchased during the 1890s had become daintier and lighter. A few discreet bar brooches and a locket, a watch or a medallion were considered stylish, while the greater supply of diamonds, together with the novelty of electric light which could set them sparkling, influenced changes in evening wear.[5] The mining brooches seen in Western Australia had their origins in a series of presentations to the royal princesses in London in the 1880s, reinterpreted in the context of the 1890s fashion for wearing 'name brooches' and the occasional piece of modish sporting jewellery.

Recognisable styles of West Australian jewellery began slowly to evolve. Ostentatious displays of wealth had never really been part of the code of the tiny and comparatively impoverished colony. With its 'genteel poor' traditions and fiercely localised, isolationist stance, it tended to look askance at the excesses of the nouveau riche in other colonies. Brooches designed for barmaids, as seen on the Victorian diggings, were not favoured in Western Australia.[6] Instead, the heraldic emblem of Western Australia – the swan (also a fashionable symbol among the 'Aesthetic' circles surrounding the artist Whistler in London in the 1890s) – was combined with nuggets and discreet mining imagery. The resultant jewellery was raised by a few makers to an art form, fit to be presented and displayed by all levels of society. Old photographs show pick-and-shovel brooches proudly pinned at the throats of not only members of the Government House set but also their governesses. The families of merchants and mining entrepreneurs can be seen wearing similar jewellery to that of miners' sweethearts.

Among the most widely worn West Australian jewellery, three types of mining brooches predominate: those which incorporate tools, those which incorporate equipment, and those which incorporate nuggets or polished quartz. A large number of brooches included the name of a goldmine or town as part of the design. Although such jewellery is often generic to colonial goldfields from Queensland to South Africa, a tutored eye can usually distinguish West Australian work. Even where there is no name or mark, it is notable that documented brooches from the west are rarely seen with the ubiquitous double-ended pick, symbol of the alluvial miner. Instead, the majority feature a one-ended pick, with a useful, flat sledge-hammer end to complete it, as a constituent of the design. This reflected the actual equipment used by local miners, as seen in contemporary photographs. The one-ended pick was perhaps developed to reduce weight on the long journey pushing a barrow into the interior. Alternatively, such picks may have been more suitable for working the hard rock of reefs where the gold was found. Whatever the reason, a one-ended pick can usually identify a goldmining brooch as West Australian.

'Swans and swags' brooches made by Charles May in Fremantle were also extremely popular among West Australian buyers (Plate 19.1). Perhaps they were seen as anti-Federation symbols, since Western Australia only joined the Commonwealth at the last moment, and then unwillingly, when the Eastern Goldfields threatened to secede from Western Australia to join the eastern colonies. Such brooches were still being made in 1906 when Western Australia attempted to secede from the Commonwealth on the grounds that membership was ruining the state's economy and stunting the growth of manufacturing. Indeed, the maker of these 'swan and swag' brooches, May, was so hostile to the economic results of Federation that he even considered emigrating to America at this time.



Coolgardie, one of the most famous of the west's desert goldfields, also featured on many mining brooches. Its distinctive icon, a gleaming white quartz reef, was shot through with attractive veins of gold. As a result, quartz pieces from there were sometimes cut and polished to be set as gems (Plate 19.2). An interesting and unusual example of quartz jewellery is the POSEIDON nugget brooch now located in the Western Australian Museum (Plate 19.5). This gold nugget embedded in white quartz has been set in an open pierced-work bar brooch, reminiscent of the neo-Gothic designs of English architect William Burges. The maker cannot be identified with any certainty. It is possible that the brooch came from the shop of well known Coolgardie jewellers Nesbit & Jerger, who advertised in *Wise's Post Office Directory* of 1895–96: 'Coolgardie quartz cut and polished' (Figure 19.1). These jewellers, it seems, suffered from two of the common hazards of the goldfields cities – fire and burglary. In 1895, their shop, known as Goldsmith's Hall, was burgled while the owners were out having supper and about £1000 worth of gold and jewellery was stolen.[7] The Hall, a corrugated iron building with an impressive sign, was then burnt down in January 1896.[8]

Coolgardie also began to generate another distinctive style of brooches when the town hosted an International Exhibition in 1899. Gold, as the source of 88 per cent of the colony's wealth, was naturally the exhibition's chief focus. The purpose-built premises opened in March, attracting 55,000 visitors by June, some of them coming up by special trains from the coast. Many souvenir brooches were made for the occasion, which is why the name 'Coolgardie' is found most frequently on surviving goldfields fretwork brooches.

Larard Brothers, jewellers of Melbourne, evidently made some of these souvenirs, and numbers of their crossed-pick, shovel brooches survive in collections around Australia (Plate 19.3). Despite the fact that the two brothers, Richard and William Larard, worked on the West Australian goldfields of Mount Magnet in 1899–1900, it is unlikely that their prolific souvenir brooches were actually made in the colony.[9] Their work, which includes some of the most attractive mining jewellery produced in Australia, seems actually to have been made to order in Victoria and delivered to Coolgardie in time for the exhibition. The Larards' pieces carry an elaborate registration mark, and are elegantly designed and immaculately made, though slight differences in filing techniques within individual pieces indicate that several hands were involved in construction.

An attractive yet more severe style of Coolgardie mining brooch was made



by Frank Piaggio, who arrived to work in Perth about 1891. By 1894 he had been joined by jeweller Robert McKinlay, though their jeweller's mark remained PIAGGIO & CO PERTH. On 10 March 1895 they registered a design for a mining brooch, 'W.A.9'. This consisted of a parallel pick and spade, with a nugget on the blade. Examples are marked with 'RdNo9' and '15 CT' in three linked rhomboids. Of the three known pieces, two have a fretted sanserif COOLGARDIE in a low curve above the handle of the shovel, and the third has HANNANS (Plate 19.3). A very similar piece was worn by Annie Campbell Loton (née Forrest) in her portrait published in *Forrest Family: Pioneers of Western Australia 1842–1982*. She was evidently proud to wear this symbol of the colony's new-found wealth. Perhaps it was a Christmas present, for Piaggio and McKinlay advertised on page 54 of the *Western Mail*'s Christmas number in 1898:

> Beautifully manufactured Kalgoorlie quartz brooches, links and charms, in all shapes and styles, are made of specially chosen stones, with veins of gold running through, and are mounted in gold. The most charming and acceptable Christmas present one could wish for.

Local manufacturing jewellers in Coolgardie were also able to take advantage of the opportunities the exhibition presented. Joseph Pearl set up a factory there in time for the exhibition. He may have originally worked for Donovan & Overland, making the finely crafted brooches characteristic of that firm from about 1898 (Plate 19.5). We can be almost certain that two surviving pieces came from his own workshop, because of a carefully worded description in a local advertisement which claimed he had a 'large and varied stock of unique and characteristic goldfields work'. More importantly, he claimed to be 'the sole maker of the Coolgardie Windlass and Five-Head Battery Design

**Figure 19.1** Nesbit & Jerger's Goldsmith's Hall, Bayley Street, Coolgardie, c. 1894. (National Library of Australia)

Brooches'. This advertisement has, however, caused some confusion among analysts, for one COOLGARDIE windlass brooch was certainly made at Donovan & Overland in Perth. It is a pair to the KALGOORLIE design registered by Donovan in 1898 and used in his LEONORA and MALCOLM brooches as well.

Only two such battery brooches are known to have survived. A delightful brooch depicting a five-head battery has ABBOTTS in serif letters applied to an arch above the stampers (Plate 19.4). Somewhat surprisingly, given the subject, the piece is aesthetically pleasing, well modelled and well balanced. A little licence has been taken with the representation, but it is undeniably a gold-stamping battery. To balance the composition, a sledge-hammer and crowbar 'out of scale' have been added. Fluted-ball finials complete the composition of what is one of the most interesting of the West Australian goldfields works. The name 'Abbotts' referred to a goldmining district which operated from around 1894 to 1910 in an area fifty miles north of Nannine, within the Murchison district of Western Australia. This brooch could well have been made to commemorate the arrival of the goldfield's first battery.

Another local, George Richard Addis, the most prominent of the Kalgoorlie–Boulder jewellers, serviced the needs of dignitaries and diggers alike. Addis was born to English parents who migrated to South Australia and then followed the gold. When Addis was about 14, the family moved to Melbourne where he may have been trained. By 1890 he was established as a goldsmith in Launceston where he remained until 1893, setting sail for Albany and the West Australian goldfields the following year.[10] There, one of his first pieces was a gold brooch with HANNAN'S arched across it (Plate 19.8). It consists of pick, spade, sledge-hammer and tap or crowbar crossed behind a sieve from which is suspended a bucket engraved with 'W.A.' It is nicely designed in a balanced arrangement, no mean feat given the requirements of using some rather prosaic tools.

Other surviving Addis mining brooches display variety and design flair. The two most notable reside in the National Gallery of Victoria and in a private collection in Western Australia (Plate 19.3). The piece in Victoria, which bears a striking resemblance to the mining brooches made by Donovan & Overland,



has a glass phial containing alluvial gold flakes, surmounted by an arch supporting the finely pierced serif letters of the name KALGOORLIE. Harmoniously arranged below are a pick, a spade and a windlass, with a rope descending through the golden hole to a bucket suspended below. Nuggets were another of Addis' passions, and he had a legendary collection which he occasionally displayed in his window. Most of his nugget brooches feature double or triple bars, joined with pierced-work rhomboids or crescents, supporting a natural nugget. Each end of the bar was usually finished with a plain or fluted ball of a type imported from France or Germany.

One of the most travelled of the West Australian goldsmiths was Charles Edward Band, who was born in 1864 in Derby, England. The son of a wealthy silk merchant, he was educated at All Saints High School, then apprenticed to R. Camp, an English watchmaker. Band was a wanderer, typical of many of his generation, who followed his fortune around the world. After moving to New York where he was employed by R. Nicholson, of Brooklyn, he journeyed to Chicago and then to Fresno, California, where he managed a jewellery business before arriving in Sydney in 1892. By 1898 he had opened his own business in Hannan Street, Kalgoorlie, only to have it destroyed by fire in 1900.[11] He is last listed in the Post Office directories in 1905 but seems to have continued in business after this, as a bracelet presented to Miss Dickson in Kalgoorlie in 1907 was purchased in a box marked 'C. E. Band'. This bracelet, given by her parents for her twenty-first birthday, is now in the collection of the Western Australian Museum.

Donovan & Overland were also the makers of an unusual brooch sold by Band. Instead of items of mining equipment, this featured a pair of axes beneath the arched sanserif lettering KURRAWANG, indicating links with the Kurrawang Woodline, a private railway run by the West Australian Goldfields

Firewood Supply Company which was controlled and managed by entrepreneur and Member of Parliament William Hedges. The piece features a large oval opal, set centrally between two axes (Plate 19.6). It is in pristine condition, in part because the imagery is unlikely to have encouraged wearing.

One important jewellery firm, Mazzucchelli's, only rose to prominence as gold was beginning to decline. It had begun in Boulder shortly after Federation as a partnership of Matthew Ernest Mazzucchelli and Sam Downes. Both had been employed by the Boulder jeweller James Robertson (1895–1909) for about five years before venturing out on their own in 1903. About 1907 they purchased the Lane Street business of T. H. Manhire (1903–07). Their shop, with its striking black paint and gold lettering, made a deep impression on Boulder residents and soon became the most important jewellers in the town. They were commissioned to make the gold key used to open the new Boulder Town Hall in 1908.

At present only three pieces marked with their M&D insignia are known to survive. One is a bar brooch of hollow tube terminated with gallery strip and scrolls. In the central section, in place of the usual nugget, is a treasured fragment of an earlier goldfields brooch. This had originally been made on the Sandhurst field during the Victorian gold rushes as a wedding present from Frederick Wheeler to Phoebe Morgan. It suffered damage and, on her death, sections were dispersed among the girls of the family. Frederica Cooke (née Wheeler), who had arrived from Victoria, sought out Mazzucchelli and had this sentimental memento of her mother incorporated into a new brooch typical of the Edwardian period (Plate 19.7).



In 1898 West Australian gold-brooch manufacture entered a new phase with the opening of a branch of the Royal Mint able to supply refined gold and ready-made sheet and rod. This proved particularly advantageous for the most prolific of the locally based wholesale firms, Donovan & Overland, as well as for some large entrepreneurial firms who moved to the west, expanding the wholesale manufacturing section of the trade from one to thirteen firms by 1901. 'Donovans' generated a considerable output from 1897 until about 1926 when, according to ex-employee Alec Lambert, the firm 'just fizzled out because not enough trained men returned from the Great War'.[12]

Before this happened, Donovans had become well known as the makers of a series of mining brooches incorporating names, such as MALCOLM, LEONORA and KALGOORLIE, in fretted letters arched above the mining equipment (Plate 19.5). These were made from about 1898, when the design was registered,[13] until around 1908, by which time the firm had instituted the use of a key as a maker's mark. The careful modelling of the constituent parts and the artistic balance to the compositions are probably why so many have survived. The series incorporated the 'golden hole' – a nugget-encrusted plate through which the rope from the windlass descends, suspending a bucket for winching up the ore. This was derived from the golden cavity of the legendary 'Londonderry' strike which had been sealed over while Lord Fingal went to London to raise the capital to work the mine. Despite the fiasco surrounding the original, the Londonderry strike symbolised fabulous wealth in West Australian goldfields iconography. Donovan & Overland also supplied many other retailers, surviving fierce competition with the larger international firms attracted to the west, perhaps because of clannish support for the local product. In 1921 the firm was still the premier West Australian wholesaler.

One fierce competitor was the London and Melbourne firm of Rosenthal Aronson. It had been represented in Western Australia by Eugenio Vanzetti, a budding mining magnate who may have registered the design 'W.A.10' on its behalf on 12 March 1896.[14] This design has a large crossed crowbar and sledgehammer surmounted by a small crossed pick and shovel. It features the one-sided geologist's pick typical of the company men who had by this time joined the prospectors and fossickers on the fields. The finished pieces are more attractive than the registered design (Plate 19.8). The goldsmith has modified the elements in the making. Two others which use the same elements plus a pan and bucket are known.[15] These have KALGOORLIE and BOULDER pierced from thin gold sheet.

In 1907, however, when the final tariff barrier was removed, such Melbourne firms had no need to keep open their smaller West Australian factories. Local subsidiaries became merely wholesale houses. By World War I, the gold boom was over and many of the goldsmiths had departed. Of those who remained, some scaled down their enterprises for the duration while quite a number went off to the war itself. As with so much in Australian history, the war marks the end of an era for West Australian jewellery production.



Overall, much of the general jewellery made in Western Australia during the late nineteenth century was little different from trade work made in England or on the east coast of Australia. It followed standard fashions, with trends being expedited by technical advances in printing and the proliferation of journals from the 1890s. It was only in the West Australian goldfields brooches, the pearling jewellery, and the 'swan and swags' brooches that a distinctive style becomes discernible. As a result, a regional genre, recognisable as West Australian, developed in concert with internationally fashionable work. These brooches are, in a sense, lost objects which deserve to be recovered and remembered as works of genuine and original art.

## Notes

1 Jewellery historian and dealer Anne Schofield has documented and discussed this. See 'Australian Jewellery in the 19th and Early 20th Century', *Australasian Antique Collector*, no. 25, pp. 72–5; and A. Schofield and K. Fahy, *Australian Jewellery: 19th and Early 20th Century*, 1990, David Ell Press, Sydney, p. 7.

2 This is how Arthur Bayley and William Ford found their riches at Coolgardie.
3 In the local Aboriginal dialect, Yilgarn referred to the beautiful white quartz outcrops in or near which much of the early gold was found. The Eastern Goldfields were later separated from the Yilgarn to make two mining districts.
4 Despite the fact that they have similar melting temperatures, only one-third the heat necessary to melt copper is required for gold.
5 See C. Gere and G. Munn, *Artists' Jewellery*, 1990, Antique Collectors Club, Woodbridge (UK), p. 99.
6 'The Flash Colonial Barmaid', in W. Coxon, *Comic Songster*, n.d., W. M. Brown, Ballarat, was proffered such an object in an attempt to win her favours.
7 *West Australian*, 13 March 1895.
8 A. Reid, *Those Were the Days*, 1933, Barclay & Sharland, Perth, p. 220.
9 *Wise's Post Office Directory 1897*, p. 56.
10 Addis is featured in a cartoon in the report on the exhibition ('Official Record of the Tasmanian International Exhibition', 1893, Exhibition Commissioners, Launceston).
11 J. S. Battye, *The Cyclopedia of Western Australia*, 1913, Cyclopedia Company, Perth, vol. 2, p. 903.
12 The impact of the expanding workshops of Levinsons, which opened its large new factory show-place that year, and the in-house work at Caris, which had just changed hands, may also have something to do with this. E. S. Lazarus also appears to have closed at this time.
13 March 1898 as 'W.A.13'.
14 Rosenthal Aronson became Aronson and Co. in the rest of Australia from 1902.
15 One is associated with Victor Streich and the Mount Lyell company in Tasmania. That piece has the name MT LYELL stamped into the head of the tap/sledge-hammer. It is in the collection of the Tasmanian Museum.

# A Broad Brush Dipped in Gold

THE EXPANSION OF AUSTRALIAN VISION

**Chapter 20**

*Anita Callaway*



> The Discovery of Gold: Two boundary riders, after their simple meal, accidentally discover a nugget. Australia rejoices, eager to shower all her wealth upon the men who love her.[1]

In the colonists' eyes at least, the discovery of gold confirmed the inevitability of Australia's progress. Surely Australia's eagerness to 'shower all her wealth upon the men who love her' was proof enough? Gold not only fuelled their material improvement but also promoted their cultural ambition. It encouraged their brazen – or, rather, gilded – confidence in themselves and their prospects. And, to represent this expansive colonial vision, art was suddenly released from the confines of the travel-artist's sketchbook or the studio artist's easel and, instead, was writ large upon canvas and, indeed, upon the stage.

Moving panoramas and pantomime transformations exemplify this new type of theatrical performance-cum-artwork. I have chosen to discuss them here not only because their popularity coincided with the economic hegemony of gold, but also because I believe that their extravagant dimensions, ambitious goals and flattering predictions figuratively express the exuberance that gold brought to Australian social, economic, political and cultural life.

Theatrical painting has been allowed little space in conventional art history. Stage scenery has been unfairly dismissed as something carelessly produced with 'gallons of distemper usually mixed in chamber pots and applied with a broom'.[2] But whatever its aesthetic qualities (and at times they were considerable), theatrical scenery was seen and appreciated by so many people that its cultural effects cannot be ignored. The colonial audience for this form of art was greater than that for traditional 'High Art' imagery.

The term 'panorama' was coined by Robert Barker to describe a personal invention – a 360° landscape painted on the inside walls of a circular building ninety feet (26.5 m) in diameter, and viewed by observers from an elevated central platform.[3] His disoriented viewers – separated from the illuminated walls of the rotunda across a dark and indefinable void – registered the images not as vertical paintings but as vistas laid out before them. With the world apparently at their feet, and with the central platform of Barker's entertainment so nicely mimicking the centre of Empire, these London observers might well be regarded as unwitting actors in a Foucauldian demonstration of the sovereign gaze – the imperial eye making them lords of all they surveyed.

The moving panorama that became so popular on the Australian goldfields was a very different device, one without the imperialist overtones of Barker's invention. It consisted of a single, continuous scene, or a series of separate scenes, painted on a length of cloth. This was wound around a vertical spindle and slowly unrolled in front of the audience onto a second spindle on the other side of the stage. The mechanism (including the stagehands who cranked the spindles at either end, and the clips that attached the canvas to a guide rope so that the panorama would not sag) was easily masked from view. As a travelling showman could cart the rolled-up canvas on its spools from place to place, the moving panorama did not require a permanent home but could be set up easily in any hall. The presentation was almost always accompanied by spoken commentary, sometimes interspersed with instrumental or vocal music. The audience sat as if in a theatre or cinema and watched as the vista unfolded before their eyes (Figure 20.1).



The passage of time was inherently factored into the moving panorama. The presentations often took two or three hours, depending on the length of the canvas, the rate at which it was unwound and whether more than one spool was used. Outrageous claims were made about the lengths of these panoramas, which were often measured in miles! These claims would be credible, however, if the advertised measurements were taken as square feet (the amount of canvas covered), rather than as a linear measurement. Thus, a panorama advertised as '5000 ft of canvas' would, if the panorama were 10 feet high, be a plausible 500 feet long.

The actual length of the panorama and the real time taken up in its presentation would have been of less significance to the audience than the illusory distance and time it covered. Today we are so familiar with the cinema's command of space and time that we no longer appreciate its remarkable nature. To a mid-nineteenth-century audience, however, the panorama's ability to telescope distance and time must have been astonishing.

The boldness of the moving panorama has misled some historians to conclude that it was an American invention.[4] This is not the case. They had originated in Britain in the early nineteenth century but, from the late 1840s, Americans enthusiastically took them up as visual expressions of their own much-vaunted Manifest Destiny. Indeed, with their reliance upon the notion of

**Figure 20.1** Diagram of the Workings of a Moving Panorama, in *Scientific American*, 16 December 1848. (National Library of Australia)

'The Course of Empire' (the belief in the inexorable westward progression of civilisation), moving panoramas became so extravagant that, to an outsider, they seemingly burlesqued American territorial and cultural aspirations. It was much the same here in the Australian colonies once gold had been discovered.

Edwin Roper Stocqueler was an exponent of the moving panorama on the Australian goldfields in the 1850s. He was the son of Joaquim Stocqueler,[5] who had written and performed the narration for the panorama of *The Route of the Overland Mail to India* at the London Gallery of Illustration in 1850, and for *The Diorama of the Ocean Mail to India and Australia* in 1853.[6] The final scene in the latter (a late addition, painted by 'Mr Danson & Sons' from a sketch by 'Mr John Calvert, the celebrated Australian Geologist') showed the 'Mount Alexander Goldfields, Forest Creek'. According to Joaquim Stocqueler's commentary:

> The picture speaks for itself. It is a graphic delineation of the country where these astounding gold-discoveries have been made, and depicts truthfully the canvass [*sic*] city of the diggers near the Golden Mount. The groups of men rocking and tossing about the golden earth, and the anxious faces ascending and descending the pits, burrowing and toiling for the coveted ore, are subjects worthy of the painter as well as description.[7]

In presenting his own 'mile-long' moving panorama of the Victorian goldfields,[8] young Edwin Roper Stocqueler might well be dismissed as merely aping his father. He was not, however, literally following in his father's footsteps: Stocqueler junior and his mother came to Victoria only after Stocqueler senior had abandoned them both in London.[9] Joaquim Stocqueler had apparently succumbed to the lure of American gold, so his son and spouse chose Australia instead. Perhaps they saw the Antipodean diggings as the next best thing to the American goldfields that Stocqueler senior had chosen over his family, or perhaps they merely saw Australia as a refuge far enough removed from the London bailiffs. Either way, young Stocqueler and his mother arrived in Victoria in 1854.[10]

Stocqueler first exhibited his moving panorama at Bendigo in August 1857. The show was presented in two parts, allowing the spools to be swapped during the intermission. The sections consisted of twenty-five and twenty-six separate paintings, respectively (Stocqueler added further pictures over the next two years), and were based upon sketches he had made while travelling about the goldfields with his mother.[11] The panorama began with a view of Port Phillip Heads as a shipload of hopeful gold diggers arrived at Victoria, followed by the landing at Port Melbourne, crossing the Yarra via the Prince's Bridge, and a view  of Melbourne as a canvas staging post for the goldfields. Then came various views of the goldfields themselves, including the miners' camps, and topographical features of the surrounding countryside.[12]

This panorama no longer exists, of course. With the requisite winding and unwinding of the canvas at every performance, the painted surface would have worn out. We can only speculate upon the content and appearance of this important work of art, but other examples of Stocqueler's painting have survived, including his *Australian Gold Diggings* (Plate 20.1); a view of the Castlemaine diggings (*Castlemaine from Ten Foot Hill*, Pioneers and Old Residents Association, Castlemaine); and a dramatic representation of a corroboree (*Corroborree*, National Library of Australia). Castlemaine was in fact the subject of the third-last scene of his panorama, a corroboree was the finale of the program's first half, and the unidentified goldfield shown in *Australian Gold Diggings* could well be one of the many diggings scenes from either half of the presentation. Even from these few examples we can glimpse what Stocqueler's panorama must have been like.

Stocqueler exhibited his work throughout the Victorian goldfields during the rest of 1857 and 1858. He often advertised it as 'Stocqueler's Diorama', for the term 'diorama' – having lost the original meaning coined by Daguerre in 1822 to define his transparent illusionary paintings – was now commonly used in the English-speaking world to describe a 'moving' panorama as opposed to a static one. But Stocqueler abandoned even this drab generic designation in June 1859 when he finally exhibited his panorama in Melbourne.[13] His bold new title – *The Golden Land of the Sunny South* – unequivocally declared the colony's confidence in its future glory. This evocative title, with its fairy-tale and mythological associations, predates by more than two decades its more familiar use in George Darrell's melodrama *The Sunny South* (Theatre Royal, Melbourne, 1883), Benjamin Farjeon's storybook *The Golden Land* (Ward Lock, London, 1886) and Tom Roberts' painting *The Sunny South* (1887, National Gallery of Victoria).

Straight after its Melbourne season, Stocqueler exhibited the panorama in the Melbourne suburb of Collingwood. He then disappears completely from colonial records. This is not too surprising. At the time, the volatility of the Victorian goldfields allowed, indeed encouraged, a cast of enigmatic figures who were as integral to the playing out of colonial history as they would have been to any melodrama. Any individual who found gold's brilliant glare too disconcerting could easily change his identity and slip back into shadowy anonymity. It is feasible that Stocqueler junior retained his father's surname only while his mother was alive, and renounced his paternal connection as soon as he could. It is also feasible – just – that Edwin Roper Stocqueler subsequently became Edward Roper, a near-contemporary fellow-artist whose parentage and early history are as yet unknown. Roper did, however, produce a novel in which the young hero's father – just like Joaquim Stocqueler – abandons his wife and son and runs off to North America under an assumed name! Roper's goldfields paintings are similar to Stocqueler's but seem to have been painted retrospectively in the 1870s and 1880s.[14]



Another panorama painter and romantic adventurer in the same mould was Augustus Baker Peirce.[15] He was born in Massachusetts, and died there, but spent most of his working life in Australia. Having run away to sea as a youth, he jumped ship at Port Phillip in 1859 and began an amazing Antipodean career as a gold digger, scene painter, snake-oil salesman, travelling photographer, riverboat captain, caricaturist and publican. Peirce's audacious willingness to try anything was echoed by many of his contemporaries, whose ready involvement in the so-called low arts has fuelled (improperly, in my opinion) the general disparagement of colonial art.

Peirce painted his first and most successful panorama in 1869. As pilot, then master, of a Murray River paddle-wheeler, Peirce had experienced his own living version of John Banvard's *Moving Panorama of the Mississippi* (1848; see Figure 20.1). Perhaps, as the *Border Post* suggested, Peirce originally intended to emulate his compatriot's panorama by substituting one mighty river for another.[16] But, taking his title from another American panorama of the past (*A Voyage round the World*, Boston, 1846–48),[17] Peirce produced instead a panorama that combined the sea voyage featured in Stocqueler senior's *The Diorama of the Ocean Mail to India and Australia* with the goldfields peregrinations of Stocqueler junior's *The Golden Land of the Sunny South*. Peirce's visual narrative of the quest for gold (familiar to all immigrant diggers) began in London with his copy of William Frith's *The Railway Station, Paddington* (1862, Royal Holloway College) and finished with goldfields scenes similar to his own *Myers Creek Rush near Sandhurst, Victoria* (Plate 20.2).

Peirce and his partner, the photographer George Kendall (who accompanied each performance with 'new songs and operatic selections'),[18] toured the *Voyage round the World* in a horse-driven cart, travelling through the Victorian goldfields and into New South Wales, adding new scenes as they went. As Kendall abandoned the enterprise somewhere en route, it was Peirce alone who brought the show to Bathurst in 1872. By this time, the panorama had grown sufficiently from its original 212 feet to allow it to be advertised at an exaggerated 5000 feet (that is, about 555 feet long, as Peirce's scenes were 9 feet high). The jocular tone of Peirce's presentation and the self-parody of his performance are clear in advertisements advising 'Heads of Families' that 'should any deaths occur in the rush to get in, a first-class undertaker will be in attendance', and warning that, although children would be admitted at half-price, 'no [o]ne over 12 stone will be considered infants'. He also announced that he had added a scene of the main street and 'local celebrities' to appeal to a Bathurst audience.[19]

One of these 'local celebrities' was the successful goldminer, Bernard Holtermann. Peirce wrote in his spirited and somewhat unreliable memoir that he had called upon Holtermann to obtain a copy of Beaufoy Merlin's photographic montage of Holtermann and his famous nugget (Holtermann Collection, State Library of New South Wales).[20] Instead of the photograph, however, the smooth-talking Peirce came away with Holtermann's commission for a new moving panorama, *The Mirror of Australia*. It was an ambitious project which celebrated all aspects of Australian colonial life, not only the gold that sustained it. Peirce exhibited this new panorama throughout central New South Wales, this time accompanied by his wife.



Theatre has always been a place of illusion and exaggeration. The very scale of the canvases on which nineteenth-century scene painters practised their art could only have encouraged extravagant ideas. Indeed, the grand themes and classical references thought to be missing from Australian colonial art can be found, after all, in theatrical scenery. Here, too, can be found landscape paintings that predicted Australia's supposedly glorious, gold-filled future as boldly and as grandiloquently as any museum picture.[21]

Although the giant moving panoramas I have already discussed were ambitious in scale, they were really very simple devices – just rolls of painted canvas on spools. They could be exhibited in any hall, anywhere, at any time. Yet those panoramas that persisted within the legitimate theatre had the benefit of sophisticated stage machinery, lighting and the whole panoply of theatrical illusion. In particular, they could use Daguerre's original dioramic effects, in which scenes painted on two or more gauze drops were rendered opaque or transparent (visible or invisible), according to whether light was reflected by or transmitted through them. By controlling incident light, the scene painted on the front cloth could be made to disappear and the second scene revealed instead. Daguerre's dioramas were static pictures that gave the illusion of the progression of time as lighting changes revealed the sequence of painted drops: day was turned into night, foggy morning into sunny noon. It would not have needed a giant conceptual leap for the diorama to depict more radical time shifts, such as Ye Olden Times to the present. Yet it was the moving panorama, and not the diorama, that presented the sublime illusion of mastery over time. Theatrical panoramas combined the long-term temporal changes of a moving panorama with the short-term temporal changes of a diorama, sometimes with dazzling effect. This type of panorama regularly featured in our colonial pantomimes, Australia being the only British colony to match – and often outdo – the spectacle of the traditional British panto.

The moving panorama painted by John Hennings for the pantomime *The House That Jack Built; or, Harlequin's Progress, and the Loves, Laughs, Laments, and Labours of Jack Melbourne and Little Victoria* (Theatre Royal, Melbourne, Christmas 1869) was itself entitled *The Progress of Melbourne.* It was presented as a vision conjured up by the Fairy Queen to show to her young protégé, Jack Melbourne, transforming the spot on which they stood ('An Aboriginal Wood in Australia Felix') – via a progressive sequence ('The "Enterprise" schooner landing the first Melbourne settlers', 'Collins Street 1838', 'A Sheep Station 1848', 'The Rush to Bendigo', and 'The Departure of Burke and Wills') – to the just-completed Melbourne Town Hall. *The House That Jack Built* was realised on stage as merely a 'Toy house, such as is drawn by school boys on their slates',[22] but the actual Town Hall was a building sufficiently grandiose to signify the high-point of Victoria's 'Course of Empire'. The whole pantomime, as well as the panoramic interlude, was in fact a complex allegory, using both young Jack and the House he built to symbolise Melbourne's growth and development.



John Hennings was born in Bremen and attended the Dusseldorf Academy.[23] He arrived in Victoria in 1855 as a young man of 20 and spent the remainder of his life in the Australian colonies as 'the *doyen* of the living galaxy of scenic artists in Australia'.[24] He was chief painter for Marcus Clarke's pantomimes *Goody Two Shoes and Little Boy Blue; or, Sing a Song of Sixpence, Harlequin Heydiddle-Diddle-'em and the Kingdom of Coins* (Theatre Royal, Melbourne, Christmas 1870) and *Twinkle, Twinkle, Little Star; or, Harlequin Jack Frost, Little Tom Tucker and the Old Woman That Lived in a Shoe* (Theatre Royal, Melbourne, Christmas 1873). Clarke afterwards used Hennings as the basis of his fictitious scene painter, 'Vandyke Brown', in his 1874 spoof article about pantomimes:

> Vandyke Brown . . . who has to paint some nineteen scenes and contrive the Transformation – the Vision of Judgment, or the Choice of Hercules, in sixteen tableaux – in addition to a Panorama consisting of the Celebrated Cities of the World ending with Fawkner's Town and the Local Pump – is apt to get cynical from tobacco and white lead. 'I don't see how to set the Thames on Fire in this last effect', he says, 'without more rosin. And, moreover seven scenes in the Panorama's plenty'. 'Oh, let's have . . . a couple of views of Metz during the siege, to melt into a pastoral picture of Geelong', suggests Buzzclack, who is a great man for 'ideas'. At which Vandyke Brown withdraws, and is heard presently in the flies, muttering like an indistinct thunderstorm.[25]

The most lavish scenic effect in any pantomime was the transformation scene. It originally formed an interlude between the fairy-tale that began the pantomime and the harlequinade that ended it, but eventually it became an entity in its own right: its structure growing longer and more complicated, its narrative becoming ever more fantastic and increasingly divorced from the plot. The mechanics were simple enough: gauzy painted cloths were hung one behind the other through the depth of the stage, and the lighting was changed so that it shone on, then through and then behind the various layers, revealing a sequence of changing scenes. These changes were like cinematic dissolves – slowly repeating through the depth of the stage, half-a-dozen times or more, and perhaps taking as long as twenty minutes to do so. Thus, in full view of the audience, a scene could be transformed slowly but convincingly from bleak winter to glorious mid-summer; from barren wasteland to fertile pasture; from the rigours of earthly existence to the delights of paradise.

Transformation scenes were perfect for illustrating temporal changes in a particular location, as if a landscape painting could be made to reveal the future. Hennings' *Progress of Melbourne* panorama would have been just as effective had it been presented as a transformation scene instead. Hennings himself demonstrated this eleven years later when he painted 'The Apotheosis of Victoria's Progress in Four Decades, from 1841 to 1881, from Chaos to Splendour' as the transformation scene for *Sinbad, the Sailor; or, The Pet of the Peri, the Old Man of the Sea, and the Dwarf of the Diamond Valley* (Theatre Royal, Melbourne, Christmas 1880). A newspaper illustration of this scene (Figure 20.2) shows an early stage of this transformation, in which a row of Aboriginal warriors is poised on a bridge over a raging torrent in front of a sublime and primitive landscape. In this picture, the transformation is about to begin. Once the Amazon queen waves her magic sword, the Aboriginal group will be flown out of view and the landscape will go through a succession of changes until it becomes Melbourne at the height of its gold prosperity in 1880, the original rocky crags now covered with grand buildings and the torrent tamed into the Yarra River.[26]



The transformation scene – as it changed wilderness into civilisation, Chaos into Splendour, or Savage State into Consummation of Empire[27] – was a powerful visual allegory of Australia's rapid progress under the stimulus of gold. The audience could see the present turn into the future before their very eyes. For twenty minutes or so, a painted landscape on a grand scale slowly changed in front of them as if they were in a time machine.

We cannot share this experience today, but we can gain some understanding of it from the exaggerated version described by Garnet Walch in his 1875 satire on pantomime production. Walch had written several local pantomimes himself, including *Trookulentos, the Tempter; or, Harlequin Cockatoo, the Demon of Discontent, the Good Fairy of Contentment, and Four-Leaved Shamrock of Australia* (Royal Victoria Theatre, Sydney, Christmas 1871) and *Australia Felix; or, Harlequin Laughing Jackass and the Magic Bat* (Opera House, Melbourne, Christmas 1873). He noted that the three 'notable individuals' in the production of a pantomime were – in descending order of importance – the manager,

**Figure 20.2** George Rossi Ashton (after John Hennings), *Sinbad (Theatre Royal),* in *Australasian Sketcher,* 1 January 1881. (La Trobe Picture Collection, State Library of Victoria)

the scene painter, and (a long way behind them) the author. Walch's fictional scenic artist, 'Christiansen', was in reality his friend and colleague Alexander Christian Habbe (another young European immigrant gold digger turned scene painter, who had arrived in 1855), and the imaginary transformation scene he painted ('The Past, Present, and Future of Australia') is a distillation of the real transformation scenes that he had produced for Walch and others:

> The scene is an Allegorical picture representing the Past, Present, and Future of Australia. First – night, and an arid waste, broken only by a few jagged rocks, and

> groups of picturesquely-posed aboriginals [*sic*] gathered round their watchfires. Gradually there steals over the scene the soft grey light of morning, the Aboriginals move as if rousing from sleep, native birds are heard, and in the distance up looms the great red summer sun. Slowly, sands, rocks, and natives sink from view, and in their place we see a headland on the coast, with the blue Southern Ocean lapping at its base, and stretching away in the distance. From the . . . waves rise shell-borne water-spirits . . . joined by wood-nymphs from the neighbouring groves . . . together [they] await the arrival of their Queen – the lovely maid Australia. She . . . gazes yearningly across the ocean. Presently, as if responsive to her prayer, the sails of a gallant ship – white-winged messenger of civilisation – mount the horizon, and as the descending gauzes hide the scene from view, the music swells into a hymn of praise and thanksgiving. The gauzy mists dispersing once again, disclose a view of busy Melbourne . . . A couple of minutes more and the huge canvas on which the city is painted lifts, and reveals the last stage of the transformation – the Future of Australia . . . [where] the noonday sun of full prosperity blazes.[28]

Even this imaginary transformation was surpassed in chauvinistic conceit by the actual transformation in the pantomime *Cinderella and the Little Glass Slipper* (Lyceum Theatre, Sydney, Christmas 1894; and Princess Theatre, Melbourne, February 1895). This transformation – 'The Wealth, Produce, and Progress of Australia' (sub-titled 'Australia's March of Triumph') – was painted by Phil Goatcher, a scene painter who had been brought to Victoria in 1890 by the actor-manager J. C. Williamson. Although he was welcomed at the time as a successful American scene painter (Goatcher and Williamson had worked together at the California Theater in San Francisco in 1870), Goatcher was in fact another Australian goldfields character. In 1865, like so many other seamen, he had at the age of 14 or 15 jumped ship in Melbourne and set off for the diggings. Finding little success at gold prospecting, he began to hang about the Ballarat theatre where John Hennings was working. Perhaps Goatcher's later memory of his meeting with Hennings was somewhat romanticised:



> The first person to notice him was Mr John Hennings, who, seeing a fair-haired, blue-eyed boy gaping at him in astonishment as he worked, jokingly asked whether he would not like to be a scene painter too. The boy's eyes nearly danced out of his head in the eagerness to say 'Yes'. Thenceforward Mr Hennings allowed his young admirer the entrée of the painting room, and noticing that the lad gave the deepest attention to the work in hand, the elder artist took an interest in him, and gave him lessons. Thus commenced Mr Goatcher's art training, and it is to the excellent grounding which he received from Mr Hennings that he ascribes his success.[29]

It is certain, however, that Hennings gave Goatcher his first lessons and first job as a scene painter in Ballarat in 1865, and that they worked together as equals in Melbourne in 1890.[30]

Goatcher's 'Wealth, Produce and Progress' transformation began in a 'primeval forest' and ended 'in an ideal sketch of federated Australia'.[31] Perhaps because of his early experience on the Australian goldfields, or his later exposure to the American faith in Manifest Destiny, or the pre-Federation exuberance of the 1890s, this transformation is outrageously smug, and yet it perfectly expresses Australia's belief in its own inexorable destiny:

1. 'Australia' wanders at first through (a) the Primeval Forest, with its weird, strange forms of fern and tree. From these spring up (b) Fair Floral Beauties;
2. Then arise (a) The Early Colonial Industries, (b) Agricultural Wealth and Wool being represented by a sturdy teamster, toiling over a hot dusty road, beside his dray, laden with The Golden Fleece of Australia;
3. Passing by a heaped-up (a) Combination of Produce, she opens (b) The Gates of Golden Wealth, inspects (c) The Interior of a Gold Mine, and takes (d) A Look at Ballarat from Black Hill;
4. Whence she surveys all The Treasures and Works of Art in Gold, and discovers, shining through the grey distance;



5. (a) The Gates of the Silver Era, through which she catches (b) A Silvery view of the Barrier Mine at Broken Hill;
6. Nor does she pause in her Triumphal March until (a) The Pearl Fisheries are reached. Here awhile she rests, gazing on (b) The Golden Splendours of the Tropical Sea, on whose warm bosom lie the slumbering fishing boats;
7. And now she turns her eyes to where (a) In Liquid Depths the mermaids sing in coral palaces, surrounded by Gardens of Anemones, a song of joy at (b) The Birth of the Pearl;
8. The journey of the Past accomplished, 'Australia' stands on the seashore and tries with hand-shaded eyes to pierce the haze, which hides the Future, and lo!
   a Glorious Vision rises – rises – rises! until she sees revealed in dazzling light –
   A Radiant Group of Beautiful Figures, allegorical of 'United Australia'![32]

What colony would dare to show itself so vainglorious, unless basking confidently in the flattering light of gold?

Perhaps it is equally vainglorious of me to have conflated the material progress that gold brought about with the colonial dream of sovereign destiny. And especially vainglorious to use theatrical bombast as my link between the two. In claiming that post-gold-rush Australian visual imagery was produced with a broad brush, both literally and figuratively, perhaps I myself am painting too broadly. Historical metaphors can be stretched only so far, after all. But I have not applied this metaphor thoughtlessly. That the relevance of the metaphor was recognised at the time is demonstrated in the final scene of the post-Federation pantomime *Little Red Riding Hood* (Bijou Theatre, Melbourne, Christmas 1903):

'The Federal Capital . . . The Gathering of the States . . . Patriotic Finale: The Shower of Gold'.[33] Like Danaë, it seems, federated Australians were the inevitable recipients of Jovian largesse.

## Notes

1 Fifth tableau in the performance given by Australian expatriate artists at the Imperial Institute, London, in 1911 (*British Australasian*, 1 June 1911, p. 15). The episodes were devised by Tom Roberts and produced by George Lambert. The boundary riders were portrayed by Fred Leist and Will Dyson.

2 P. J. Zeplin Waite, ' "Three Shillings for a Bolt of Hessian": Stage Design in Melbourne 1919–1953', *Australasian Drama Studies*, vol. 2, no. 1, October 1983, p. 48.

3 Patent for the device was issued 3 July 1787. Barker first used the term 'panorama' in 1792 when he built a permanent exhibit in London.



4 Claims for the uniqueness of the American moving panorama have been made (inter alia) by K. H. McGinnis, 'Moving Right Along: Nineteenth Century Panorama Painting in the United States', University of Minnesota thesis, 1983, quoted in M. Colligan, 'Canvas and Wax: Images of Information in Australian Panoramas and Waxworks, with Particular Reference to Melbourne 1849–1920', Monash University thesis, 1987, p. 85; and E. J. Fruitema and P. A. Zoetmulder (eds), *The Panorama Phenomenon*, 1981, Mesdag Panorama catalogue, The Hague, p. 74.

5 Also recorded as Joachim Hayward Stocqueler, Joachim Heyward Stocqueler, and J. H. Siddons.

6 The former was painted by Thomas Grieve and William Telbin; the latter by Messrs Grieve, Telbin, Callcott, Cuthbert, and Gordon.

7 J. Stocqueler, in *A Descriptive Guide to the Diorama of the Ocean Mail to India and Australia*, 1853, Gallery of Illustration, London, pp. 65–6.

8 *Bendigo Advertiser*, 8 August 1857; quoted in Colligan, 'Canvas and Wax', p. 105.

9 See F. Cusack, 'Edwin Roper Loftus Stocqueler', in J. Kerr (ed.), *The Dictionary of Australian Artists: Painters, Sketchers, Photographers and Engravers to 1870*, 1992, Oxford University Press, Melbourne, pp. 758–60.

10 When Stocqueler first showed his panorama in Bendigo in August 1857, he claimed that the scenes of Melbourne and Bendigo were taken from sketches he had made 'four years ago' (*Bendigo Advertiser*, 1 August 1857; quoted in National Library of Australia's artist file).

11 *Bendigo Advertiser*, 29 June 1857; quoted in Colligan, 'Canvas and Wax', p. 105. Mother and son claimed to have seen a bunyip during these travels. Alas, Stocqueler did not record this incident for his show!

12 Frank Cusack, ms held in the Dictionary of Australian Artists files.

13 He called his Bourke Street exhibition hall The Gallery of Illustration, after the London venue where his father had appeared.

14 E. Roper, *Fred Seagood: His Travels and His Triumphs*, 1904, Swan Sonnenscheim & Co., London. So far, the evidence that Stocqueler and Roper are one and the same is only circumstantial. However, their individual fragmented biographies blend together nicely into one comprehensive artistic career. For Stocqueler (1829–?), see Cusack, 'Edwin Roper Loftus Stocqueler', pp. 758–60; for Roper (c. 1832–1909), see S. Jones and J. Kerr, 'Edward Roper', also in Kerr, *Dictionary of Australian Artists*, pp. 683–4.

15 He variously signed himself Peirce, Pierce, or Pearce, using the initials A. B. for Augustus Baker, or G. for Gus.

16 *Border Post*, 15 September 1869, p. 2.

17 J. F. McDermott, *The Lost Panoramas of the Mississippi*, 1958, University of Chicago Press, Chicago, p. 68.

18 *Beechworth, Ovens and Murray Advertiser*, 28 March 1871, p. 2; Information, Pictorial Collection, State Library of Victoria.

19 A. Callaway, 'Augustus Baker Peirce', in Kerr, *Dictionary of Australian Artists*, pp. 614–16.



20 A. B. Peirce, *Knocking About: Being Some Adventures of Augustus Baker Peirce in Australia*, 1924, Yale University Press, New Haven.

21 See A. Callaway, *Visual Ephemera: Theatrical Art in Nineteenth-Century Australia*, 2000, University of New South Wales Press, Sydney, chs 8–9.

22 W. M. Akhurst, *The House That Jack Built; or, Harlequin's Progress, and the Loves, Laughs, Laments and Labours, of Jack Melbourne, and Little Victoria*, 1869, pantomime libretto, Theatre Royal, Melbourne.

23 A. Callaway, 'John (Johann Friedrich) Hennings', in Kerr, *Dictionary of Australian Artists*, pp. 35–59.

24 Old Stager, 'Some Bygone Pantomimes', *Australasian*, 6 January 1894, p. 23.

25 M. Clarke, 'Rehearsing a Pantomime', *Weekly Times*, 31 January 1874, p. 9.

26 M. Williams, *Australia on the Popular Stage, 1829–1929*, 1983, Oxford University Press, Melbourne, pp. 76–7; Colligan, 'Canvas and Wax', p. 209.

27 Stages in the 'Course of Empire' sequence, as depicted by the American painter Thomas Cole.

28 G. Walch, *On the Cards; or, a Motley Pack*, 1875, Baillière, Melbourne, pp. 1–4.

29 *Table Talk*, 31 October 1890, p. 4.

30 *Table Talk*, 21 November 1890, p. 17.

31 *Australasian*, 9 February 1895, p. 260.

32 T. Taylor, *Cinderella and the Little Glass Slipper*, pantomime libretto, Lyceum Theatre, Sydney, 1894.

33 This particular scene was designed by John Little (Collection of Melbourne Theatre and Concert Programmes 1885–1913, 12 vols, Mitchell Library).

# Index

Durham University

Durham University Library

Please return or renew this item by the last date shown below or the date shown on your Self-Issue receipt.

If this item is requested by another user it may be recalled and a revised due date will be given. The item MUST be returned by the revised due date. Fines are incurred on overdue books.

www.durham.ac.uk/library/

DURHAM UNIVERSITY LIBRARY
3 0104 01388742 1